Wavelet Methods for Solving Partial Differential Equations and Fractional Differential Equations

Wavelet Methods for Solving Partial Differential Equations and Fractional Differential Equations

Dr. Santanu Saha Ray
Department of Mathematics
National Institute of Technology Rourkela
Rourkela, Odisha, India

Dr. Arun Kumar Gupta
School of Applied Sciences
KIIT University
Bhubaneswar, Odisha, India

CRC Press
Taylor & Francis Group
Boca Raton London New York

CRC Press is an imprint of the
Taylor & Francis Group, an **informa** business
A CHAPMAN & HALL BOOK

CRC Press
Taylor & Francis Group
6000 Broken Sound Parkway NW, Suite 300
Boca Raton, FL 33487-2742

Printed on acid-free paper

International Standard Book Number-13: 978-1-138-05381-6 (Hardback)
International Standard Book Number-13: 978-1-315-16718-3 (eBook)

Library of Congress Cataloging-in-Publication Data

Names: Saha Ray, Santanu, author. | Gupta, Arun Kumar (Mathematician), author.
Title: Wavelet methods for solving partial differential equations and
fractional differential equations / Santanu Saha Ray and Arun Kumar Gupta.
Description: Boca Raton : CRC Press, 2018. | Includes bibliographical
references and index.
Identifiers: LCCN 2017039698| ISBN 9781138053816 (hardback : alk. paper) |
ISBN 9781315167183 (ebook)
Subjects: LCSH: Differential equations, Partial. | Wavelets (Mathematics) |
Fractional differential equations. | Differential equations--Numerical
solutions.
Classification: LCC QA374 .S24 2018 | DDC 515/.353--dc23
LC record available at https://lccn.loc.gov/2017039698

Visit the Taylor & Francis Web site at
http://www.taylorandfrancis.com

and the CRC Press Web site at
http://www.crcpress.com

Contents

List of Figures

List of Tables

Preface

The subject of fractional calculus has gained considerable popularity and importance during the past three decades or so, mainly due to its demonstrated applications in numerous seemingly diverse and widespread fields of science and engineering. It deals with derivatives and integrals of arbitrary orders. The fractional derivative has been occurring in many physical problems, such as frequency-dependent damping behavior of materials, motion of a large, thin plate in a Newtonian fluid, creep and relaxation functions for viscoelastic materials, the $PI^{\lambda}D^{\mu}$ controller for the control of dynamical systems, etc. Phenomena in electromagnetics, acoustics, viscoelasticity, electrochemistry, control theory, neutron point kinetic model, anomalous diffusion, Brownian motion, signal and image processing, fluid dynamics, and material science are well described by differential equations of fractional order.

Generally, nonlinear partial differential equations (PDEs) of fractional order are difficult to solve. So for the last few decades, a great deal of attention has been directed toward the solution (both exact and numerical) of these problems. The aim of this book is to present an extensive study of different wavelet methods for obtaining numerical solutions of mathematical problems occurring in the disciplines of science and engineering. This present work also provides a comprehensive foundation of different wavelet methods comprising the Haar wavelet method, Legendre wavelet method, Legendre multiwavelet methods, Chebyshev wavelet method, Hermite wavelet method, and Petrov–Galerkin method. The intention is to examine the accuracy of various wavelet methods and their efficiency for solving nonlinear fractional differential equations.

With the widespread applications of wavelet methods for solving difficult problems in diverse fields of science and engineering such as wave propagation, data compression, image processing, pattern recognition, computer graphics, and medical technology, these methods have been implemented to develop accurate and fast algorithms for solving integral, differential, and integrodifferential equations, especially those whose solutions are highly localized in position and scale. The main feature of wavelets is their ability to convert the given differential and integral equations to a system of linear or nonlinear algebraic equations, which can be solved by numerical methods. Therefore, our main focus in the present work is to analyze the application of wavelet-based methods for solving the problem of fractional order PDEs.

The introductory concepts of wavelet, wavelet transform, and multiresolution analysis have been discussed in mathematical preliminary. The basic idea of various analytical and numerical methods, namely, variational iteration method (VIM), homotopy perturbation method (HPM), homotopy analysis method, first integral method (FIM), optimal homotopy asymptotic method (OHAM), Haar wavelet method, Legendre wavelet method, Chebyshev wavelet method, and Hermite wavelet method have been presented in Chapter 1.

In Chapter 2, we have considered both analytical and numerical approach for solving some particular nonlinear PDEs like Burgers' equation, modified Burgers' equation, the Huxley equation, the Burgers–Huxley equation, and the modified Korteweg–de Vries (mKdV) equation, which have a wide variety of applications in physical models. VIM and the Haar wavelet method are applied to obtain the analytical and numerical approximate solution of Huxley and Burgers–Huxley equations. Comparisons between analytical solution and numerical solution have been cited in tables and also graphically. The Haar wavelet method has also been applied to solve Burgers', modified Burgers', and mKdV equations numerically. The results thus obtained are compared with exact solutions as well as solutions available in open literature. Error of collocation method is presented in this chapter.

Methods like HPM and OHAM are very powerful and efficient techniques for solving nonlinear PDEs. Using these methods, many functional equations such as ordinary differential equations, PDEs, and integral equations have been solved. We have implemented HPM and OHAM in Chapter 3 in order to obtain the analytical approximate solutions of system of nonlinear PDE, namely, the Boussinesq–Burgers' equations. Also, the Haar wavelet method has been applied to obtain the numerical solution of Boussinesq–Burgers' equations. Moreover, the convergence of HPM and OHAM is discussed in this chapter.

The mathematical modeling and simulation of systems and processes, based on the description of their properties in terms of fractional derivatives, naturally leads to differential equations of fractional order and the necessity to solve such equations. The mathematical preliminaries of fractional calculus, definitions, and theorems are presented in Chapter 4. Next, in this chapter, the Haar wavelet method is analyzed for solving fractional differential equations. The time-fractional Burgers–Fisher, generalized Fisher-type equations, and nonlinear time- and space-fractional Fokker–Planck equations are solved by using two-dimensional Haar wavelet method. The obtained results are compared with the OHAM, the exact solutions' and results available in open literature. Comparison of obtained results with OHAM, Adomian decomposition method, VIM, and operational Tau method (OTM) are demonstrated in order to justify the accuracy and efficiency of the proposed schemes. The convergence of the two-dimensional Haar wavelet technique is provided at the end of this chapter.

In Chapter 5, the fractional differential equations such as KdV–Burger–Kuramoto (KBK) equation, seventh-order KdV (sKdV) equation, and Kaup–Kupershmidt equation are solved by using two-dimensional Legendre wavelet and Legendre multiwavelet methods. The main focus of this chapter is the application of the two-dimensional Legendre wavelet method for solving nonlinear fractional differential equations like the time-fractional KBK equation and time-fractional sKdV equation in order to demonstrate the efficiency and accuracy of the proposed wavelet method. Similarly in Chapter 6, two-dimensional Chebyshev wavelet method has been implemented to obtain the numerical solutions of the time-fractional Sawada–Kotera equation, fractional order Camassa–Holm equation, and Riesz space-fractional sine-Gordon equations. Convergence analysis is also done for these wavelet methods.

In Chapter 7, the solitary wave solution of fractional modified Fornberg–Whitham equation is attained by using FIM, and the approximate solutions obtained by OHAM are also compared with the exact solutions acquired by FIM. Moreover, the Hermite wavelet method is implemented to obtain approximate solutions of the fractional modified Fornberg–Whitham equation. The Hermite wavelet method is implemented to a system of nonlinear fractional differential equations, namely, the fractional Jaulent–Miodek equations. The convergence of this wavelet method is discussed in this chapter. Chapter 8 places emphasis on the application of the Petrov–Galerkin method for solving the fractional differential equations, such as the fractional KdV-Burgers' equation and the fractional Sharma–Tasso–Olver equation, with the aim to exhibit the capabilities of this method in handling nonlinear equations. The main objective of this chapter is to establish the efficiency and accuracy of the Petrov–Galerkin method in solving fractional differential equations numerically by implementing a linear hat function as the trial function and a quintic B-spline function as the test function.

Various wavelet methods have been successfully employed to numerous partial and fractional differential equations in order to demonstrate the validity and accuracy of these procedures. Analyzing the numerical results, it can be concluded that the wavelet methods provide worthy numerical solutions for both classical and fractional order PDEs. Finally, it is worthwhile to mention that the proposed wavelet methods are promising and powerful methods for solving fractional differential equations in mathematical physics. This work also aims at making this subject popular and acceptable to the engineering and science communities to appreciate the universe of wonderful mathematics, which is in between classical integer order differentiation and integration, which even today is not much acknowledged and is hidden from scientists and engineers. Therefore, our goal is to encourage the reader to appreciate the beauty as well as the usefulness of these numerical wavelet-based techniques in the study of nonlinear physical systems.

Acknowledgements

Foremost, I express my deepest sense of gratitude to my supervisor Dr. S. Saha Ray for his insightful guidance, lively discussion, expert advice, constant encouragement, and constructive and honest criticism. I am greatly indebted to him for his invaluable and sustained supervision during my PhD. He is an excellent mentor and was very supportive throughout the work. I feel extremely fortunate to have had excellent supervision from him during the past several years. I am also thankful to his family for their affection and hospitality.

I also express my sincere gratitude to Dr. Achyuta Samanta, founder of Kalinga Institute of Industrial Technology (KIIT) and Kalinga Institute of Social Sciences, Bhubaneswar, Odisha, India, for his kind cooperation and support. The moral support received from my colleagues at KIIT is also acknowledged.

I am extremely thankful to all the supporting and technical staff of the Department of Mathematics for their kind help and moral support. I would like to gratefully acknowledge the incredible moments I spent with some special friends: Subhrakanta, Prakash, Manas, and Subha. I am also thankful to my childhood friends Radha Mohan, Anil, and Pintu for their selfless help and encouragement. My sincere thanks to everyone who has given me new ideas and useful criticism. I am truly indebted.

Last but not the least, I extend my heartfelt gratitude to my beloved parents Rameswar Gupta and Bidusi Gupta, and I am greatly indebted to them for bearing the inconvenience during the work. I owe a deep sense of indebtedness to my family members Santosh Gupta, Sanju Gupta, Anita Gupta, and Girish Kumar Gupta for their care, affection, immeasurable support, and invariable source of motivation.

Arun Kumar Gupta

Mathematical Preliminary

M.1 Introduction

Partial differential equations (PDEs) are of widespread interest because of their connection with phenomena in the physical world. These are useful tools for describing the natural phenomena of science and engineering models. For instance, in physics, the heat flow and wave propagation phenomena are well described via PDEs. Many engineering applications are simulated mathematically as PDEs with initial and boundary conditions. The diffusion of neutrons in nuclear reactor dynamics, population models, the dispersion of a chemically reactive material, and many physical phenomena of fluid dynamics, quantum mechanics, electricity, and so on are governed by PDEs.

PDEs originated from the study of solutions to a wide variety of problems in mechanics. Even though the foundation of nonlinear PDEs is ancient, they have undergone remarkable new developments during the second half of the twentieth century. Scientists and mathematicians have become actively involved in the study of countless problems offered by PDEs. The primary reason for this research was that it plays a vital role in modern mathematical sciences, mainly in applied physics, mathematical modeling, and engineering. With the development of PDEs, several methods such as the characteristics method, spectral methods, and perturbation techniques have been employed to evaluate the solution of nonlinear problems. However, there is no general method of finding analytical solutions of nonlinear PDEs. Hence, new numerical techniques are required for finding solutions of nonlinear equations. Therefore, it becomes increasingly important to be familiar with all traditional and recently developed methods for solving PDEs and the implementations of these methods. In this context, a relatively new and emerging area in mathematical research with a variety of applications in engineering disciplines, namely, wavelets theory, has attracted the focus of researchers in the field of science and engineering. Wavelets are very successfully used in signal analysis for wave form demonstration and segmentations, time-frequency analysis, medical diagnostics, geophysical signal processing, statistical analysis, pattern recognition, and fast algorithms for easy execution. This work particularly deals with the development of various wavelet methods for the solution of PDEs.

M.2 Wavelets

The term "wavelet" was derived from the French word *ondelette*, which means "small wave." An oscillatory function $\psi(x) \in L^2(\Re)$ with zero mean and compact support is a wavelet if it has the following desirable characteristics:

i. Smoothness: $\psi(x)$ is n times differentiable and its derivatives are continuous.
ii. Localization: $\psi(x)$ is well localized both in time and frequency domains, that is, $\psi(x)$, and its derivatives must decay rapidly. For frequency localization, $\hat{\Psi}(\omega)$ must decay sufficiently fast as $\omega \to \infty$ and $\hat{\Psi}(\omega)$ becomes flat in the neighborhood of $\omega = 0$.

The flatness is associated with the number of vanishing moments of $\psi(x)$, that is, $\int_{-\infty}^{\infty} x^k \psi(x)\,dx = 0$ or equivalently $\frac{d^k}{d\omega^k}\hat{\Psi}(\omega) = 0$ for $k = 0,1,\ldots,n$ in the sense that the larger the number of vanishing moments, the greater is the flatness when ω is small.

iii. The admissibility condition

$$\int_{-\infty}^{\infty} \frac{\left|\hat{\Psi}(\omega)\right|}{|\omega|} d\omega < \infty$$

suggests that $\left|\hat{\Psi}(\omega)\right|$ decay at least as $|\omega|^{-1}$ or $|x|^{\varepsilon-1}$ for $\varepsilon>0$.

Although most of the numerical methods have been successfully applied for many linear and nonlinear differential equations, they have also some drawbacks in regions where singularities or sharp transitions occur. In those cases, the solutions may be oscillating, and for accurate representation of the results, adaptive numerical schemes must be used, which complicates the solution. To overcome the preceding difficulty, wavelet transform methods are quite useful.

M.3 Wavelet Transform [1]

Morlet and Grossmann [2,3] first introduced the concept of wavelets in the early 1980s. Since then, a lot of researchers were involved in the development of wavelets. Some notable contributors include Morlet and Grossmann [3] for the formulation of continuous wavelet transform (CWT), Stromberg [4] for early works on discrete wavelet transform (DWT), Meyer [5] and Mallat [6] for multiresolution analysis (MRA) using wavelet transform, and Daubechies [7] for the proposal of orthogonal compactly supported wavelets. Thereafter, a lot of work has been done both on the development and application of wavelet analysis on a wide variety of problems like signal and image processing, data condensation, and solution of differential equations.

In 1982, Jean Morlet, a French geophysical engineer, first introduced the concept of wavelets as a family of functions constructed from dilation and translation of a single function known as the "mother wavelet" $\psi(t)$. They are defined by

$$\psi_{a,b}(t) = \frac{1}{\sqrt{|a|}}\psi\left(\frac{t-b}{a}\right), \quad a,b \in \Re, \quad a \neq 0 \tag{M.1}$$

where

a is called a scaling parameter that measures the degree of compression or scale
b is a translation or shifting parameter that determines the location of the wavelet

If $|a|<1$, the wavelet (M.1) is the compressed version of the mother wavelet and corresponds mainly to higher frequencies. On the other hand, when $|a|>1$, $\psi_{a,b}(t)$ has a larger time width than $\psi(t)$ and corresponds to lower frequencies. Thus, wavelets have time widths adapted to their frequencies, which is the main reason for the success of the Morlet wavelets in signal processing and time-frequency signal analysis. It can be noted that the resolution of wavelets at different scales varies in the time and frequency domains as governed by the Heisenberg uncertainty principle. At a large scale, the solution is coarse in the time domain and fine in the frequency domain. As the scale a decreases, the resolution in the time domain becomes finer, while that in the frequency domain becomes coarser.

The success of Morlet's numerical algorithms encouraged Grossmann, a French theoretical physicist, to make an extensive study of the Morlet wavelet transform, which led to the recognition that wavelets $\psi_{a,b}(t)$ correspond to a square integrable representation of the affine group. Grossmann was concerned with the wavelet transform of $f \in L^2(\Re)$ defined by

$$\mathcal{W}_{\psi}\left[f\right](a,b) = \left(f, \psi_{a,b}\right) = \frac{1}{\sqrt{|a|}} \int_{-\infty}^{\infty} f(t) \overline{\psi\left(\frac{t-b}{a}\right)} dt, \tag{M.2}$$

where $\psi_{a,b}(t)$ plays the same role as the kernel $e^{i\omega t}$ in the Fourier transform. The CWT $\mathcal{W}_{\psi}$ is linear. The inverse wavelet transform can be defined so that f can be reconstructed by means of the formula

$$f(t) = C_{\psi}^{-1} \int_{-\infty}^{\infty} \int_{-\infty}^{\infty} \mathcal{W}_{\psi}\left[f\right](a,b) \psi_{a,b}(t) \left(a^{-2} da\right) db, \tag{M.3}$$

provided C_{ψ} satisfies the so-called admissibility condition, that is,

$$C_{\psi} = 2\pi \int_{-\infty}^{\infty} \frac{\left|\hat{\Psi}(\omega)\right|^2}{|\omega|} d\omega < \infty, \tag{M.4}$$

where $\hat{\Psi}(\omega)$ is the Fourier transform of the mother wavelet $\psi(t)$.

Grossmann's ingenious work revealed that certain algorithms that decompose a signal on the whole family of scales can be utilized as an efficient tool for multiscale analysis. In practical applications, the continuous wavelet can be computed at discrete grid points. For this, a general wavelet, ψ, can be defined by replacing a with a_0^m $(a_0 \neq 0,1)$, b with $nb_0a_0^m$ $(b_0 \neq 0)$, where m and n are integers, making

$$\psi_{m,n}(t) = a_0^{-m/2} \psi\left(a_0^{-m} t - nb_0\right). \tag{M.5}$$

The DWT of f is defined as

$$\bar{f}(m,n) = \mathcal{W}\left[f\right](m,n) = \left(f, \psi_{m,n}\right) = \int_{-\infty}^{\infty} f(t) \bar{\psi}_{m,n}(t) dt \tag{M.6}$$

where $\psi_{m,n}(t)$ is given in Equation M.5.

The series

$$\sum_{m,n=-\infty}^{\infty} \bar{f}(m,n) \psi_{m,n}(t) \tag{M.7}$$

is called the wavelet series (WS) of f, and the functions $\{\psi_{m,n}(t)\}$ are called the discrete wavelets or simply wavelets. To compute the wavelet transform of a function at some point in the time-scale plane, we do not need to know the function values for the entire time axis. All we need is the function at those values of time at which the wavelet is nonzero. Consequently, the evaluation of the wavelet transform can be done almost in real time.

It is known that the CWT is a two-parameter representation of a function. In many practical applications, particularly in signal processing, data are represented by a finite number of values, so it is essential and often expedient to consider the discrete version of CWT. From a mathematical perspective, a continuous representation of a function of two continuous parameters a, b in Equation M.2 can be converted into a discrete one by assuming that a and b take only integral values as given in Equation M.5. In general, the function f belonging to the Hilbert space, $L^2(\mathfrak{R})$ can be completely determined by its DWT if the wavelets form a complete system in $L^2(\mathfrak{R})$. In other words, if the wavelets form an orthonormal basis of $L^2(\mathfrak{R})$, then they are complete and f can be reconstructed from its DWT $\left\{\bar{f}(m,n)=(f,\psi_{m,n})\right\}$ by means of the formula

$$f(t)=\sum_{m,n=-\infty}^{\infty}(f,\psi_{m,n})\psi_{m,n}(t), \tag{M.8}$$

provided the wavelets form an orthonormal basis.

Alternatively, the function f can be determined by the formula

$$f(t)=\sum_{m,n=-\infty}^{\infty}(f,\psi_{m,n})\bar{\psi}_{m,n}(t), \tag{M.9}$$

provided the wavelets form a basis and $\{\bar{\psi}_{m,n}(t)\}$ is the dual basis.

For some particular choice of ψ and a_0, b_0, the $\psi_{m,n}$ constitute an orthonormal basis for $L^2(\mathfrak{R})$. If $a_0=2$ and $b_0=1$, then there exists a function, ψ, with good time-frequency localization properties such that

$$\psi_{m,n}(t)=2^{-m/2}\psi\left(2^{-m}t-n\right) \tag{M.10}$$

form an orthonormal basis for $L^2(\mathfrak{R})$. These $\{\psi_{m,n}(t)\}$ are known as the Littlewood–Paley wavelets. They have a good space-frequency localization, given in the following representation of f as

$$f(t)=\sum_{m,n}(f,\psi_{m,n})\psi_{m,n}(t). \tag{M.11}$$

An orthodox example of a wavelet, ψ, for which $\psi_{m,n}$ constitute an orthonormal basis of $L^2(\mathfrak{R})$ is the Haar wavelet

$$\psi(t)=\begin{cases}1, & 0\le t<\frac{1}{2}\\ -1, & \frac{1}{2}\le t<1\ .\\ 0, & \text{otherwise}\end{cases} \tag{M.12}$$

Historically, the Haar basis is the first orthonormal wavelet basis that was invented long before the concept of wavelet was introduced. The joint venture of Morlet and Grossmann led to a detailed mathematical study of the wavelet transforms, their properties, and applications.

M.3.1 Properties of CWTs

Let $\psi, \varphi \in L^2(\mathfrak{R})$ be wavelets and g and h be functions that belong to $L^2(\mathfrak{R})$, then

i. Linearity property

$$\mathcal{W}_{\psi}\left(k_1 g + k_2 h\right)(\xi,\eta) = k_1\left(\mathcal{W}_{\psi} g\right)(\xi,\eta) + k_2\left(\mathcal{W}_{\psi} h\right)(\xi,\eta)$$

where k_1 and k_2 are any two scalars

ii. Translation property

$$\left(\mathcal{W}_{\psi}\left(T_c g\right)\right)(\xi,\eta) = \left(\mathcal{W}_{\psi} g\right)(\xi,\eta - c)$$

where T_c denotes the translation operator defined by $T_c f(t) = f(t-c)$

iii. Dilation property

$$\left(\mathcal{W}_{\psi}\left(D_c g\right)\right)(\xi,\eta) = \frac{1}{\sqrt{c}}\left(\mathcal{W}_{\psi} g\right)\left(\frac{\xi}{c}, \frac{\eta}{c}\right), \quad c > 0$$

where D_c is a dilation operator defined by $D_c g(t) = \frac{1}{c} g\left(\frac{t}{c}\right), \quad c > 0$

iv. Symmetry property

$$\left(\mathcal{W}_{\psi} g\right)(\xi,\eta) = \overline{\left(\mathcal{W}_{g} \psi\right)\left(\frac{1}{\xi}, -\frac{b}{\eta}\right)}, \quad \xi \neq 0$$

v. Parity property

$$\left(\mathcal{W}_{P\psi} P g\right)(\xi,\eta) = \left(\mathcal{W}_{\psi} g\right)(\xi, -\eta)$$

where P is the parity operator defined by $Pg(t) = g(-t)$

vi. Antilinearity property

$$\left(\mathcal{W}_{k\psi + l\phi} g\right)(\xi,\eta) = \bar{k}\left(\mathcal{W}_{\psi} g\right)(\xi,\eta) + \bar{l}\left(\mathcal{W}_{\phi} g\right)(\xi,\eta)$$

for any scalars k and l

vii. $\left(\mathcal{W}_{T_c \psi} g\right)(\xi,\eta) = \left(\mathcal{W}_{\psi} g\right)(\xi, \eta + c\xi)$

viii. $\left(\mathcal{W}_{D_c \psi} g\right)(\xi,\eta) = \frac{1}{\sqrt{c}}\left(\mathcal{W}_{\psi} g\right)(c\xi, \eta), \quad c > 0$

Hence, the integral wavelet transform (IWT) is defined to be the convolution with respect to the dilation of the reflection of some function, called a "basic wavelet", while the WS is expressed in terms of a single function, called a "wavelet" by means of two very simple operations: binary dilations and integral translations. In wavelet analysis, WS and IWT are intimately related. The IWT of a function on the real line evaluated at certain points in the time-scale domain gives the coefficients for its WS representation. Wavelet techniques enable us to divide a complicated function into several simpler ones and study them separately. This property, along with fast wavelet algorithm, makes these techniques very attractive for analysis and synthesis. Unlike Fourier-based analyses that use global (nonlocal) sine and cosine functions as bases, wavelet analysis uses bases that are localized in time and frequency to more effectively represent nonstationary signals. As a result, a wavelet representation is much more compact and easier for implementation. Using the powerful MRA, one can represent a function by a finite sum of components at different resolutions so that each component can be adaptively processed based on the objectives of the application. This capability of representing functions compactly and in several levels of resolution is the major strength of the wavelet analysis.

M.4 Orthonormal Wavelets

The orthonormal wavelets with good time-frequency localization are found to play a significant role in wavelet theory and have a great variety of applications. In general, the theory of wavelets begins with a single function $\psi \in L^2(\Re)$, and a family of functions, $\psi_{m,n}$, is constructed from this single function ψ by the operation of binary dilation (i.e., dilation by 2^m) and dyadic translation of $n2^{-m}$ so that

$$\psi_{m,n}(t) = 2^{\frac{m}{2}} \psi\left(2^m\left(t - \frac{n}{2^m}\right)\right), \quad m,n \in Z$$

$$= 2^{\frac{m}{2}} \psi\left(2^m t - n\right), \tag{M.13}$$

where the factor $2^{\frac{m}{2}}$ is introduced to ensure orthonormality.

Definition of Orthonormal Wavelet

A wavelet $\psi \in L^2(\Re)$ is called an orthonormal if the family of functions $\psi_{m,n}$, generated from ψ, is an orthonormal basis of $L^2(\Re)$, that is,

$$\langle \psi_{i,j}, \psi_{m,n} \rangle = \delta_{i,m}\delta_{j,n}, \quad i,j,m,n \in Z.$$

Definition of Semi-Orthogonal Wavelet

A wavelet $\psi \in L^2(\Re)$ is called an semi-orthogonal wavelet [8] if the family $\{\psi_{m,n}\}$ satisfies the following condition:

$$\langle \psi_{i,j}, \psi_{m,n} \rangle = 0, \quad i \neq m, \quad i,j,m,n \in Z.$$

Obviously, every semi-orthogonal wavelet generates an orthogonal decomposition of $L^2(\mathfrak{R})$, and every orthonormal wavelet is also a semi-orthogonal wavelet.

M.4.1 Construction of Orthonormal Wavelets

The construction of an orthonormal wavelet, namely, the Haar wavelet, is discussed by using the properties of scaling functions and filters. The scaling function (ϕ) satisfies the dilation equation as

$$\phi(t) = \sqrt{2}\sum_{n=-\infty}^{\infty} c_n\phi(2t-n) \tag{M.14}$$

where the coefficients c_n are given by

$$c_n = \sqrt{2}\int_{-\infty}^{\infty}\phi(t)\phi(2t-n)\,dt. \tag{M.15}$$

Evaluating the integral given in Equation M.15 with $\phi = \chi_{[0,1)}$ gives c_n as follows:

$$c_0 = c_1 = \frac{1}{\sqrt{2}} \quad \text{and} \quad c_n = 0 \quad \text{for } n \neq 0,1.$$

Here $\chi_{[0,1)}$ denotes the characteristic function given by

$$\chi_{[0,1)}(t) = \begin{cases} 1, & 0 \le t < 1 \\ 0, & \text{otherwise.} \end{cases}$$

Hence, the dilation equation becomes

$$\phi(t) = \phi(2t) + \phi(2t-1). \tag{M.16}$$

This means that $\phi(t)$ is a linear combination of the even and odd translates of $\phi(2t)$ and satisfies a very simple two-scale relation (M.16), which is shown in Figure M.1.

Thus, the Haar mother wavelet is obtained as a simple two-scale relation.

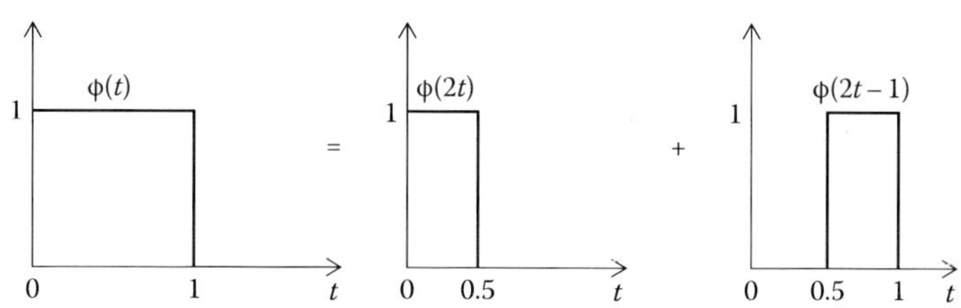

FIGURE M.1
Two-scale relation of $\phi(t) = \phi(2t) + \phi(2t-1)$.

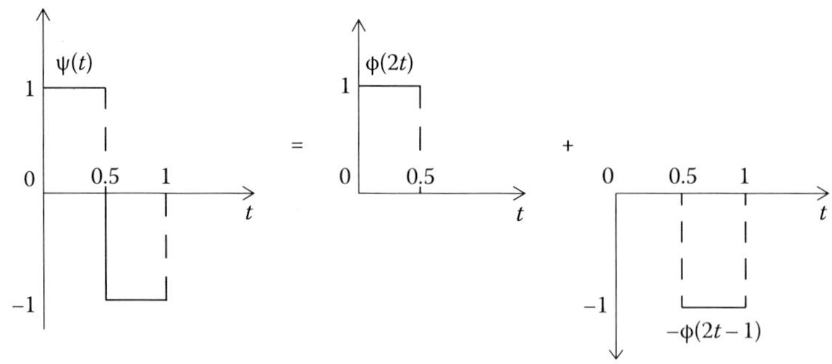

FIGURE M.2
Two-scale relation of $\psi(t) = \phi(2t) - \phi(2t-1)$.

$$\psi(t) = \phi(2t) - \phi(2t-1) \tag{M.17}$$

$$= \chi_{[0,0.5)} - \chi_{[0.5,1)}$$

$$= \begin{cases} 1, & 0 \le t < \dfrac{1}{2} \\ -1, & \dfrac{1}{2} \le t < 1 \\ 0, & \text{otherwise} \end{cases} \tag{M.18}$$

This two-scale relation (M.17) of ψ is represented in Figure M.2.

M.5 Multiresolution Analysis

In 1989, Stephane Mallat and Yves Meyer introduced the idea of multiresolution analysis (MRA) [8]. The fundamental idea of MRA is to represent a function as a limit of successive approximations, each of which is a "smoother" version of the original function. The successive approximations correspond to different resolutions, which leads to the name MRA as a formal approach to construct orthogonal wavelet bases utilizing a definite set of rules. It also provides the existence of so-called scaling functions and scaling filters that are then used for the construction of wavelets and fast numerical algorithms. In applications, it is an effective mathematical framework for hierarchical decomposition of a signal or an image into components of different scales represented by a sequence of function spaces on $\mathfrak{R}$.

Any wavelet, orthogonal or semi-orthogonal, generates a direct sum decomposition of $L^2(\mathfrak{R})$. For each $j \in Z$, let us consider the closed subspaces

$$V_j = \ldots \oplus W_{j-2} \oplus W_{j-1}, \quad j \in Z,$$

of $L^2(\mathfrak{R})$. A set of subspaces $\{V_j\}_{j \in Z}$ is said to be MRA of $L^2(\mathfrak{R})$ if it possesses the following properties:

1. $V_j \subset V_{j+1}, \quad \forall j \in Z$
2. $\bigcup_{j \in Z} V_j$ is dense in $L^2(\mathfrak{R})$

3. $\bigcap_{j \in Z} V_j = \{0\}$
4. $V_{j+1} = V_j \oplus W_j$
5. $f(t) \in V_j \Leftrightarrow f(2t) \in V_{j+1}, \quad \forall j \in Z$

Properties (2)–(5) state that $\{V_j\}_{j \in Z}$ is a nested sequence of subspaces that effectively covers $L^2(\mathfrak{R})$. That is, every square integrable function can be approximated as closely as desired by a function that belongs to at least one of the subspaces V_j. A function $\varphi \in L^2(\mathfrak{R})$ is called a scaling function if it generates the nested sequence of subspaces V_j and satisfies the dilation equation, namely,

$$\varphi(t) = \sum_k p_k \varphi(at - k), \tag{M.19}$$

with $p_k \in l^2$ and a being any rational number.

For each scale j, since $V_j \subset V_{j+1}$, there exists a unique orthogonal complementary subspace W_j of V_j in V_{j+1}. This subspace W_j is called wavelet subspace and is generated by $\psi_{j,k} = \psi(2^j t - k)$, where $\psi \in L^2$ is called the wavelet. From the preceding discussion, these results follow easily:

- $V_{j_1} \cap V_{j_2} = V_{j_2}, \quad j_1 > j_2$
- $W_{j_1} \cap W_{j_2} = 0, \quad j_1 \neq j_2$
- $V_{j_1} \cap W_{j_2} = 0, \quad j_1 < j_2$

Mathematically, the fundamental idea of MRA is to represent a function or signal as a limit of successive approximations, each of which is a finer version of the function. These successive approximations correspond to different levels of resolutions. Thus, MRA is a formal approach to constructing orthonormal wavelet bases using a definite set of rules and procedures. The key feature of this analysis is to describe mathematically the process of studying signals or images at different time scales. From the point of view of practical applications, MRA is a very effective mathematical framework for hierarchical decomposition of an image or signal into components of different scales or frequencies.

In recent years, there have been many developments and new applications of wavelet analysis for describing complex algebraic functions and analyzing empirical continuous data obtained from many kinds of signals at different scales of resolutions. The wavelet-based approximations of ordinary differential equations and PDEs, have been attracting attention since the contribution of orthonormal bases of compactly supported wavelet by Daubechies and MRA-based fast wavelet transform algorithm by Beylkin et al. [9] gained momentum.

In order to solve PDEs by numerical methods, the unknown solution can be represented by wavelets of different resolutions, resulting in a multigrid representation. The dense matrix resulting from an integral operator can be sparsified using wavelet-based thresholding techniques to attain an arbitrary degree of solution accuracy. The main feature of wavelets is their ability to convert the given differential and integral equations to a system of linear or nonlinear algebraic equations that can be solved by numerical methods. The goal of this chapter is to convey a general idea about wavelets.

1

Numerous Analytical and Numerical Methods

1.1 Introduction

The purpose of this chapter is to deliver a brief description of various analytical and numerical methods, namely, variational iteration method (VIM), first integral method (FIM), homotopy perturbation method (HPM), homotopy analysis method (HAM), optimal homotopy asymptotic method (OHAM), Haar wavelet method, Legendre wavelet method (LWM), Chebyshev wavelet method (CWM), Hermite wavelet method (HWM), and Petrov–Galerkin method.

The entire chapter can be divided into two parts. In the first part, the basic ideas of some well-known analytical techniques such as VIM, FIM, HPM, HAM, and OHAM have been discussed, whereas the second part is devoted to study the elemental concept of various methods based on wavelet functions. The applicability of these proposed methods have been examined for solving nonlinear partial differential equations (PDEs) and fractional partial differential equations (FPDEs). Our goal is to encourage the reader to appreciate the beauty as well as the effectiveness of these analytical and numerical techniques in the study of nonlinear physical phenomena.

1.2 Part I: Fundamental Idea of Various Analytical Methods

1.2.1 Variational Iteration Method (VIM)

The concept of VIM was first developed by Ji-Huan He in the year 1997 [10]. The method has been favorably applied to various kinds of nonlinear problems by many researchers in a variety of scientific fields. The key advantage of the method is its flexibility and potential to solve nonlinear equations accurately and conveniently. The method gives the solution in the form of rapidly convergent successive approximations that may give the exact solution if such a solution exists. The method has been proved by many authors to be reliable and efficient for a variety of scientific applications, linear and nonlinear equations as well.

To illustrate the elemental concept of VIM [11–13], we consider the general differential equation as follows:

$$Lu + Nu = g(x,t) \tag{1.1}$$

where

L is a linear operator
N is a nonlinear operator
$g(x,t)$ is a known analytical function

According to the VIM, a correction functional can be constructed as follows:

$$u_{n+1}(x,t) = u_n(x,t) + \int_0^t \lambda\left(Lu_n(x,\xi) + N\tilde{u}_n(x,\xi) - g(x,\xi)\right)d\xi, \quad n \geq 0 \tag{1.2}$$

where

λ is a general Lagrange multiplier that can be identified optimally by the variational theory
Subscript n denotes the nth order approximation
$\tilde{u}_n$ is regarded as a restricted variation, that is, $\delta\tilde{u}_n = 0$

The Lagrange multiplier λ can be determined from the stationary condition of the correction functional $\delta u_{n+1} = 0$.

The main advantages of this method are as follows:

1. The correction functions can be constructed easily by the general Lagrange multipliers that can be optimally determined by the variational theory. The application of restricted variations in correction functional makes it much easier to determine the multiplier.
2. The initial approximation can be freely selected with possible unknown constants that can be identified by various methods.
3. The approximations acquired with the aid of this method are valid not only for a small parameter but also for a very large parameter.

Being different from the other analytical methods, such as perturbation methods, this method does not depend on small or large parameters, it could possibly in finding wide application in nonlinear problems without linearization, discretization or small perturbations.

1.2.2 First Integral Method

The FIM is one of the powerful mathematical techniques for finding exact solutions of partial as well as fractional differential equations. This method is based on the ring theory of commutative algebra. It was first proposed by Feng [14] and was further developed by the same author in [15,16]. Many authors have used FIM effectively to establish the exact solutions of various nonlinear PDEs and FPDEs arising in mathematical physics [17–19]. In order to apply the FIM over FPDEs, the fractional differential equations may be transformed into classical ordinary differential equations (ODEs) through fractional complex transform with the help of local fractional derivative. The section below demonstrates the fundamental concept of FIM via algorithm.

1.2.2.1 Algorithm of FIM

In this section, the fundamental concept of FIM has been established. The main steps of this proposed method are described as follows:

Step 1: Consider the following general nonlinear fractional order PDE:

$$F\left(u, u_x, D_t^{\alpha} u, u_{xx}, D_t^{\alpha} u_{xx}, u_{xxx}, \ldots\right) = 0, \quad 0 < \alpha \leq 1 \tag{1.3}$$

where $u = u(x,t)$ is an unknown function, $D_t^{\alpha} u$ is a local fractional derivative of u, and F is a polynomial in u and its numerous partial derivatives in which the highest-order derivatives and nonlinear terms are involved

Step 2: Using the fractional complex transform [20,21]

$$u(x,t) = \Psi(\xi), \quad \xi = kx - \frac{\lambda t^{\alpha}}{\Gamma(\alpha+1)} \tag{1.4}$$

where k and λ are constants, the FPDE (1.3) is transformed to a nonlinear ODE for $u(x,t) = \Psi(\xi)$ of the following form:

$$F\left(\Psi, k\Psi_{\xi}, \lambda\Psi_{\xi}, k^2\Psi_{\xi\xi}, \lambda k^2\Psi_{\xi\xi\xi}, k^3\Psi_{\xi\xi\xi}, \ldots\right) = 0. \tag{1.5}$$

Step 3: Suppose Equation 1.5 has a solution of the form

$$\Psi(\xi) = X(\xi), \tag{1.6}$$

and introducing a new variable $Y(\xi) = \Psi_{\xi}(\zeta)$ leads to a system of ODEs of the form

$$\frac{dX(\xi)}{d\xi} = Y(\xi), \quad \frac{dY(\xi)}{d\xi} = H\left(X(\xi), Y(\xi)\right). \tag{1.7}$$

In general, it is quite difficult to solve a two-dimensional autonomous planar system of ODEs, such as Equation 1.7.

Step 4: Utilizing the qualitative notion of differential equations [22], if we are able to evaluate the integrals to Equation 1.7 under the same conditions, then the general solutions to Equation 1.7 may also be derived directly. With the aid of the division theorem for two variables in the complex domain C, which is based on the Hilbert-Nullstellensatz theorem [23], one can obtain the first integral to Equation 1.7. This first integral can reduce Equation 1.5 to a first-order integrable ODE. Then by solving this equation directly, the exact solution to Equation 1.3 is obtained.

Theorem 1.1: Division Theorem

Let $Q(x,y)$ and $R(x,y)$ be polynomials in $C[x,y]$, and $Q(x,y)$ is irreducible in $C[x,y]$. If $R(x,y)$ vanishes at all zero points of $Q(x,y)$, then there exists a polynomial $H(x,y)$ in $C[x,y]$ such that

$$R(x,y)=Q(x,y)H(x,y). \tag{1.8}$$

The division theorem follows immediately from the Hilbert-Nullstellensatz theorem from the ring theory of commutative algebra [23,24]. The elementary idea of this procedure is to construct a first integral with polynomial coefficients of an explicit form to an equivalent autonomous planar system by utilizing the division theorem.

1.2.3 Homotopy Perturbation Method (HPM)

In the last three decades with the rapid development of nonlinear sciences, there has appeared increasing interest of scientists and engineers in the analytical techniques for nonlinear problems. Various perturbation methods have been widely applied to solve nonlinear problems in science and engineering. But, most of the perturbation techniques require the existence of a small parameter in the equation. An unsuitable choice of such parameter would lead to very bad results. The solutions obtained through perturbation methods can be valid only when a small value of the parameter is used. Hence, it is necessary to check the validity of the approximations through numerical processes.

In contrast to the traditional perturbation methods, the HPM technique is independent of small or large physical parameters and provides a simple way to ensure the convergence of solution.

To illustrate the basic ideas of HPM [25], we consider the following nonlinear differential equation:

$$\mathcal{A}(u)-f(\boldsymbol{r})=0, \quad \boldsymbol{r}\in\Omega \tag{1.9}$$

with the boundary conditions

$$B\left(u,\frac{\partial u}{\partial n}\right)=0, \quad \boldsymbol{r}\in\Gamma \tag{1.10}$$

where

- $\mathcal{A}$ is a general differential operator
- B is a boundary operator
- $f(\boldsymbol{r})$ is a known analytic function
- Γ is the boundary of the domain Ω

The operator $\mathcal{A}$ can be divided into two parts: linear $\mathcal{L}$ and nonlinear $\mathcal{N}$. Therefore, Equation 1.9 can be rewritten as follows:

$$\mathcal{L}(u)+\mathcal{N}(u)-f(\boldsymbol{r})=0. \tag{1.11}$$

We construct a homotopy $v(\boldsymbol{r},p)$ of Equation 1.9 as follows $v(\boldsymbol{r},p):\Omega\times[0,1]\rightarrow\mathfrak{R}$, which satisfies [26]

$$H(v,p)\equiv(1-p)\left[\mathcal{L}(v)-\mathcal{L}(u_0)\right]+p\left[\mathcal{A}(v)-f(\boldsymbol{r})\right]=0, \tag{1.12}$$

or

$$H(v,p)\equiv\mathcal{L}(v)-\mathcal{L}(u_0)+p\mathcal{L}(u_0)+p\left[\mathcal{N}(v)-f(\boldsymbol{r})\right]=0 \tag{1.13}$$

where
$p\in[0,1]$ is an embedding parameter
u_0 is an initial approximation of Equation 1.9, which satisfies the boundary conditions

It follows from (1.12) and (1.13) that

$$H(v,0)=\mathcal{L}(v)-\mathcal{L}(u_0)=0 \tag{1.14}$$

or

$$H(v,1)=\mathcal{A}(v)-f(\boldsymbol{r})=0. \tag{1.15}$$

The changing process of p from zero to unity is just that of $v(\boldsymbol{r},p)$ from $u_0(\boldsymbol{r})$ to $u(\boldsymbol{r})$. In topology, this is called deformation, and $\mathcal{L}(v)-\mathcal{L}(u_0)$, $\mathcal{A}(v)-f(\boldsymbol{r})$ are called homotopic.

We assume that the solution of Equation 1.13 can be written as a power series in p:

$$v=v_0+pv_1+p^2v_2+\cdots. \tag{1.16}$$

The approximate solution of Equation 1.9 can be obtained by setting $p=1$:

$$u=\lim_{p\rightarrow 1}v=v_0+v_1+v_2+\cdots. \tag{1.17}$$

The series in Equation 1.17 is convergent for most cases, and the convergence rate depends upon the nonlinear operator $\mathcal{A}(v)$.

The nonlinear term $\mathcal{N}(u)$ can be expressed in He polynomials [27] as

$$\mathcal{N}(u)=\sum_{m=0}^{\infty}p^mH_m(v_0,v_1,\ldots,v_m) \tag{1.18}$$

where

$$H_m(v_0,v_1,\ldots,v_m)=\frac{1}{m!}\frac{\partial^m}{\partial p^m}\left(N\left(\sum_{k=0}^{m}p^kv_k\right)\right)\Bigg|,\quad m=0,1,2,\ldots. \tag{1.19}$$

1.2.4 Optimal Homotopy Asymptotic Method (OHAM)

Unlike perturbation methods, the OHAM is independent of small or large physical parameters and provides a simple way to ensure the convergence of solution series.

The method was first devised by Marinca et al. [28–30]. Recently, many researchers have successfully applied this method to various nonlinear problems in science and engineering. This is an effective and powerful method to find the approximate solution of nonlinear problems. The advantage of OHAM is built in convergence criteria, which is controllable. In OHAM, the control and adjustment of the convergence region are provided in a convenient way.

In HPM and OHAM, the concept of homotopy from topology and conventional perturbation technique were merged to propose a general analytic procedure for the solution of nonlinear problems. Thus, these methods are independent of the existence of a small parameter in the problem at hand and thereby overcome the limitations of conventional perturbation technique. OHAM, however, is the most generalized form of HPM as it employs a more general auxiliary function, $H(p)$, in place of HPM's p.

To illustrate the basic ideas of OHAM [31,32], we consider the following nonlinear differential equation:

$$A\big(u(x,t)\big)+g(x,t)=0,\quad x\in\Omega \tag{1.20}$$

with the boundary conditions

$$B\left(u,\frac{\partial u}{\partial t}\right)=0,\quad x\in\Gamma, \tag{1.21}$$

where

A is a differential operator
B is a boundary operator
$u(x,t)$ is an unknown function
Γ is the boundary of the domain Ω
$g(x,t)$ is a known analytic function

The operator A can be decomposed as

$$A=L+N, \tag{1.22}$$

where

L is a linear operator
N is a nonlinear operator

We construct a homotopy $\varphi(x,t;p):\Omega\times[0,1]\rightarrow\Re$ that satisfies

$$H\big(\varphi(x,t;p),p\big)\equiv(1-p)\Big[L\big(\varphi(x,t;p)\big)+g(x,t)\Big]-H(p)\Big[A\big(\varphi(x,t;p)\big)+g(x,t)\Big]=0, \tag{1.23}$$

where

$p\in[0,1]$ is an embedding parameter
$H(p)$ is a nonzero auxiliary function for $p\neq 0$ and $H(0)=0$

When $p=0$ and $p=1$, we have $\varphi(x,t;0)=u_0(x,t)$ and $\varphi(x,t;1)=u(x,t)$, respectively.

Thus as p varies from 0 to 1, the solution $\varphi(x,t;p)$ approaches from $u_0(x,t)$ to $u(x,t)$.

Here, $u_0(x,t)$ is obtained from Equations 1.23 and 1.21 with $p=0$ yields

$$L\left(\varphi\left(x,t;0\right)\right)+g\left(x,t\right)=0,\quad B\left(u_0,\frac{\partial u_0}{\partial t}\right)=0. \tag{1.24}$$

The auxiliary function $H(p)$ is chosen in the form

$$H\left(p\right)=C_1p+C_2p^2+C_3p^3+\cdots \tag{1.25}$$

where $C_1,C_2,C_3,\ldots$ are convergence control parameters to be determined. To get an approximate solution, $\tilde{\varphi}\left(x,t;C_1,C_2,C_3,\ldots\right)$ is expanded in a series about p as

$$\tilde{\varphi}\left(x,t;p,C_1,C_2,C_3,\ldots\right)=u_0\left(x,t\right)+\sum_{i=1}^{\infty}u_i\left(x,t,C_1,C_2,C_3,\ldots\right)p^i. \tag{1.26}$$

Substituting Equation 1.26 in Equation 1.23 and equating the coefficients of like powers of p, we will have the following equations:

$$L\left(u_1\left(x,t\right)+g\left(x,t\right)\right)=C_1N_0\left(u_0\left(x,t\right)\right),\quad B\left(u_1,\frac{\partial u_1}{\partial t}\right)=0, \tag{1.27}$$

$$L\left(u_2\left(x,t\right)\right)-L\left(u_1\left(x,t\right)\right)=C_2N_0\left(u_0\left(x,t\right)\right)+C_1\left(L\left(u_1\left(x,t\right)\right)+N_1\left(u_0\left(x,t\right),u_1(x,t)\right)\right),$$
$$B\left(u_2,\frac{\partial u_2}{\partial t}\right)=0. \tag{1.28}$$

And hence, the general governing equations for $u_j(x,t)$ is given by

$$\begin{aligned}L\left(u_j\left(x,t\right)\right)&=L\left(u_{j-1}\left(x,t\right)\right)+C_jN_0\left(u_0\left(x,t\right)\right)\\&+\sum_{i=1}^{j-1}C_i\left[L\left(u_{j-i}\left(x,t\right)\right)+N_{j-i}\left(u_0\left(x,t\right),\ldots,u_{j-1}\left(x,t\right)\right)\right];\quad j=2,3,\ldots\end{aligned} \tag{1.29}$$

where $N_j(u_0(x,t),\ldots,u_j(x,t))$ is the coefficient of p^j in the expansion of $N(\varphi(x,t;p))$ about the embedding parameter p and

$$N\left(\varphi\left(x,t;p,C_1,C_2,C_3,\ldots\right)\right)=N_0\left(u_0\left(x,t\right)\right)+\sum_{j=1}^{\infty}N_j\left(u_0,u_1,\ldots,u_j\right)p^j. \tag{1.30}$$

It is observed that the convergence of the series (1.26) depends upon the convergence control parameters $C_1,C_2,C_3,\ldots$.

The approximate solution of Equation 1.20 can be written in the following form:

$$\tilde{u}\left(x,t;C_1,C_2,C_3,\ldots\right)=u_0\left(x,t\right)+\sum_{j=1}^{n-1}u_j\left(x,t,C_1,C_2,C_3,\ldots\right). \tag{1.31}$$

Substituting Equation 1.31 in Equation 1.20, we get the following expression for the residual:

$$R_n\left(x,t;C_1,C_2,C_3,\ldots\right)=L\left(\tilde{u}\left(x,t;C_1,C_2,C_3,\ldots\right)\right)+N\left(\tilde{u}\left(x,t;C_1,C_2,C_3,\ldots\right)\right)+g\left(x,t\right). \quad (1.32)$$

If $R_n(x,t;C_1,C_2,C_3,\ldots)=0$, then $\tilde{u}\left(x,t;C_1,C_2,C_3,\ldots\right)$ is the exact solution. Generally, such a case does not arise for nonlinear problems. The nth-order approximate solution given by Equation 1.31 depends on the convergence control parameters $C_1,C_2,C_3,\ldots$, and these parameters can be optimally determined by various methods such as weighted residual least square method, Galerkin method, collocation method, and so on.

Case I

According to the collocation method, the optimal values of the parameters $C_1,C_2,C_3,\ldots$ can be obtained by solving the following system of equations:

$$R_n\left(x_i,t_j;C_1,C_2,C_3,\ldots,C_{k^2}\right)=0 \quad \text{for } i=1,2,\ldots,k \text{ and } j=1,2,\ldots,k. \quad (1.33)$$

Case II

According to weighted residual least square method, the optimal values of the convergence control parameters $C_1,C_2,C_3,\ldots$ can be obtained by solving the following functional:

$$J\left(C_1,C_2,C_3,\ldots,C_k\right)=\int_{t_1}^{t_2}\int_{a}^{b}R_n^2\left(x,t;C_1,C_2,C_3,\ldots,C_k\right)dx\,dt, \quad (1.34)$$

where a and b are two values depending on the given problem. The unknown parameters $C_1,C_2,C_3,\ldots$ can be identified from the conditions

$$\frac{\partial J}{\partial C_1}=\frac{\partial J}{\partial C_2}=\cdots=\frac{\partial J}{\partial C_k}=0. \quad (1.35)$$

The convergence of the nth approximate solution depends upon unknown parameters $C_1,C_2,C_3,\ldots$. When the convergence control parameters $C_1,C_2,C_3,\ldots$ are known by the preceding method, the approximate solution of (1.20) is well determined.

1.2.5 Homotopy Analysis Method (HAM)

The HAM was introduced by Liao [33] and is an effective and powerful method to find the approximate solution of nonlinear problems. To illustrate the basic ideas of HAM, we consider the following nonlinear differential equation [34,35]:

$$\mathcal{N}\left[u\left(x,t\right)\right]=0, \quad (1.36)$$

where

$\mathcal{N}$ is a nonlinear operator
x and t denote the independent variables
$u(x,t)$ is an unknown function

For simplicity, we ignore all boundary or initial conditions, which can be treated in a similar way. By means of generalizing HAM, we first construct the zeroth-order deformation equation as follows:

$$(1-p)\mathscr{L}\left[\phi(x,t;p)-u_0(x,t)\right]=p\hbar H(x,t)\mathcal{N}\left[\phi(x,t;p)\right], \tag{1.37}$$

where
$p\in[0,1]$ is the embedding parameter
$\hbar\neq 0$ is an auxiliary parameter
$\mathscr{L}$ is an auxiliary linear operator
$\phi(x,t;p)$ is an unknown function
$u_0(x,t)$ is an initial guess of $u(x,t)$
$H(x,t)$ is a nonzero auxiliary function

For $p=0$ and $p=1$, the zeroth-order deformation equation given by Equation 1.37 leads to

$$\phi(x,t;0)=u_0(x,t) \quad \text{and} \quad \phi(x,t;1)=u(x,t). \tag{1.38}$$

Thus as p increases from 0 to 1, the solution $\phi(x,t;p)$ varies from the initial guess $u_0(x,t)$ to the solution $u(x,t)$. Expanding $\phi(x,t;p)$ in Taylor's series with respect to the embedding parameter p, we have

$$\phi(x,t;p)=u_0(x,t)+\sum_{m=1}^{\infty}p^m u_m(x,t), \tag{1.39}$$

where

$$u_m(x,t)=\frac{1}{m!}\frac{\partial^m\phi(x,t;p)}{\partial p^m}\bigg|_{p=0}. \tag{1.40}$$

The convergence of the series (1.39) depends upon the auxiliary parameter $\hbar$. If it is convergent at $p=1$, we have

$$u(x,t)=u_0(x,t)+\sum_{m=1}^{\infty}u_m(x,t), \tag{1.41}$$

which must be one of the solutions of the original nonlinear differential equation. Differentiating the zeroth-order deformation Equation 1.37 for m-times with respect to p then dividing them by $m!$ and finally setting $p=0$, we obtain the following mth order deformation equation:

$$\mathscr{L}\left[u_m(x,t)-\chi_m u_{m-1}(x,t)\right]=\hbar H(x,t)\Re_m\left(u_0,u_1,\ldots,u_{m-1}\right), \tag{1.42}$$

where

$$\Re_m\left(u_0,u_1,\ldots,u_{m-1}\right)=\frac{1}{(m-1)!}\frac{\partial^{m-1}\mathcal{N}\left[\phi(x,t,p)\right]}{\partial p^{m-1}}\bigg|_{p=0} \tag{1.43}$$

and

$$\chi_m = \begin{cases} 1, & m > 1 \\ 0, & m \le 1. \end{cases} \tag{1.44}$$

Now, the solution for mth order deformation Equation 1.42 can be obtained by applying $\mathcal{L}^{-1}$ on both sides, we get

$$u_m(x,t) = \chi_m u_{m-1}(x,t) + \mathcal{L}^{-1}\left[\hbar H(x,t)\Re_m(u_0,u_1,\ldots,u_{m-1})\right]. \tag{1.45}$$

In this way, it is easy to obtain $u_m(x, t)$ for $m > 1$, and at Mth order we have

$$u(x,t) = \sum_{m=0}^{M} u_m(x,t). \tag{1.46}$$

When $M \to \infty$, we obtain an accurate approximation of the original Equation 1.36.

1.3 Part II: Fundamental Idea of Various Numerical Methods

"Wavelets" has been a very popular topic of conversations in many scientific and engineering gatherings these days. The subject of wavelet analysis has recently drawn a great deal of attention from mathematical scientists in various disciplines. The integral wavelet transform (IWT) is defined to be the convolution with respect to the dilation of the reflection of some function called a "basic wavelet," while the wavelet series (WS) is expressed in terms of a single function called a "wavelet" by means of two very simple operations: binary dilations and integral translations. Analogous to Fourier analysis, there are WSs and IWTs. In wavelet analysis, WS and IWT are intimately related. The IWT of a function on the real line evaluated at certain points in the time-scale domain gives the coefficients for its WS representation. As the polynomial spline functions are the simplest functions for both computational and implementation purposes, they are most attractive for analyzing and constructing wavelets.

Some view wavelets as a new basis for representing functions, some consider it as a technique for time-frequency analysis, and others think of it as a new mathematical subject. Of course, all of them are right, since "wavelets" are a versatile tool with very rich mathematical content and great potential for applications. However, as this subject is still in the midst of rapid development, it is definitely too early to give a unified presentation.

Wavelets are very effectively used in signal analysis for wave form demonstration and segmentations, time-frequency analysis, medical diagnostics, geophysical signal processing, statistical analysis, pattern recognition, and fast algorithms for easy execution. The wavelet analysis could be a promising tool for solving various difficulties in physics, engineering, and image processing [1]. The wavelet method is an exciting method for solving difficult problems in mathematics, physics, and engineering, with modern applications in diverse fields such as wave propagation, data compression, image processing, pattern recognition, computer graphics, detection of aircraft and submarines, and improvement in

CAT scans and other medical technologies. Also, wavelet methods have been used to develop accurate and fast algorithms for solving integral and differential equations of fractional order, especially those whose solutions are highly localized in position and scale. While wavelets have gained popularity in these areas, new applications are continually being investigated.

Wavelet techniques enable us to divide a complicated function into several simpler ones and study them separately. This property, along with fast wavelet algorithm, makes these techniques very attractive for analysis and synthesis. Unlike Fourier-based analyses that use global (nonlocal) sine and cosine functions as bases, wavelet analysis uses bases that are localized in time and frequency to more effectively represent nonstationary signals. As a result, a wavelet representation is much more compact and easier for implementation. Using the powerful multiresolution analysis, one can represent a function by a finite sum of components at different resolutions so that each component can be adaptively processed based on the objectives of the application. This capability of representing functions compactly and in several levels of resolutions is the major strength of the wavelet analysis.

In the case of solving PDEs by numerical methods, the unknown solution can be represented by wavelets of different resolutions, resulting in a multigrid representation. The dense matrix resulting from an integral operator can be sparsified using wavelet-based thresholding techniques to attain an arbitrary degree of solution accuracy. Wavelets allow accurate depiction of a variety of functions and operators. The main feature of wavelets is their ability to convert the given differential and integral equations to a system of linear or nonlinear algebraic equations that can be solved by numerical methods. The goal of this chapter is to convey a general idea about wavelets and to describe different wavelet methods in detail.

Orthogonal functions and polynomial series have received considerable attention in dealing with various problems of dynamic systems. The main characteristic of this technique is that it reduces these problems to those of solving a system of algebraic equations, thus greatly simplifying the problem. Special attention has been given to applications of Haar wavelets, Legendre wavelets, Chebyshev wavelets, and Hermite wavelets.

In the present chapter, we introduce different wavelet-based methods, namely, the Haar wavelet method, the LWM, the CWM, and the HWM. These methods are studied in detail in subsequent chapters.

1.3.1 Haar Wavelets and the Operational Matrices

In 1982, Morlet [8] first introduced the idea of wavelets as a family of functions constructed from dilation and translation of a single function called the "mother wavelet." Haar wavelet functions have been used from 1910 and were introduced by the Hungarian mathematician Alfred Haar [36]. Haar wavelets (which are Daubechies wavelets of order 1) consist of piecewise constant functions on the real line that can take only three values, that is, 0, 1, and −1, and are therefore the simplest orthonormal wavelets with a compact support. Haar wavelet method is used for the following features: simpler and faster, flexible, convenient, small computational costs, and computationally attractive. The Haar functions are a family of switched rectangular wave forms where amplitudes can differ from one function to another.

The Haar wavelet family for $x \in [0,1)$ is defined as follows [37]:

$$h_i(x) = \begin{cases} 1 & x \in [\xi_1, \xi_2) \\ -1 & x \in [\xi_2, \xi_3) \\ 0 & \text{elsewhere} \end{cases} \tag{1.47}$$

where

$$\xi_1 = \frac{k}{m}$$
$$\xi_2 = \frac{k+0.5}{m}$$
$$\xi_3 = \frac{k+1}{m}.$$

In these formulae, integer $m=2^j$ and $j=0,1,2,\ldots$, where j indicates the level of the wavelet; $k=0,1,2,\ldots,m-1$ is the translation parameter. Maximum level of resolution is J. The index i is calculated from the formula $i=m+k+1$; in the case of minimal values $m=1$, $k=0$, we have $i=2$. The maximal value of $i=2M=2^{J+1}$. It is assumed that the value $i=1$ corresponds to the scaling function for which

$$h_i(x)=\begin{cases}1 & \text{for } x\in[0,1)\\ 0 & \text{elsewhere.}\end{cases} \tag{1.48}$$

In the following analysis, integrals of the wavelets are defined as

$$p_i(x)=\int_0^x h_i(x)dx,\quad q_i(x)=\int_0^x p_i(x)dx,\quad r_i(x)=\int_0^x q_i(x)dx.$$

This can be done with the aid of (1.47)

$$p_i(x)=\begin{cases}x-\xi_1 & \text{for } x\in[\xi_1,\xi_2)\\ \xi_3-x & \text{for } x\in[\xi_2,\xi_3)\\ 0 & \text{elsewhere}\end{cases} \tag{1.49}$$

$$q_i(x)=\begin{cases}0 & \text{for } x\in[0,\xi_1)\\ \frac{1}{2}(x-\xi_1)^2 & \text{for } x\in[\xi_1,\xi_2)\\ \frac{1}{4m^2}-\frac{1}{2}(\xi_3-x)^2 & \text{for } x\in[\xi_2,\xi_3)\\ \frac{1}{4m^2} & \text{for } x\in[\xi_3,1]\end{cases} \tag{1.50}$$

$$r_i(x)=\begin{cases}\frac{1}{6}(x-\xi_1)^3 & \text{for } x\in[\xi_1,\xi_2)\\ \frac{1}{4m^2}(x-\xi_2)+\frac{1}{6}(\xi_3-x)^3 & \text{for } x\in[\xi_2,\xi_3)\\ \frac{1}{4m^2}(x-\xi_2) & \text{for } x\in[\xi_3,1)\\ 0 & \text{elsewhere.}\end{cases} \tag{1.51}$$

The collocation points are defined as

$$x_l = \frac{l-0.5}{2M}, \quad l = 1,2,\ldots,2M.$$

It is expedient to introduce the $2M \times 2M$ matrices $\boldsymbol{H}$, $\boldsymbol{P}$, $\boldsymbol{Q}$, and $\boldsymbol{R}$ with the elements $H(i,l) = h_i(x_l)$, $P(i,l) = p_i(x_l)$, $Q(i,l) = q_i(x_l)$, and $R(i,l) = r_i(x_l)$, respectively.

In 2012, the generalized Haar wavelet operational matrix of integration has been derived by the learned researcher Saha Ray [38]. Usually, the Haar wavelets are defined for the interval $t \in [0,1)$, but in the general case $t \in [A,B]$, we divide the interval $[A, B]$ into m equal subintervals, each of width $\Delta t = (B-A)/m$. In this case, the orthogonal set of Haar functions are defined in the interval $[A, B]$ by [38]

$$h_0(t) = \begin{cases} 1 & t \in [A,B] \\ 0 & \text{elsewhere} \end{cases}, \tag{1.52}$$

and

$$h_i(t) = \begin{cases} 1, & \zeta_1(i) \le t < \zeta_2(i) \\ -1, & \zeta_2(i) \le t < \zeta_3(i) \\ 0, & \text{otherwise} \end{cases} \tag{1.53}$$

where

$$\zeta_1(i) = A + \left(\frac{k-1}{2^j}\right)(B-A) = A + \left(\frac{k-1}{2^j}\right) m\Delta t$$

$$\zeta_2(i) = A + \left(\frac{k-(1/2)}{2^j}\right)(B-A) = A + \left(\frac{k-(1/2)}{2^j}\right) m\Delta t$$

$$\zeta_3(i) = A + \left(\frac{k}{2^j}\right)(B-A) = A + \left(\frac{k}{2^j}\right) m\Delta t$$

for $i = 1, 2, \ldots, m$, $m = 2^J$, and J is a positive integer that is called the maximum level of resolution. Here, j and k represent the integer decomposition of the index i, that is, $i = k + 2^j - 1$, $0 \le j < i$, and $1 \le k < 2^j + 1$.

The mutual orthogonalities of all Haar wavelets can be expressed as

$$\int_a^b h_m(t) h_n(t)\,dt = (b-a)2^{-j}\delta_{mn} = \begin{cases} (b-a)2^{-j}, & m = n = 2^j + k \\ 0, & m \ne n. \end{cases}$$

1.3.1.1 Function Approximation

Any function $y(t) \in L^2([0,1))$ can be expanded into Haar wavelets by [39]

$$y(t) = c_0 h_0(t) + c_1 h_1(t) + c_2 h_2(t) + \cdots, \quad \text{where } c_j = \int_0^1 y(t) h_j(t)\,dt. \tag{1.54}$$

If $y(t)$ is approximated as a piecewise constant in each subinterval, the sum in Equation 1.54 may be terminated after m terms, and consequently, we can write a discrete version in the matrix form as

$$\boldsymbol{Y} \approx \sum_{i=0}^{m-1} c_i h_i(t_l) = \boldsymbol{C}_m^T H_m, \tag{1.55}$$

where $\boldsymbol{Y}$ and $\boldsymbol{C}_m^T$ are the m-dimensional row vectors.

Here, $\boldsymbol{H}$ is the Haar wavelet matrix of order m defined by $\boldsymbol{H}=[\boldsymbol{h}_0, \boldsymbol{h}_1, \dots, \boldsymbol{h}_{m-1}]^T$, that is,

$$\boldsymbol{H} = \begin{bmatrix} \boldsymbol{h}_0 \\ \boldsymbol{h}_1 \\ \vdots \\ \boldsymbol{h}_{m-1} \end{bmatrix} = \begin{bmatrix} h_{0,0} & h_{0,1} & \cdots & h_{0,m-1} \\ h_{1,0} & h_{1,1} & \cdots & h_{1,m-1} \\ \vdots & \vdots & & \vdots \\ h_{m-1,0} & h_{m-1,1} & \cdots & h_{m-1,m-1} \end{bmatrix} \tag{1.56}$$

where $\boldsymbol{h}_0, \boldsymbol{h}_1, \dots, \boldsymbol{h}_{m-1}$ are the discrete forms of the Haar wavelet bases.

The collocation points are given by

$$t_l = A + (l - 0.5)\Delta t, \quad l = 1, 2, \dots, m. \tag{1.57}$$

1.3.1.2 Operational Matrix of the General Order Integration

The integration of the $H_m(t) = [h_0(t), h_1(t), \dots, h_{m-1}(t)]^T$ can be approximated by [40]

$$\int_0^t H_m(\tau)\, d\tau \cong \boldsymbol{Q} H_m(t) \tag{1.58}$$

where $\boldsymbol{Q}$ is called the Haar wavelet operational matrix of integration, which is a square matrix of m-dimension. To derive the Haar wavelet operational matrix of the general order of integration, we recall the fractional integral of order $\alpha(>0)$, which is defined by Podlubny [41]:

$$J^\alpha f(t) = \frac{1}{\Gamma(\alpha)} \int_0^t (t - \tau)^{\alpha - 1} f(\tau)\, d\tau, \quad \alpha > 0, \ \alpha \in \Re^+ \tag{1.59}$$

where $\Re^+$ is the set of positive real numbers.

The operational matrix for general order was first derived by learned researcher Saha Ray [38]. The Haar wavelet operational matrix $\boldsymbol{Q}^\alpha$ for integration of the general order α is given by [38]

$$\begin{aligned} \boldsymbol{Q}^\alpha H_m(t) = J^\alpha H_m(t) &= \left[J^\alpha h_0(t), J^\alpha h_1(t), \dots, J^\alpha h_{m-1}(t) \right]^T \\ &= \left[Q h_0(t), Q h_1(t), \dots, Q h_{m-1}(t) \right]^T \end{aligned} \tag{1.60}$$

where

$$Qh_0(t)=\begin{cases}\dfrac{t^{\alpha}}{\Gamma(1+\alpha)}, & t\in[A,B],\\ 0, & \text{elsewhere}\end{cases} \tag{1.61}$$

and

$$Qh_i(t)=\begin{cases}0, & A\le t<\zeta_1(i),\\ \phi_1, & \zeta_1(i)\le t<\zeta_2(i),\\ \phi_2, & \zeta_2(i)\le t<\zeta_3(i),\\ \phi_3, & \zeta_3(i)\le t<B,\end{cases} \tag{1.62}$$

where

$$\phi_1=\frac{(t-\zeta_1(i))^{\alpha}}{\Gamma(\alpha+1)},$$

$$\phi_2=\frac{(t-\zeta_1(i))^{\alpha}}{\Gamma(\alpha+1)}-2\frac{(t-\zeta_2(i))^{\alpha}}{\Gamma(\alpha+1)},$$

$$\phi_3=\frac{(t-\zeta_1(i))^{\alpha}}{\Gamma(\alpha+1)}-2\frac{(t-\zeta_2(i))^{\alpha}}{\Gamma(\alpha+1)}+\frac{(t-\zeta_3(i))^{\alpha}}{\Gamma(\alpha+1)}$$

for $i=1,2,\ldots,m$, $m=2^J$, and J is a positive integer, called the maximum level of resolution. Here, j and k represent the integer decomposition of the index i, that is, $i=k+2^j-1$, $0\le j<i$, and $1\le k<2^j+1$.

1.3.2 Legendre Wavelets

The application of Legendre wavelets for solving differential and integral equations is thoroughly considered by many researchers in [42–44] and references therein. Both initial and boundary-value problems can be solved efficiently by using the LWM. Here, the basic idea of Legendre wavelets are introduced; the operational matrix of integration is then derived. The derived operational matrix of fractional order integration is then applied to solve fractional differential equations. The method reduces the fractional initial or boundary value problem to a system of algebraic equations. The large systems of algebraic equations may lead to greater computational complexity and large storage requirements. However, the operational matrix for the Legendre wavelets is structurally sparse. This reduces the computational complexity of the resulting algebraic system.

Wavelets constitute a family of functions constructed from dilation and translation of a single function called the mother wavelet $\psi(t)$. They are defined by

$$\psi_{a,b}(t)=\frac{1}{\sqrt{|a|}}\psi\left(\frac{t-b}{a}\right),\quad a,b\in\Re \tag{1.63}$$

where
a is dilation parameter
b is translation parameter

By restricting a, b to discrete values as $a=a_0^{-j}$, $b=kb_0a_0^{-j}$, where $a_0>1$, $b_0>0$, and $n,k\in N$.

The Legendre polynomials of order m, denoted by $L_m(t)$, are defined on the interval [−1, 1] and can be determined with the help of the following recurrence formulae [42]:

$$\begin{aligned} L_0(t) &= 1, \\ L_1(t) &= t, \\ L_{m+1}(t) &= \frac{2m+1}{m+1} t L_m(t) - \frac{m}{m+1} L_{m-1}(t), \quad m = 1,2,3,\dots. \end{aligned} \tag{1.64}$$

Legendre wavelets $\psi_{n,m}(t) = \psi(k,\hat{n},m,t)$ have four arguments, defined on interval [0, 1) by

$$\psi_{n,m}(t) = \begin{cases} \left(m+\frac{1}{2}\right)^{1/2} 2^{k/2} L_m\left(2^k t - \hat{n}\right), & \frac{\hat{n}-1}{2^k} \le t < \frac{\hat{n}+1}{2^k}, \\ 0, & \text{elsewhere} \end{cases} \tag{1.65}$$

where

$k = 2,3,\dots$, $\hat{n} = 2n-1$, $n = 1,2,3,\dots,2^{k-1}$, $m = 0,1,\dots,M-1$ is the order of the Legendre polynomials

M is a fixed positive integer

The set of Legendre wavelets form an orthogonal basis of $L^2(\mathfrak{R})$.

The two-dimensional Legendre wavelets are defined as

$$\psi_{n_1,m_1,n_2,m_2}(x,t) = \begin{cases} A L_{m_1}\left(2^{k_1}x - \hat{n}_1\right) L_{m_2}\left(2^{k_2}t - \hat{n}_2\right), & \frac{\hat{n}_1-1}{2^{k_1}} \le x < \frac{\hat{n}_1+1}{2^{k_1}}, \ \frac{\hat{n}_2-1}{2^{k_2}} \le t < \frac{\hat{n}_2+1}{2^{k_2}} \\ 0, & \text{elsewhere} \end{cases} \tag{1.66}$$

where

$A = \sqrt{\left(m_1+\frac{1}{2}\right)\left(m_2+\frac{1}{2}\right) 2^{(k_1+k_2)/2}}$

$\hat{n}_1$ and $\hat{n}_2$ are defined similarly to $\hat{n}$

k_1 and k_2 are any positive integers

m_1 and m_2 are the orders for Legendre polynomials

The two-dimensional Legendre $\psi_{n_1,m_1,n_2,m_2}(x,t)$ forms a basis for $L^2([0,1)\times[0,1))$.

1.3.2.1 *Function Approximation*

A function, $f(x,t)$, defined over $[0,1)\times[0,1)$ can be expanded in terms of Legendre wavelet as [43]

$$f(x,t) = \sum_{n=1}^{\infty}\sum_{i=0}^{\infty}\sum_{l=1}^{\infty}\sum_{j=0}^{\infty} c_{n,i,l,j}\,\psi_{n,i,l,j}(x,t). \tag{1.67}$$

If the infinite series in Equation 1.67 is truncated, then it can be written as

$$f(x,t) \cong \sum_{n=1}^{2^{k1-1}} \sum_{i=0}^{M_1-1} \sum_{l=1}^{2^{k2-1}} \sum_{j=0}^{M_2-1} c_{n,i,l,j} \psi_{n,i,l,j}(x,t) = \Psi^T(x) C \Psi(t), \tag{1.68}$$

where $\Psi(x)$ and $\Psi(t)$ are $2^{k_1-1}M_1 \times 1$ and $2^{k_2-1}M_2 \times 1$ matrices, respectively, given by

$$\Psi(x) \equiv \left[\psi_{1,0}(x),\ldots,\psi_{1,M_1-1}(x),\psi_{2,0}(x),\ldots,\psi_{2,M_1-1}(x),\ldots,\psi_{2^{k1-1},0}(x),\ldots,\psi_{2^{k1-1},M_1-1}(x)\right]^T,$$

$$\Psi(t) \equiv \left[\psi_{1,0}(t),\ldots,\psi_{1,M_2-1}(t),\psi_{2,0}(t),\ldots,\psi_{2,M_2-1}(t),\ldots,\psi_{2^{k2-1},0}(t),\ldots,\psi_{2^{k2-1},M_2-1}(t)\right]^T.$$

Also, C is a $2^{k_1-1}M_1 \times 2^{k_2-1}M_2$ matrix whose elements can be calculated from the formula

$$c_{n,i,l,j} = \int_0^1 \int_0^1 \psi_{n,i}(x) \psi_{l,j}(t) f(x,t) dt\ dx, \tag{1.69}$$

with $n=1,\ldots,2^{k_1-1}$, $i=0,\ldots,M_1-1$, $l=1,\ldots,2^{k_2-1}$, $j=0,\ldots,M_2-1$.

1.3.2.2 Operational Matrix of the General Order Integration

The integration of the Legendre wavelet function $\Psi(t)$, $\Psi(t) \equiv [\psi_{1,0}(t),\ldots,\psi_{1,M-1}(t),\psi_{2,0}(t),\ldots,\psi_{2,M-1}(t),\ldots,\psi_{2^{k-1},0}(t),\ldots,\psi_{2^{k-1},M-1}(t)]^T$ can be approximated by

$$\int_0^t \Psi(\tau) d\tau \cong Q\Psi(t), \tag{1.70}$$

where Q is called the Legendre wavelet operational matrix of integration. To derive the Legendre wavelet operational matrix of the general order of integration, we recall the fractional integral of order $\alpha(>0)$, which is defined by Podlubny [41]:

$$J^\alpha f(t) = \frac{1}{\Gamma(\alpha)} \int_0^t (t-\tau)^{\alpha-1} f(\tau) d\tau, \quad \alpha > 0,\ \alpha \in \mathfrak{R}^+ \tag{1.71}$$

where $\mathfrak{R}^+$ is the set of positive real numbers.

The Legendre wavelet operational matrix Q^α for integration of the general order α is given by

$$\begin{aligned} Q^\alpha \Psi(t) &= J^\alpha \Psi(t) \\ &= \left[J^\alpha \psi_{1,0}(t),\ldots,J^\alpha \psi_{1,M-1}(t),J^\alpha \psi_{2,0}(t),\ldots,J^\alpha \psi_{2,M-1}(t),J^\alpha \psi_{2^{k-1},0}(t),\ldots,J^\alpha \psi_{2^{k-1},M-1}(t)\right]^T \end{aligned}$$

where

$$J^\alpha \psi_{n,m}(x) = \begin{cases} \left(m+\dfrac{1}{2}\right)^{1/2} 2^{k/2} J^\alpha L_m\left(2^k x - \hat{n}\right), & \dfrac{\hat{n}-1}{2^k} \le x < \dfrac{\hat{n}+1}{2^k} \\ 0, & \text{elsewhere} \end{cases} \tag{1.72}$$

for $k = 2,3,\ldots,\ \hat{n} = 2n-1,\ n = 1,2,3,\ldots,2^{k-1},\ m = 0,1,\ldots,M-1$ is the order of the Legendre polynomials, and M is a fixed positive integer.

1.3.3 Chebyshev Wavelets

The Chebyshev wavelets are competent for solving some fractional and integral equations [45,46]. Nowadays, Chebyshev polynomials have become more significant in numerical evaluation. Among the four forms of Chebyshev polynomials, the first and second kinds are certain cases of the symmetric Jacobi polynomials, whereas the third and fourth kinds are unique instances of the nonsymmetric Jacobi polynomials. Great attention has been focused on the first and second kinds of Chebyshev polynomials $T_n(x)$ and $U_n(x)$ and their various uses in numerous applications. Nevertheless, there are very few articles that concentrate on the wavelets shaped through these two types of Chebyshev polynomials for application in FPDEs. This motivates our curiosity in such wavelets.

There are several advantages of using Chebyshev wavelets approximations based on collocation spectral method. First, unlike most numerical methods, it is now conventional that they are characterized by the use of exponentially decaying errors. Second, various numerical methods do not perform well near singularities, whereas approximations through wavelets effectively handle singularities in the problem. In the end, due to their fast convergence, the Chebyshev wavelets method does not undergo from the instability problems related with other numerical methods.

The Chebyshev wavelets $\psi_{n,m}(t) = \psi(k,n,m,t)$ have four arguments, defined on interval [0, 1) by [45]

$$\psi_{n,m}(t) = \begin{cases} 2^{k/2}\bar{U}_m\left(2^k t - 2n + 1\right), & \dfrac{n-1}{2^{k-1}} \le t < \dfrac{n}{2^{k-1}}, \\ 0, & \text{elsewhere} \end{cases} \tag{1.73}$$

where
- $n = 1,2,3,\ldots,2^{k-1}$
- k is assumed to be any positive integer
- m is the degree of the second kind Chebyshev polynomials
- t is the normalized time

Here, $\bar{U}_m(t) = \sqrt{\dfrac{2}{\pi}}U_m(t)$, $U_m(t)$, $m = 0,1,2,\ldots,M$ are the second kind of Chebyshev polynomials of degree m defined on the interval $[-1,1]$ and satisfy the following recursive formula:

$$\begin{aligned} U_0(t) &= 1, \\ U_1(t) &= 2t, \\ U_{m+1}(t) &= 2tU_m(t) - U_{m-1}(t), \quad m = 1,2,3,\ldots. \end{aligned} \tag{1.74}$$

The two-dimensional Chebyshev wavelets are defined as

$$\psi_{n_1,m_1,n_2,m_2}(x,t) = \begin{cases} 2^{(k_1+k_2)/2}\bar{U}_{m_1}\left(2^{k_1}x - 2n_1 + 1\right)\bar{U}_{m_2}\left(2^{k_2}t - 2n_2 + 1\right), & \dfrac{n_1-1}{2^{k_1-1}} \le x < \dfrac{n_1}{2^{k_1-1}},\ \dfrac{n_2-1}{2^{k_2-1}} \le t < \dfrac{n_2}{2^{k_2-1}} \\ 0, & \text{elsewhere} \end{cases} \tag{1.75}$$

where
n_1 and n_2 are defined similarly to n
k_1 and k_2 are any positive integers
m_1 and m_2 are the orders for the second kind Chebyshev polynomials

1.3.3.1 Function Approximation

A function, $f(x,t)$, defined over $[0,1)\times[0,1)$ may be expanded in terms of Chebyshev wavelets as [46]

$$f(x,t)=\sum_{n=1}^{\infty}\sum_{i=0}^{\infty}\sum_{l=1}^{\infty}\sum_{j=0}^{\infty}c_{n,i,l,j}\psi_{n,i,l,j}(x,t). \tag{1.76}$$

If the infinite series in Equation 1.76 is truncated, then Equation 1.76 can be written as

$$f(x,t)\cong\sum_{n=1}^{2^{k_1-1}}\sum_{i=0}^{M_1-1}\sum_{l=1}^{2^{k_2-1}}\sum_{j=0}^{M_2-1}c_{n,i,l,j}\psi_{n,i,l,j}(x,t)=\Psi^T(x)C\Psi(t), \tag{1.77}$$

where $\Psi(x)$ and $\Psi(t)$ are $2^{k_1-1}M_1\times1$ and $2^{k_2-1}M_2\times1$ matrices, respectively, given by

$$\Psi(x)\equiv\left[\psi_{1,0}(x),\psi_{1,1}(x),\ldots,\psi_{1,M_1-1}(x),\psi_{2,0}(x),\ldots,\psi_{2,M_1-1}(x),\ldots,\psi_{2^{k1-1},0}(x),\ldots,\psi_{2^{k1-1},M_1-1}(x)\right]^T,$$

$$\Psi(t)\equiv\left[\psi_{1,0}(t),\psi_{1,1}(t),\ldots,\psi_{1,M_2-1}(t),\psi_{2,0}(t),\ldots,\psi_{2,M_2-1}(t),\ldots,\psi_{2^{k2-1},0}(t),\ldots,\psi_{2^{k2-1},M_2-1}(t)\right]^T.$$

Also, $\boldsymbol{C}$ is a $2^{k_1-1}M_1\times2^{k_2-1}M_2$ matrix whose elements can be calculated from the formula

$$c_{n,i,l,j}=\int_0^1\int_0^1\psi_{n,i}(x)\psi_{l,j}(t)f(x,t)dt\ dx,$$

with $n=1,\ldots,2^{k_1-1}$, $i=0,\ldots,M_1-1$, $l=1,\ldots,2^{k_2-1}$, $j=0,\ldots,M_2-1$.

1.3.3.2 Operational Matrix of the General Order Integration

The integration of the Chebyshev wavelet function $\Psi(t)$, $\Psi(t)\equiv[\psi_{1,0}(t),\ldots,\psi_{1,M-1}(t),\psi_{2,0}(t),\ldots,\psi_{2,M-1}(t),\ldots,\psi_{2^{k-1},0}(t),\ldots,\psi_{2^{k-1},M-1}(t)]^T$ can be approximated by

$$\int_0^t\Psi(\tau)d\tau\approx P\Psi(t) \tag{1.78}$$

where $\boldsymbol{P}$ is called the Chebyshev wavelet operational matrix of integration. To derive the Chebyshev wavelet operational matrix of the general order of integration, we recall the fractional integral of order $\alpha(>0)$, which is defined by Podlubny [41]:

$$J^\alpha f(t)=\frac{1}{\Gamma(\alpha)}\int_0^t(t-\tau)^{\alpha-1}f(\tau)d\tau,\quad \alpha>0,\ \alpha\in\Re^+ \tag{1.79}$$

where $\Re^+$ is the set of positive real numbers.

The Chebyshev wavelet operational matrix P^α for integration of the general order α is given by

$$P^\alpha\Psi(t) = J^\alpha\Psi(t)$$
$$= \left[J^\alpha\psi_{1,0}(t),\ldots,J^\alpha\psi_{1,M-1}(t),J^\alpha\psi_{2,0}(t),\ldots,J^\alpha\psi_{2,M-1}(t),J^\alpha\psi_{2^{k-1},0}(t),\ldots,J^\alpha\psi_{2^{k-1},M-1}(t)\right]^T$$

where

$$J^\alpha\psi_{n,m}(x) = \begin{cases} 2^{k/2}J^\alpha\bar{U}_m\left(2^k x - 2n + 1\right), & \frac{n-1}{2^{k-1}} \le x < \frac{n}{2^{k-1}} \\ 0, & \text{elsewhere} \end{cases}$$

for $n=1,2,3,\ldots,2^{k-1}$, $m=0,1,\ldots,M-1$ is the order of the Chebyshev polynomials, and M is a fixed positive integer.

1.3.4 Hermite Wavelets

The Hermite polynomials $H_m(x)$ of order m are defined on the interval $(-\infty,\infty)$ and can be deduced with the assistance of the following recurrence formulae:

$$\begin{aligned} H_0(x) &= 1 \\ H_1(x) &= 2x, \\ H_{m+1}(x) &= 2xH_m(x) - 2mH_{m-1}(x), \quad m = 1,2,3,\ldots. \end{aligned} \tag{1.80}$$

The Hermite polynomials $H_m(x)$ are orthogonal with respect to the weight function e^{-x^2}.

The Hermite wavelets are defined on interval [0, 1) by [47]

$$\psi_{n,m}(x) = \begin{cases} 2^{k/2}\sqrt{\frac{1}{n!2^n\sqrt{\pi}}}H_m\left(2^k x - \hat{n}\right), & \text{for } \frac{\hat{n}-1}{2^k} \le x < \frac{\hat{n}+1}{2^k} \\ 0, & \text{otherwise} \end{cases} \tag{1.81}$$

where

$k=1,2,\ldots$ is the level of resolution

$n = 1,2,\ldots,2^{k-1}$, $\hat{n} = 2n-1$ is the translation parameter

$m=1,2,\ldots,M-1$ is the order of Hermite polynomials

The two-dimensional Hermite wavelets are outlined as

$$\psi_{n_1,m_1,n_2,m_2}(x,t) = \begin{cases} AH_{m_1}\left(2^{k_1}x - \hat{n}_1\right)H_{m_2}\left(2^{k_2}t - \hat{n}_2\right), & \frac{\hat{n}_1-1}{2^{k_1}} \le x < \frac{\hat{n}_1+1}{2^{k_1}}, \ \frac{\hat{n}_2-1}{2^{k_2}} \le t < \frac{\hat{n}_2+1}{2^{k_2}} \\ 0, & \text{elsewhere} \end{cases} \tag{1.82}$$

where

$$A=\sqrt{\frac{1}{n_1!2^{n_1}\sqrt{\pi}}}\sqrt{\frac{1}{n_2!2^{n_2}\sqrt{\pi}}}2^{(k_1+k_2)/2}$$

n_1 and n_2 defined similarly to n
k_1 and k_2 are any positive integers
m_1 and m_2 are the orders for Hermite polynomials

1.3.4.1 Function Approximation

A function, $f(x,t)$, defined over $[0,1)\times[0,1)$ can be expanded in terms of Hermite wavelet as [48]

$$f(x,t)=\sum_{n=1}^{\infty}\sum_{i=0}^{\infty}\sum_{l=1}^{\infty}\sum_{j=0}^{\infty}c_{n,i,l,j}\psi_{n,i,l,j}(x,t). \tag{1.83}$$

If the infinite series in Equation 1.83 is truncated, then it can be written as

$$f(x,t)\cong\sum_{n=1}^{2^{k_1-1}}\sum_{i=0}^{M_1-1}\sum_{l=1}^{2^{k_2-1}}\sum_{j=0}^{M_2-1}c_{n,i,l,j}\psi_{n,i,l,j}(x,t)=\Psi^T(x)C\Psi(t), \tag{1.84}$$

where $\Psi(x)$ and $\Psi(t)$ are $2^{k_1-1}M_1\times 1$ and $2^{k_2-1}M_2\times 1$ matrices, respectively, given by

$$\Psi(x)\equiv\left[\psi_{1,0}(x),\psi_{1,1}(x),\ldots,\psi_{1,M_1-1}(x),\psi_{2,0}(x),\ldots,\psi_{2,M_1-1}(x),\ldots,\psi_{2^{k_1-1},0}(x),\ldots,\psi_{2^{k_1-1},M_1-1}(x)\right]^T,$$

$$\Psi(t)\equiv\left[\psi_{1,0}(t),\psi_{1,1}(t),\ldots,\psi_{1,M_2-1}(t),\psi_{2,0}(t),\ldots,\psi_{2,M_2-1}(t),\ldots,\psi_{2^{k_2-1},0}(t),\ldots,\psi_{2^{k_2-1},M_2-1}(t)\right]^T.$$

Also, C is a $2^{k_1-1}M_1\times 2^{k_2-1}M_2$ matrix whose elements can be calculated from the formula

$$c_{n,i,l,j}=\int_0^1\int_0^1\psi_{n,i}(x)\psi_{l,j}(t)f(x,t)dt\ dx, \tag{1.85}$$

with $n=1,\ldots,2^{k_1-1}$, $i=0,\ldots,M_1-1$, $l=1,\ldots,2^{k_2-1}$, $j=0,\ldots,M_2-1$.

1.3.4.2 Operational Matrix of the General-Order Integration

The integration of the Hermite wavelet function $\Psi(t)$, $\Psi(t)\equiv[\psi_{1,0}(t),\ldots,\psi_{1,M-1}(t),\psi_{2,0}(t),\ldots,\psi_{2,M-1}(t),\ldots,\psi_{2^{k-1},0}(t),\ldots,\psi_{2^{k-1},M-1}(t)]^T$ can be approximated by

$$\int_0^t\Psi(\tau)d\tau\cong Q\Psi(t), \tag{1.86}$$

where Q is called the Hermite wavelet operational matrix of integration. To derive the Hermite wavelet operational matrix of the general order of integration, we recall the fractional integral of order α(>0), which is defined by Podlubny [41]:

$$J^{\alpha} f(t) = \frac{1}{\Gamma(\alpha)} \int_0^t (t-\tau)^{\alpha-1} f(\tau) d\tau, \quad \alpha > 0, \ \alpha \in \Re^{+} \tag{1.87}$$

where $\Re^+$ is the set of positive real numbers.

The Hermite wavelet operational matrix $\mathbf{Q}^{\alpha}$ for integration of the general order α is given by

$$\begin{aligned} \mathbf{Q}^{\alpha}\Psi(t) &= J^{\alpha}\Psi(t) \\ &= \left[J^{\alpha}\psi_{1,0}(t),\ldots,J^{\alpha}\psi_{1,M-1}(t),J^{\alpha}\psi_{2,0}(t),\ldots,J^{\alpha}\psi_{2,M-1}(t),J^{\alpha}\psi_{2^{k-1},0}(t),\ldots,J^{\alpha}\psi_{2^{k-1},M-1}(t)\right]^{T} \end{aligned}$$

where

$$J^{\alpha}\psi_{n,m}(t) = \begin{cases} 2^{k/2}\sqrt{\dfrac{1}{n!2^{n}\sqrt{\pi}}} J^{\alpha} H_m\left(2^k t - \hat{n}\right), & \text{for } \dfrac{\hat{n}-1}{2^k} \le t < \dfrac{\hat{n}+1}{2^k} \\ 0, & \text{elsewhere} \end{cases} \tag{1.88}$$

where

$k = 1,2,\ldots$ is the level of resolution

$n = 1,2,\ldots,2^{k-1}$, $\hat{n} = 2n-1$ is the translation parameter

$m = 1,2,\ldots,M-1$ is the order of Hermite polynomials

2

Numerical Solution of Partial Differential Equations by Haar Wavelet Method

2.1 Introduction

Historically, partial differential equations originated from the study of surfaces in geometry and for solving a wide variety of problems in mechanics. Although the origin of nonlinear partial differential equations is very old, they have undergone remarkable new developments during the last half of the twentieth century. A large number of mathematicians became actively involved in the investigation of numerous problems presented by partial differential equations. The primary reason for this research was that partial differential equations both express many fundamental laws of nature and frequently arise in the mathematical analysis of diverse problems in science and engineering. Partial differential equations arise frequently in the formulation of fundamental laws of nature and in the mathematical analysis of a wide variety of problems in applied mathematics, mathematical physics, and engineering sciences including fluid dynamics, nonlinear optics, solid mechanics, plasma physics, quantum field theory, and condensed-matter physics. This subject plays a central role in modern mathematical sciences, especially in applied physics, mathematical modeling, and engineering. In fact, partial differential equations have been found to be essential to develop the theory of surfaces on the one hand and to the solution of physical problems on the other.

The development of linear partial differential equations is characterized by the efforts to develop the general theory and various methods of solutions of linear equations. Several methods such as the characteristics method, spectral methods, and perturbation techniques have been utilized to study these problems. But, as most solution methods for linear equations cannot be applied to nonlinear equations, there is no general method of finding analytical solutions of nonlinear partial differential equations. New numerical techniques are usually required for finding solutions of nonlinear equations. Methods of solution for nonlinear equations represent only one aspect of the theory of nonlinear partial differential equations. Like linear equations, questions of existence, uniqueness, and stability of solutions of nonlinear partial differential equations are of fundamental importance. These and other aspects of nonlinear equations have led the subject into one of the most diverse and active areas of modern mathematics.

Many problems of physical interest are described by partial differential equations with appropriate initial and boundary conditions. These problems are usually formulated as initial-value problems, boundary-value problems, or initial boundary-value problems. Indeed, the theory of nonlinear waves and solitons has experienced a revolution over the past three decades. During this revolution, many remarkable and unexpected phenomena have also been observed in physical, chemical, and biological systems.

2.2 Outline of Present Study

In this chapter, we will focus our study on the nonlinear partial differential equations those have particular applications appearing in applied sciences and engineering. We have considered both analytical and numerical approach for solving some specific nonlinear partial differential equations like Burgers' equation, modified Burgers' equation, the Huxley equation, the Burgers–Huxley equation, and the modified Korteweg–de Vries (mKdV) equation, which have a wide variety of applications in physical models.

2.2.1 Burgers' Equation

The one-dimensional Burgers' equation [49]

$$u_t + uu_x - \nu u_{xx} = 0, \quad 0 \le x \le 1 \tag{2.1}$$

is a nonlinear homogeneous parabolic partial differential equation. Here ν (>0) can be interpreted as viscosity. The Burgers' equation is considered as a model equation that describes the interaction of convection and diffusion. It arises in many physical problems including one-dimensional turbulence, sound waves in viscous medium, shock waves in a viscous medium, waves in fluid-filled viscous elastic tubes, and magnetohydrodynamic waves in a medium with finite electrical conductivity.

Various mathematical methods such as the Galerkin finite element method [50], spectral collocation method [51], quartic B-spline differential quadrature method [52], quartic B-spline collocation method [53], finite element method [54], fourth-order finite difference method [55], explicit and exact explicit finite difference method [56], and least-squares quadratic B-splines finite element method [57] have been used in attempting to solve Burgers' equations. Our aim in the present work is to implement the Haar wavelet method to stress its power in handling nonlinear equations so that one can apply it to various types of nonlinearity.

2.2.2 Modified Burgers' Equation

Modifying the nonlinear term uu_x in Equation 2.1 to $u^p u_x$, the generalized modified Burgers' equation [58] can be obtained in the following form:

$$u_t + u^p u_x - \nu u_{xx} = 0, \quad 0 \le x \le 1, \tag{2.2}$$

where p is a positive constant and ν (>0) can be interpreted as viscosity.

The modified Burgers' equation [59] has the strong nonlinear aspects of the governing equation in many practical transport problems such as nonlinear waves in medium with low-frequency pumping or absorption, ion reflection at quasi-perpendicular shocks, turbulence transport, wave processes in thermoelastic medium, transport and dispersion of pollutants in rivers, and sediment transport. Numerous mathematical methods such as the Petrov–Galerkin method [60], quintic spline method [61], Sextic B-spline collocation method [58], local discontinuous Galerkin method [62], and Lattice Boltzmann model [63] have been used in attempting to solve modified Burgers' equations.

2.2.3 Burgers–Huxley and Huxley Equations

The generalized Burgers–Huxley equation [64–66] is a nonlinear partial differential equation of the form

$$u_t + \alpha u^{\delta} u_x - u_{xx} = \beta u\left(1-u^{\delta}\right)\left(u^{\delta}-\gamma\right), \quad 0 \le x \le 1,\ t \ge 0, \tag{2.3}$$

where α, β, γ, and δ are parameters, $\beta \ge 0$, γ, $\delta > 0$. When $\alpha = 0$, $\delta = 1$, Equation 2.3 reduces to the Huxley equation. The Huxley equation [67,68] is a nonlinear partial differential equation of second order of the form

$$u_t = u_{xx} + u\left(k-u\right)\left(u-1\right), \quad k \ne 0. \tag{2.4}$$

This equation is an evolution equation that describes nerve pulse propagation in biology from which molecular CB properties can be calculated. The generalized Burgers–Huxley equation is of high importance for describing the interaction between reaction mechanisms, convection effects, and diffusion transport.

Various powerful mathematical methods such as the Adomian decomposition method [64,69], spectral collocation method [65], tanh–coth method [66], homotopy perturbation method [67], Exp-Function method [68], variational iteration method (VIM) [70], and differential quadrature method [71] have been used in attempting to solve the Burgers–Huxley and Huxley equations. The solitary wave solutions of the generalized Burgers–Huxley equation have been studied by the learned researchers Wang et al. [72] and El-Danaf [73].

2.2.4 Modified Korteweg–de Vries (mKdV) Equation

Next, we consider the generalized mKdV equation, [49] which is a nonlinear partial differential equation of the form

$$u_t + qu^2 u_x + ru_{xxx} = 0, \quad 0 \le x \le 1,\ t \ge 0 \tag{2.5}$$

where q and r are parameters.

The mKdV equations are the most popular soliton equations and have been extensively investigated. The mKdV equation is of high significance in many branches of the nonlinear science field. The mKdV equation appears in many fields such as acoustic waves in certain anharmonic lattices, Alfvén waves in collisionless plasma, transmission lines in a Schottky barrier, models of traffic congestion, ion acoustic soliton, elastic media, and so on [74].

2.3 Application of the Haar Wavelet Method to Obtain the Numerical Solution of Burgers' Equation

Haar wavelet collocation method is used for solving generalized Burgers' equation. This method consists of reducing the problem to a set of algebraic equations by expanding the term, which has maximum derivatives, given in the equation as Haar functions with unknown coefficients. The operational matrix of integration is utilized to evaluate the coefficients of Haar functions. This method gives us the implicit form of the approximate solutions of the problems.

Consider the one-dimensional Burgers' equation [49]

$$u_t + uu_x - \nu u_{xx} = 0, \quad 0 \le x \le 1, \tag{2.6}$$

with the following associated initial and boundary conditions:

$$u(x,t_0) = f(x), \quad 0 \le x \le 1$$

and

$$u(0,t) = u(1,t) = 0, \quad t \ge t_0. \tag{2.7}$$

It is assumed that $\dot{u}''(x,t)$ can be expanded in terms of Haar wavelets as

$$\dot{u}''(x,t) = \sum_{i=1}^{2M} a_s(i) h_i(x) \quad \text{for } t \in [t_s, t_{s+1}] \tag{2.8}$$

where "·" and "′" stand for differentiation with respect to t and x, respectively.

Integrating Equation 2.8 with respect to t from t_s to t and twice with respect to x from 0 to x, the following equations are obtained:

$$u''(x,t) = (t - t_s) \sum_{i=1}^{2M} a_s(i) h_i(x) + u''(x,t_s), \tag{2.9}$$

$$u'(x,t) = (t - t_s) \sum_{i=1}^{2M} a_s(i) p_i(x) + u'(x,t_s) - u'(0,t_s) + u'(0,t), \tag{2.10}$$

$$u(x,t) = (t - t_s) \sum_{i=1}^{2M} a_s(i) q_i(x) + u(x,t_s) - u(0,t_s) + x\left[u'(0,t) - u'(0,t_s)\right] + u(0,t), \tag{2.11}$$

$$\dot{u}(x,t) = \sum_{i=1}^{2M} a_s(i) q_i(x) + x\dot{u}'(0,t) + \dot{u}(0,t). \tag{2.12}$$

By using the boundary conditions at $x = 1$ and from Equations 2.12 and 2.11, respectively, we have

$$\dot{u}'(0,t) = -\sum_{i=1}^{2M} a_s(i) q_i(1) \tag{2.13}$$

and

$$u'(0,t) - u'(0,t_s) = -(t - t_s) \sum_{i=1}^{2M} a_s(i) q_i(1). \tag{2.14}$$

From Equation 1.50, it is obtained that

$$q_i(1) = \begin{cases} 0.5 & \text{if } i = 1 \\ \dfrac{1}{4m^2} & \text{if } i > 1. \end{cases} \tag{2.15}$$

Substituting Equations 2.13 through 2.15 in Equations 2.10 through 2.12 and discretizing the results by assuming $x \to x_l$, $t \to t_{s+1}$, the following equations are obtained:

$$u''(x_l, t_{s+1}) = (t_{s+1} - t_s)\sum_{i=1}^{2M} a_s(i) h_i(x_l) + u''(x_l, t_s), \tag{2.16}$$

$$u'(x_l, t_{s+1}) = (t_{s+1} - t_s)\sum_{i=1}^{2M} a_s(i)\left[p_i(x_l) - q_i(1)\right] + u'(x_l, t_s), \tag{2.17}$$

$$u(x_l, t_{s+1}) = (t_{s+1} - t_s)\sum_{i=1}^{2M} a_s(i)\left[q_i(x_l) - x_l q_i(1)\right] + u(x_l, t_s), \tag{2.18}$$

$$\dot{u}(x_l, t_{s+1}) = \sum_{i=1}^{2M} a_s(i)\left[q_i(x_l) - x_l q_i(1)\right]. \tag{2.19}$$

Substituting Equations 2.16 through 2.19 in Equation 2.6, we have

$$\begin{aligned}\sum_{i=1}^{2M} a_s(i)\left[q_i(x_l) - x_l q_i(1)\right] = \nu &\left[(t_{s+1} - t_s)\sum_{i=1}^{2M} a_s(i) h_i(x_l) + u''(x_l, t_s)\right] \\ &- \left[(t_{s+1} - t_s)\sum_{i=1}^{2M} a_s(i)\left[q_i(x_l) - x_l q_i(1)\right] + u(x_l, t_s)\right] \\ &\times \left[(t_{s+1} - t_s)\sum_{i=1}^{2M} a_s(i)\left[p_i(x_l) - q_i(1)\right] + u'(x_l, t_s)\right].\end{aligned} \tag{2.20}$$

From Equation 2.20, the wavelet coefficients $a_s(i)$ can be successively calculated using mathematical software. This process starts with

$$u(x_l, t_0) = f(x_l),$$

$$u'(x_l, t_0) = f'(x_l),$$

$$u''(x_l, t_0) = f''(x_l).$$

To show the effectiveness and accuracy of proposed scheme, we consider two test examples. The numerical solutions thus obtained are compared with the analytical solutions as well as available numerical results.

Example 2.1

Consider Burgers' equation with the following initial and boundary conditions [56]:

$$\begin{aligned} u(x,0) &= \sin(\pi x), \quad 0 \le x \le 1 \\ u(0,t) &= u(1,t) = 0, \quad t > 0. \end{aligned} \tag{2.21}$$

The exact solution of Equation 2.6 is given by [56]

$$u(x,t) = \frac{2\pi\nu \sum_{n=1}^{\infty} A_n n \sin(n\pi x)\exp(-n^2\pi^2\nu t)}{A_0 + \sum_{n=1}^{\infty} A_n \cos(n\pi x)\exp(-n^2\pi^2\nu t)}, \tag{2.22}$$

where

$$A_0 = \int_0^1 \exp\left(\frac{-1}{2\pi\nu}(1-\cos(\pi x))\right) dx,$$

$$A_n = 2\int_0^1 \exp\left(\frac{-1}{2\pi\nu}(1-\cos(\pi x))\right) dx.$$

The numerical solutions of Example 2.1 are presented for $\nu=0.01$ with $\Delta t=0.001$, taking $M=64$ in Table 2.1 and Figures 2.1 and 2.2. The obtained results are compared with explicit finite difference method (EFDM), exact explicit finite difference method (EEFDM), least square quadratic B-spline finite element method and Crank–Nicolson method as given in References [56,57,75]. Consequently it is found that the present method provides much better results than that presented in [56,57,75]. Figures 2.1 and 2.2 are in good agreement with the results obtained by learned researcher Jiwari [76].

TABLE 2.1

Comparison of Haar Wavelet Solution with Other Numerical Methods for Burgers' Equation (Example 2.1) at Different Values of t with $a=2$, $\nu=0.01$, and $\Delta t=0.001$

x	t	EFDM [56] $\Delta t=0.001$	EEFDM [56] $\Delta t=0.001$	Least-Squares Quadratic B-Splines FEM [57] $\Delta t=0.0001$	Crank–Nicolson Method [75] $\Delta t=0.01$	Present Method $\Delta t=0.001$	Exact Solution
0.25	0.4	0.34244	0.34164	0.34244	0.34229	0.34224	0.34191
	0.6	0.26905	0.26890	0.27536	0.26902	0.26924	0.26896
	0.8	0.22145	0.22150	0.22752	—	0.22170	0.22148
	1.0	0.18813	0.18825	0.19375	0.18817	0.18837	0.18819
	3.0	0.07509	0.07515	0.07754	0.07511	0.07516	0.07511
0.5	0.4	0.67152	0.65606	0.66543	0.66797	0.66106	0.66071
	0.6	0.53406	0.52658	0.53525	0.53211	0.52984	0.52942
	0.8	0.44143	0.43743	0.44526	—	0.43953	0.43914
	1.0	0.37568	0.37336	0.38047	0.37500	0.37476	0.37442
	3.0	0.15020	0.15015	0.15362	0.15018	0.15027	0.15018
0.75	0.4	0.94675	0.90111	0.91201	0.93680	0.90980	0.91026
	0.6	0.78474	0.75862	0.77132	0.77724	0.76745	0.76724
	0.8	0.65659	0.64129	0.65254	—	0.64778	0.64740
	1.0	0.56135	0.55187	0.56157	0.55833	0.55647	0.55605
	3.0	0.22502	0.22454	0.22874	0.22485	0.22497	0.22481

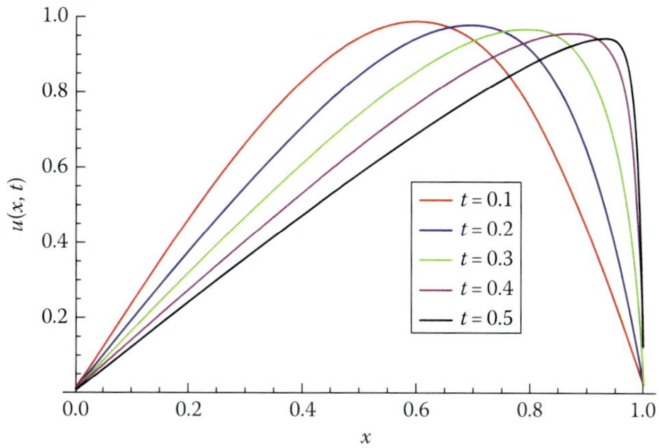

FIGURE 2.1
Behavior of numerical solutions for Burgers' equation (Example 2.1) when $\nu = 0.01$ and $\Delta t = 0.001$ at times $t = 0.1, 0.2, 0.3, 0.4$, and 0.5.

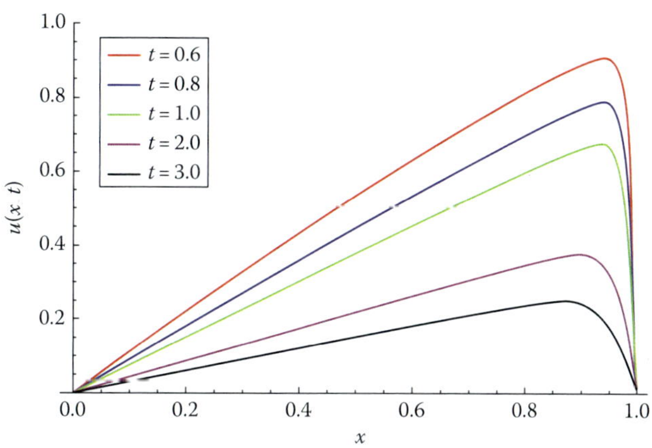

FIGURE 2.2
Behavior of numerical solutions for Burgers' equation (Example 2.1) when $\nu = 0.01$ and $\Delta t = 0.001$ at times $t = 0.6, 0.8, 1.0, 2.0$, and 3.0.

Example 2.2

In this example, we consider Burgers' equation with an initial condition in the following form:

$$u(x,0) = \frac{2\pi\nu \sin(\pi x)}{a + \cos(\pi x)}, \quad a > 1. \tag{2.23}$$

The exact solution of Equation 2.6 is given by [77]

$$u(x,t) = \frac{2\pi\nu \exp(-\pi^2 \nu t)\sin(\pi x)}{a + \exp(-\pi^2 \nu t)\cos(\pi x)}, \quad a > 1. \tag{2.24}$$

TABLE 2.2

Comparison of L_2 and L_∞ Errors with Other Numerical Methods for Burgers' Equation (Example 2.2), Taking $a=100$, $\nu=0.01$ at $t=1$

	Rahman [78]		Mittal and Jain [79]			Present Method $\Delta t=0.01$		Present Method $\Delta t=0.001$	
N	L_2	L_∞	L_2	L_∞	M	L_2	L_∞	L_2	L_∞
10	3.455E−7	4.881E−7	3.284E−7	4.628E−7	4	3.267E−8	4.634E−8	1.498E−8	2.157E−8
20	1.013E−7	1.431E−7	8.192E−8	1.164E−7	8	2.288E−8	3.239E−8	5.235E−9	7.452E−9
40	4.003E−8	5.668E−8	2.047E−8	2.907E−8	16	2.042E−8	2.889E−8	2.779E−9	3.939E−9
80	4.003E−8	3.499E−8	5.119E−9	7.271E−9	32	1.981E−8	2.802E−8	2.165E−9	3.064E−9

TABLE 2.3

Comparison of L_2 and L_∞ Errors with Other Numerical Methods for Burgers' Equation (Example 2.2), Taking $a=100, \nu=0.005$ at $t=1$

	Rahman [78]		Mittal and Jain [79]			Present Method $\Delta t=0.01$		Present Method $\Delta t=0.001$	
N	L_2	L_∞	L_2	L_∞	M	L_2	L_∞	L_2	L_∞
10	8.819E−8	1.246E−7	8.631E−8	1.215E−7	4	4.266E−9	6.0565E−9	1.9418E−9	2.7999E−9
20	2.403E−8	3.394E−8	2.153E−8	3.062E−8	8	2.9996E−9	4.2463E−9	6.8086E−10	9.7005E−10
40	7.942E−9	1.125E−8	5.378E−9	7.644E−9	16	2.681E−9	3.7933E−9	3.6323E−10	5.1519E−10
80	3.918E−9	5.549E−9	1.345E−9	7.644E−9	32	2.6013E−9	3.6797E−9	2.8393E−10	4.0182E−10

In the case of Example 2.2, Tables 2.2 and 2.3 show the L_2 and L_∞ errors at different values of a, ν, and M. Moreover, the results are compared with semi-implicit finite difference method and cubic B-splines collocation method as given in References [78,79], respectively. It has been observed that the present method is more accurate and efficient than the other numerical solutions. The physical behaviors of solutions at different time stages are shown in Figures 2.3 and 2.4.

2.3.1 Numerical Results and Discussion for Burgers' Equation

Table 2.1 shows the comparison of exact solutions with the approximate solutions of different numerical methods for Burgers' equation. Agreement between present numerical results and exact solutions appears very satisfactory through illustration in Table 2.1. In Table 2.1, J has been taken as 6, that is, $M=64$ with $\nu=0.01$ and different values of t. Similarly, Tables 2.2 and 2.3 show the comparison of L_2 and L_∞ errors with other numerical methods for $\nu=0.01$ and 0.005 with $a=100$ and $t=1$. From Tables 2.2 and 2.3, it has been observed that the present method is more accurate and efficient than the other numerical methods presented in References [78,79].

Figures 2.1 through 2.4 cite the behavior of numerical solutions obtained for Burgers' equation at different time stages, taking $\nu=0.01$.

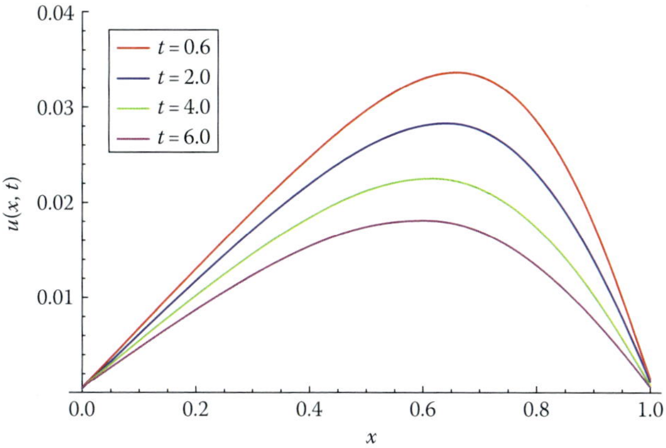

FIGURE 2.3
Behavior of numerical solutions for Burgers' equation (Example 2.2) when $\nu=0.01$ and $\Delta t=0.001$ at times $t = 0.6, 2.0, 4.0$, and 6.0.

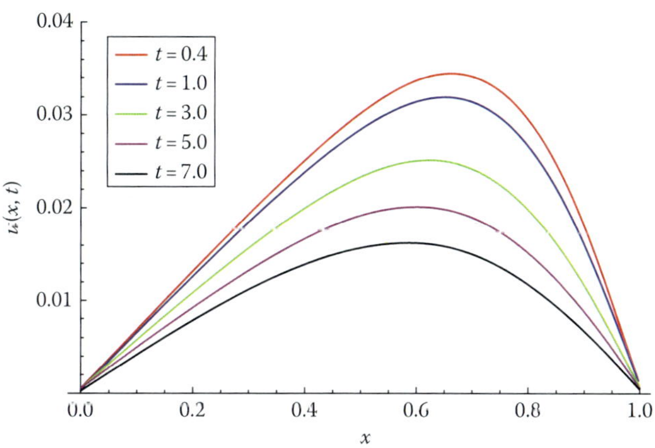

FIGURE 2.4
Behavior of numerical solutions for Burgers' equation (Example 2.2) when $\nu=0.01$ and $\Delta t=0.001$ at times $t = 0.4, 1.0, 3.0, 5.0$, and 7.0.

2.4 Haar Wavelet-Based Scheme for Modified Burgers' Equation

Taking $p = 2$ in Equation 2.2, the generalized modified Burgers' equation [58] can be obtained as follows:

$$u_t + u^2 u_x - \nu u_{xx} = 0, \quad 0 \le x \le 1, \tag{2.25}$$

where ν is a parameter.

The initial condition associated with Equation 2.25 will be

$$u(x, t_0) = f(x), \quad 0 \le x \le 1 \tag{2.26}$$

with boundary conditions

$$u(0,t) = u(1,t) = 0, \quad t \geq t_0.$$

Let us suppose that $\dot{u}''(x,t)$ can be expanded in terms of Haar wavelets as

$$\dot{u}''(x,t) = \sum_{i=1}^{2M} a_s(i)h_i(x) \quad \text{for } t \in [t_s, t_{s+1}] \tag{2.27}$$

where "$\cdot$" and "$'$" stand for differentiation with respect to t and x, respectively.

Now, integrating Equation 2.27 with respect to t from t_s to t and twice with respect to x from 0 to x, the following equations are obtained:

$$u''(x,t) = (t - t_s)\sum_{i=1}^{2M} a_s(i)h_i(x) + u''(x,t_s), \tag{2.28}$$

$$u'(x,t) = (t - t_s)\sum_{i=1}^{2M} a_s(i)p_i(x) + u'(x,t_s) + u'(0,t) - u'(0,t_s), \tag{2.29}$$

$$u(x,t) = (t - t_s)\sum_{i=1}^{2M} a_s(i)q_i(x) + u(x,t_s) + x\left[u'(0,t) - u'(0,t_s)\right] + u(0,t) - u(0,t_s), \tag{2.30}$$

$$\dot{u}(x,t) = \sum_{i=1}^{2M} a_s(i)q_i(x) + x\dot{u}'(0,t) + \dot{u}(0,t). \tag{2.31}$$

Using the boundary condition at $x = 1$, from Equation 2.31 we have

$$\dot{u}'(0,t) = -\sum_{i=1}^{2M} a_s(i)q_i(1), \tag{2.32}$$

and from Equation 2.30, we obtain

$$u'(0,t) - u'(0,t_s) = -(t - t_s)\sum_{i=1}^{2M} a_s(i)q_i(1). \tag{2.33}$$

Substituting Equations 2.32 and 2.33 in Equations 2.29 through 2.31 and discretizing the above results by assuming $x \to x_l$, $t \to t_{s+1}$, we obtain

$$u''(x_l,t_{s+1}) = (t_{s+1} - t_s)\sum_{i=1}^{2M} a_s(i)h_i(x_l) + u''(x_l,t_s), \tag{2.34}$$

$$u'(x_l,t_{s+1}) = (t_{s+1} - t_s)\sum_{i=1}^{2M} a_s(i)\left[p_i(x_l) - q_i(1)\right] + u'(x_l,t_s), \tag{2.35}$$

$$u(x_l,t_{s+1}) = (t_{s+1}-t_s)\sum_{i=1}^{2M} a_s(i)\left[q_i(x_l) - x_l q_i(1)\right] + u(x_l,t_s), \tag{2.36}$$

$$\dot{u}(x_l,t_{s+1}) = \sum_{i=1}^{2M} a_s(i)\left[q_i(x_l) - x_l q_i(1)\right]. \tag{2.37}$$

Substituting Equations 2.34 through 2.37 in Equation 2.25, we have

$$\begin{aligned}\sum_{i=1}^{2M} a_s(i)\left[q_i(x_l) - x_l q_i(1)\right] = \nu\left[(t_{s+1}-t_s)\sum_{i=1}^{2M} a_s(i) h_i(x_l) + u''(x_l,t_s)\right] \\ -\left[(t_{s+1}-t_s)\sum_{i=1}^{2M} a_s(i)\left[q_i(x_l) - x_l q_i(1)\right] + u(x_l,t_s)\right]^2 \\ \times\left[(t_{s+1}-t_s)\sum_{i=1}^{2M} a_s(i)\left[p_i(x_l) - q_i(1)\right] + u'(x_l,t_s)\right].\end{aligned} \tag{2.38}$$

From Equation 2.38, the wavelet coefficients $a_s(i)$ can be successively calculated. This process starts with

$$u(x_l,t_0) = f(x_l),$$

$$u'(x_l,t_0) = f'(x_l),$$

$$u''(x_l,t_0) = f''(x_l).$$

To show the efficiency and accuracy of the proposed scheme, two test examples have been considered taking $p=2$. The numerical solutions thus acquired are compared with the analytical solutions as well as available numerical results.

Example 2.3

Consider modified Burgers' equation with the following initial and boundary conditions [60,61]:

$$\begin{aligned} u(x,1) &= \frac{x}{1+\frac{1}{c_0}e^{x^2/4\nu}} \\ u(0,t) &= u(1,t) = 0, \quad t \geq 1 \end{aligned} \tag{2.39}$$

where $c_0 = e^{1/8\nu}$.

The exact solution of Equation 2.25 is given by [60,61]

$$u(x,t) = \frac{x/t}{1+\frac{\sqrt{t}}{c_0}e^{x^2/4\nu t}}, \quad t \geq 1. \tag{2.40}$$

Example 2.4

In this example, we consider modified Burgers' equation with initial and boundary conditions in the following form:

$$u(x,0) = \sin(\pi x), \quad 0 \le x \le 1,$$
$$u(0,t) = u(1,t) = 0, \quad t > 0. \tag{2.41}$$

In the case of Example 2.3, the Haar wavelet numerical solutions have been compared with the results obtained by Ramadan et al. [61], using the collocation method with quintic splines as demonstrated in Table 2.4. In the case of Example 2.4, the solutions have been compared with the results obtained by Duan et al. [63], using the 2-bit lattice Boltzmann method (LBM). Tables 2.5 and 2.6 cite the comparison of Haar wavelet solutions with LBM and quintic splines numerical solutions at $t=0.4$ and $t=2$, and hence the numerical solutions at different time stages for Examples 2.3 and 2.4 are shown in Figures 2.5 through 2.10.

TABLE 2.4

L_2 and L_∞ Error Norm for Modified Burgers' Equation (Example 2.3) at Different Values of t with $\nu=0.001$ and $\Delta t=0.001$

	Present Method		Quintic Spline	
Time (sec)	$L_2 \times 10^{-3}$	$L_\infty \times 10^{-3}$	$L_2 \times 10^{-3}$ [61]	$L_\infty \times 10^{-3}$ [61]
2	0.0755325	0.289254	0.0670395601	0.2796704002
3	0.0711446	0.256515	0.0689577701	0.2514379353
4	0.065229	0.214577	0.0666974605	0.2185661439
5	0.0604007	0.182222	0.0636023977	0.1923643818
6	0.0565458	0.157428	0.0604622308	0.1717652452
7	0.0534117	0.138249	0.0575085655	0.1553318123
8	0.0508078	0.123651	0.0548010376	0.1418932282

TABLE 2.5

Comparison of Haar Wavelet Solutions with the LBM Solutions and Quintic-Splines Solution of Modified Burgers' Equation (Example 2.4) at $t=0.4$ and $\nu=0.01$

x	Approximate Solution Using Haar Wavelet Method (u_{approx})	Approximate Solution Using Lattice Boltzmann Method [63]	Approximate Solution Using Quintic Spline Method [61]
0.10	0.221423	0.22177116	0.22033034
0.20	0.396841	0.39414890	0.39460783
0.30	0.531256	0.53134565	0.53244922
0.40	0.64835	0.64627793	0.64763455
0.50	0.744936	0.74511632	0.74643231
0.60	0.831235	0.83048713	0.83133318
0.70	0.902641	0.90235089	0.90195203
0.80	0.95132	0.95495434	0.95119837
0.90	0.825329	0.83737688	0.82794559
0.99	0.0623064	0.06214261	0.04674614

TABLE 2.6

Comparison of Haar Wavelet Solutions with the LBM Solutions and Quintic-Splines Solution of Modified Burgers' Equation (Example 2.4) at $t=2.0$ and $\nu=0.01$

x	Approximate Solution Using Haar Wavelet Method (u_{approx})	Approximate Solution Using Lattice Boltzmann Method [63]	Approximate Solution Using Quintic Spline Method [61]
0.10	0.111789	0.11194772	0.11013979
0.20	0.208539	0.20710153	0.20614825
0.30	0.284853	0.28512152	0.28477813
0.40	0.351297	0.35038171	0.35045112
0.50	0.406404	0.40665374	0.40700602
0.60	0.457189	0.45649486	0.45704614
0.70	0.501339	0.50155303	0.50224419
0.80	0.542602	0.54199420	0.54265295
0.90	0.536499	0.53547356	0.53225529
0.99	0.0790367	0.08046491	0.05693884

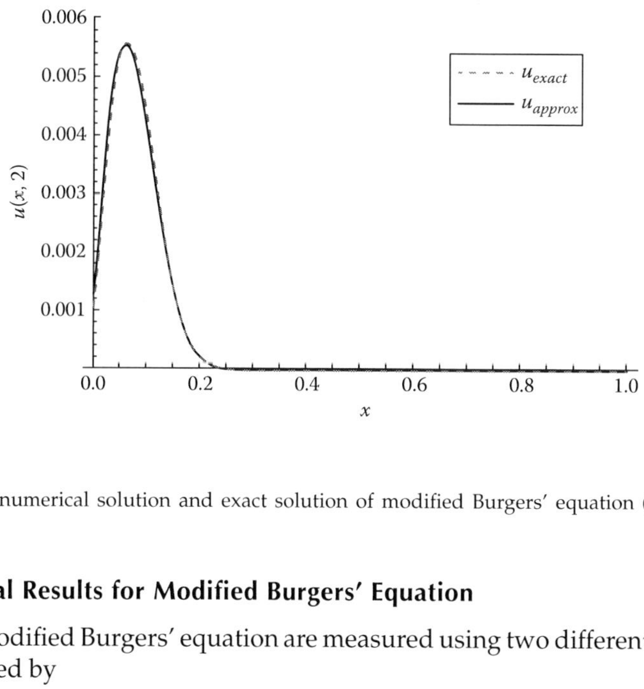

FIGURE 2.5
Comparison of the numerical solution and exact solution of modified Burgers' equation (Example 2.3) when $t=2.0$ and $\nu=0.001$.

2.4.1 Numerical Results for Modified Burgers' Equation

The errors for modified Burgers' equation are measured using two different norms, namely, L_2 and L_∞, defined by

$$L_2 = R.M.S.\ error = \frac{1}{\sqrt{2M}}\sqrt{\sum_{l=1}^{2M}\left(\left|u_{approx}\left(x_l,t\right)-u_{exact}\left(x_l,t\right)\right|\right)^2} \tag{2.42}$$

$$L_\infty = \max\left|u_{approx}\left(x_l,t\right)-u_{exact}\left(x_l,t\right)\right|. \tag{2.43}$$

Table 2.4 exhibits the L_2 and L_∞ error norms for modified Burgers' equation taking $p=2$, $\nu=0.001$ and different values of t. In Table 2.4, J is taken as 5, that is, $M=32$ and Δt is taken as 0.001.

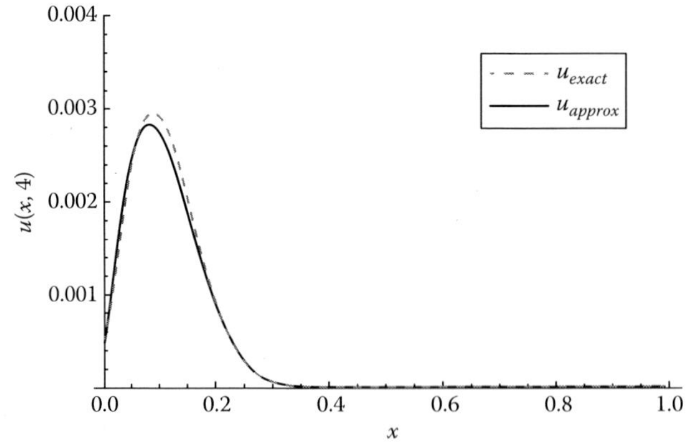

FIGURE 2.6
Comparison of the numerical solution and exact solution of modified Burgers' equation (Example 2.3) when $t=4$ and $\nu=0.001$.

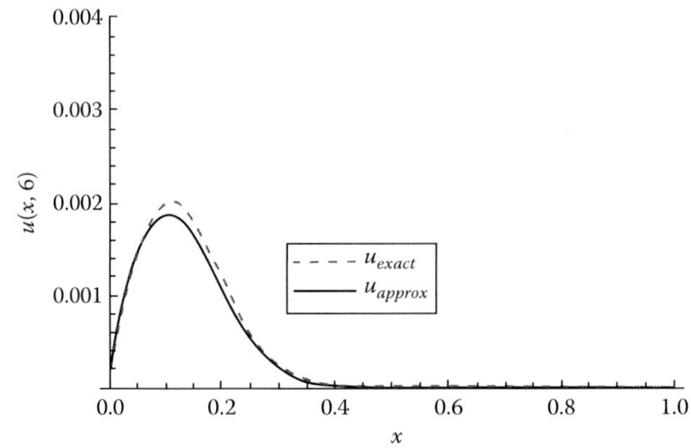

FIGURE 2.7
Comparison of the numerical solution and exact solution of modified Burgers' equation (Example 2.3) when $t=6$ and $\nu=0.001$.

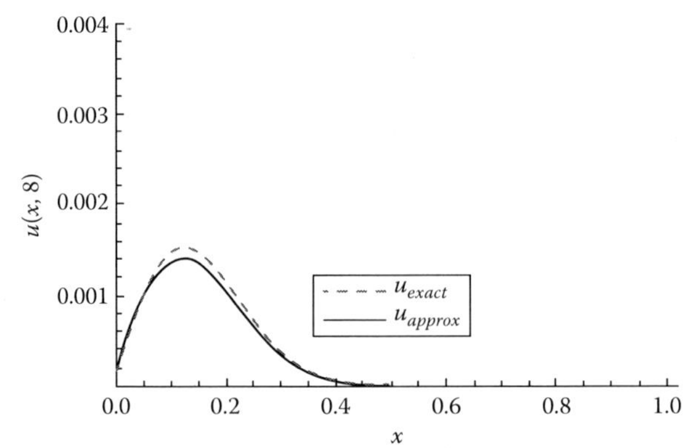

FIGURE 2.8
Comparison of the numerical solution and exact solution of modified Burgers' equation (Example 2.3) when $t=8$ and $\nu=0.001$.

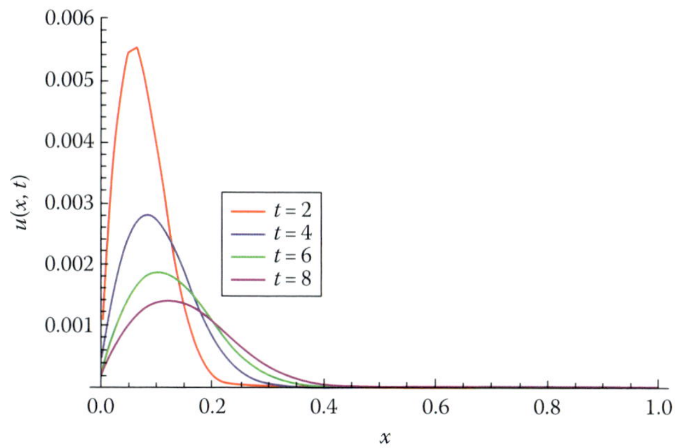

FIGURE 2.9
Behavior of numerical solutions for modified Burgers' equation (Example 2.3) when $\nu = 0.001$ and $\Delta t = 0.001$ at times $t = 2, 4, 6,$ and 8.

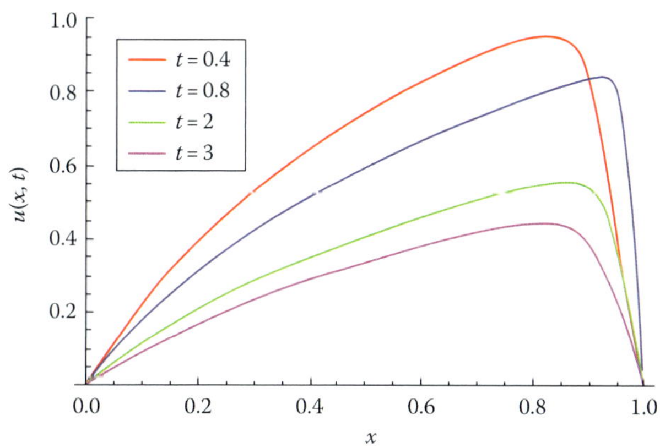

FIGURE 2.10
Behavior of the numerical solutions for modified Burgers' equation (Example 2.4) when $\nu = 0.01$ and $\Delta t = 0.001$ at times t = 0.4, 0.8, 2, and 3.

2.5 Analytical and Numerical Methods for Solving the Burgers–Huxley Equation

The generalized Burgers–Huxley equation [64–66] is a nonlinear partial differential equation of the form

$$u_t + \alpha u^\delta u_x - u_{xx} = \beta u\left(1 - u^\delta\right)\left(u^\delta - \gamma\right), \quad 0 \le x \le 1, \ t \ge 0, \tag{2.44}$$

where α, β, γ, and δ are parameters, $\beta \ge 0$, $\gamma, \delta > 0$.

The initial condition associated with Equation 2.44 is given by

$$u(x,0)=\left(\frac{\gamma}{2}+\frac{\gamma}{2}\tanh\left[A_1x\right]\right)^{1/\delta}. \tag{2.45}$$

The exact solution of Equation 2.44 is given by [65,72]

$$u(x,t)=\left(\frac{\gamma}{2}+\frac{\gamma}{2}\tanh\left[A_1\left(x-A_2t\right)\right]\right)^{1/\delta}, \tag{2.46}$$

where

$$A_1=\frac{-\alpha\delta+\delta\sqrt{\alpha^2+4\beta(1+\delta)}}{4(1+\delta)}\gamma$$

$$A_2=\frac{\gamma\alpha}{(1+\delta)}-\frac{(1+\delta-\gamma)\left(-\alpha+\sqrt{\alpha^2+4\beta(1+\delta)}\right)}{2(1+\delta)}$$

where α, β, γ, and δ are parameters with $\beta\geq 0$ and $\delta\geq 0$.

This exact solution satisfies the following boundary conditions:

$$\begin{aligned} u(0,t)&=\left[\frac{\gamma}{2}+\frac{\gamma}{2}\tanh(-A_1A_2t)\right]^{1/\delta}, \quad t\geq 0\\ u(1,t)&=\left[\frac{\gamma}{2}+\frac{\gamma}{2}\tanh\left(A_1(1-A_2t)\right)\right]^{1/\delta}, \quad t\geq 0. \end{aligned} \tag{2.47}$$

In our present study, $\delta=1$ has been considered.

2.5.1 Application of Variational Iteration Method for Solving the Burgers–Huxley Equation

Construct a correctional functional for Equation 2.44 as follows:

$$u_{n+1}(x,t)=u_n(x,t)+\int_0^t\lambda\left[\left(u_n\right)_\tau-\widetilde{\left(u_n^\delta\right)\left(u_n\right)_x}-\left(\tilde{u}_n\right)_{xx}-u_n\widetilde{(1-u_n)(u_n}-0.001)\right]d\tau, \tag{2.48}$$

where

λ is a general Lagrangian multiplier whose optimal value can be found using variational theory

$\tilde{u}_n(x,t)$ is the restricted variation, that is, $\delta\tilde{u}_n(x,t)=0$.

$$\delta u_{n+1}(x,t)=\delta u_n(x,t)+\delta\int_0^t\lambda\left[\left(u_n\right)_\tau-\widetilde{\left(u_n^\delta\right)\left(u_n\right)_x}-\left(\tilde{u}_n\right)_{xx}-u_n\widetilde{(1-u_n)(u_n}-0.001)\right]d\tau. \tag{2.49}$$

From Equation 2.49, we have

$$\delta u_{n+1}(x,t) = \delta u_n(x,t) + \delta \int_0^t \lambda \left(\frac{\partial u_n}{\partial \tau} \right) d\tau \tag{2.50}$$

since $\delta\left(\widetilde{(u_n^\delta)(u_n)}_x \right) = 0$, $\delta\left(\widetilde{u_n} \right)_{xx} = 0$ and $\delta\left(u_n \widetilde{(1-u_n)(u_n-0.001)} \right) = 0$.

Integrating the right-hand side of Equation 2.50 yields

$$\delta u_{n+1}(x,t) = \delta u_n(x,t)(1+\lambda) - \int_0^t \delta u_n(x,\tau)\lambda' d\tau. \tag{2.51}$$

From stationary condition, we know

$$\delta u_{n+1}(x,t) = 0,$$

which yields that

$$\delta u_n(x,t)(1+\lambda) - \int_0^t \delta u_n(x,\tau)\lambda' d\tau = 0.$$

This implies $1+\lambda=0$ and $\lambda'(t)=0$.

Hence

$$\lambda = -1. \tag{2.52}$$

Therefore, λ can be identified as -1, and the following variational iteration formula can be obtained as

$$u_{n+1}(x,t) = u_n(x,t) - \int_0^t \left[(u_n)_\tau - \widetilde{(u_n)(u_n)}_x - \left(\widetilde{u_n}\right)_{xx} - u_n \widetilde{(1-u_n)(u_n-0.001)} \right] d\tau. \tag{2.53}$$

From the preceding iteration formula, Equation 2.53, we can obtain

$$u_1(x,t) = 0.0005 + 0.0005\tanh(0.00025x) - 0.000000249938t\,\text{sech}^2(0.00025x)$$

with the initial condition

$$u_0(x,t) = 0.0005 + 0.0005\tanh(0.00025x).$$

2.5.2 Application of Haar Wavelet Method for Solving the Burgers–Huxley Equation

The Haar wavelet solution of $u(x,t)$ is sought by assuming that $\dot{u}''(x,t)$ can be expanded in terms of Haar wavelets as

$$\dot{u}''(x,t) = \sum_{i=1}^{2M} a_s(i) h_i(x) \quad \text{for } t \in [t_s, t_{s+1}]. \tag{2.54}$$

Integrating Equation 2.54 with respect to t from t_s to t and twice with respect to x from 0 to x, the following equations are obtained:

$$u''(x,t)=(t-t_s)\sum_{i=1}^{2M}a_s(i)h_i(x)+u''(x,t_s), \tag{2.55}$$

$$u'(x,t)=(t-t_s)\sum_{i=1}^{2M}a_s(i)p_i(x)+u'(x,t_s)-u'(0,t_s)+u'(0,t), \tag{2.56}$$

$$u(x,t)=(t-t_s)\sum_{i=1}^{2M}a_s(i)q_i(x)+u(x,t_s)-u(0,t_s)+x\left[u'(0,t)-u'(0,t_s)\right]+u(0,t), \tag{2.57}$$

$$\dot{u}(x,t)=\sum_{i=1}^{2M}a_s(i)q_i(x)+x\dot{u}'(0,t)+\dot{u}(0,t). \tag{2.58}$$

By using the boundary conditions, at $x=1$, we have

$$\dot{u}'(0,t)=-\sum_{i=1}^{2M}a_s(i)q_i(1)+\dot{u}(1,t)-\dot{u}(0,t). \tag{2.59}$$

Discretizing the results by assuming $x\to x_l$, $t\to t_{s+1}$, we obtain

$$u''(x_l,t_{s+1})=(t_{s+1}-t_s)\sum_{i=1}^{2M}a_s(i)h_i(x_l)+u''(x_l,t_s), \tag{2.60}$$

$$u'(x_l,t_{s+1})=(t_{s+1}-t_s)\sum_{i=1}^{2M}a_s(i)p_i(x_l)+u'(x_l,t_s)-u'(0,t_s)+u'(0,t_{s+1}), \tag{2.61}$$

$$\begin{aligned}u(x_l,t_{s+1})&=(t_{s+1}-t_s)\sum_{i=1}^{2M}a_s(i)q_i(x_l)+u(x_l,t_s)-u(0,t_s)\\&\quad+x_l\left[u'(0,t_{s+1})-u'(0,t_s)\right]+u(0,t_{s+1}),\end{aligned} \tag{2.62}$$

$$\dot{u}(x_l,t_{s+1})=\sum_{i=1}^{2M}a_s(i)q_i(x_l)+x_l\dot{u}'(0,t_{s+1})+\dot{u}(0,t_{s+1}). \tag{2.63}$$

Substituting Equations 2.63 and 2.59 in Equation 2.44, we have

$$\begin{aligned}\sum_{i=1}^{2M}a_s(i)\left[q_i(x_l)-x_lq_i(1)\right]&=u''(x_l,t_s)+u(x_l,t_s)\left[1-u(x_l,t_s)\right]\left[u(x_l,t_s)-0.001\right]\\&\quad-u(x_l,t_s)u'(x_l,t_s)-\dot{u}(0,t_{s+1})-x_l\left[\dot{u}(1,t_{s+1})-\dot{u}(0,t_{s+1})\right].\end{aligned} \tag{2.64}$$

From the preceding equation, the wavelet coefficients $a_s(i)$ can be successively calculated. This process started with

$$u(x_l, t_0) = 1 + \tanh\left(\frac{x_l}{2}\right),$$

$$u'(x_l, t_0) = \frac{1}{2}\operatorname{sech}^2\left(\frac{x_l}{2}\right),$$

$$u''(x_l, t_0) = -\frac{1}{2}\operatorname{sech}^2\left(\frac{x_l}{2}\right)\tanh\left(\frac{x_l}{2}\right).$$

2.5.3 Numerical Results for the Burgers–Huxley Equation

Tables 2.7 through 2.9 show the comparisons of the exact solutions with the approximate solutions of the Burgers–Huxley equation taking $\alpha=1$, $\beta=1$, $\gamma=0.001$, $\delta=1$, and different values of t. In Tables 2.7 through 2.9, J is taken as 3, that is, $M=8$ and Δt is taken as 0.0001.

In the case of $\gamma=0.001$, the *R.M.S. error* between the Haar wavelet solutions and the exact solutions of Burgers–Huxley equations for $t=0.4, 0.6$, and 1 are 3.00204E−6, 4.50295E−6, and 7.50449E−6, respectively, and that of the VIM solutions and the exact solutions are 1.49925E−7, 2.24888E−7, and 3.74813E−7, respectively. In the case of the Burgers–Huxley equation, Figures 2.11 through 2.13 cite the comparison graphically between the numerical solutions obtained by Haar wavelet method, VIM, and exact solutions for different values of t and γ.

TABLE 2.7

Absolute Errors for the Solutions of the Burgers–Huxley Equation Using the Haar Wavelet Method and One Iteration of VIM at Various Collocation Points for x with $t=0.4$ and $\gamma=0.001$

x	Approximate Solutions Using Haar Wavelet Method (u_{approx})	Approximate Solutions Using VIM (u_{approx})	Exact Solutions (u_{exact})	Absolute Errors Using Haar Wavelet Method	Absolute Errors Using VIM
0.03125	0.00050006	0.00049994	0.000500054	6.56610E−9	1.49925E−7
0.09375	0.000500121	0.000499912	0.000500062	5.90949E−8	1.49925E−7
0.15625	0.000500234	0.00049992	0.000500069	1.64153E−7	1.49925E−7
0.21875	0.000500399	0.000499927	0.000500077	3.21739E−7	1.49925E−7
0.28125	0.000500617	0.000499935	0.000500085	5.31854E−7	1.49925E−7
0.34375	0.000500887	0.000499943	0.000500093	7.94498E−7	1.49925E−7
0.40625	0.000501210	0.000499951	0.000500101	1.10967E−6	1.49925E−7
0.46875	0.000501586	0.000499959	0.000500109	1.47737E−6	1.49925E−7
0.53125	0.000502014	0.000499966	0.000500116	1.89760E−6	1.49925E−7
0.59375	0.000502495	0.000499974	0.000500124	2.37036E−6	1.49925E−7
0.65625	0.000503028	0.000499982	0.000500132	2.89565E−6	1.49925E−7
0.71875	0.000503613	0.00049999	0.00050014	3.47347E−6	1.49925E−7
0.78125	0.000504251	0.000499998	0.000500148	4.10381E−6	1.49925E−7
0.84375	0.000504942	0.000500005	0.000500155	4.78669E−6	1.49925E−7
0.90625	0.000505685	0.000500013	0.000500163	5.52210E−6	1.49925E−7
0.96875	0.000506481	0.000500021	0.000500171	6.31009E−6	1.49925E−7

TABLE 2.8

Absolute Errors for the Solutions of the Burgers–Huxley Equation Using the Haar Wavelet Method and One Iteration of VIM at Various Collocation Points for x with $t=0.6$ and $\gamma=0.001$

x	Approximate Solutions Using Haar Wavelet Method (u_{approx})	Approximate Solutions Using VIM (u_{approx})	Exact Solutions (u_{exact})	Absolute Errors Using Haar Wavelet Method	Absolute Errors Using VIM
0.03125	0.000500089	0.000499854	0.000500079	9.84903E−9	2.24888E−7
0.09375	0.000500175	0.000499862	0.000500087	8.86412E−8	2.24888E−7
0.15625	0.000500341	0.00049987	0.000500094	2.46226E−7	2.24888E−7
0.21875	0.000500585	0.000499877	0.000500102	4.82602E−7	2.24888E−7
0.28125	0.000500908	0.000499885	0.00050011	7.97771E−7	2.24888E−7
0.34375	0.00050131	0.000499893	0.000500118	1.19173E−6	2.24888E−7
0.40625	0.00050179	0.000499901	0.000500126	1.66400E−6	2.24888E−7
0.46875	0.00050235	0.000499909	0.000500134	2.21600E−6	2.24888E−7
0.53125	0.000502988	0.000499916	0.000500141	2.84700E−6	2.24888E−7
0.59375	0.000503705	0.000499924	0.000500149	3.55600E−6	2.24888E−7
0.65625	0.0005045	0.000499932	0.000500157	4.34300E−6	2.24888E−7
0.71875	0.000505375	0.00049994	0.000500165	5.21000E−6	2.24888E−7
0.78125	0.000506328	0.000499948	0.000500173	6.15500E−6	2.24888E−7
0.84375	0.00050736	0.000499956	0.00050018	7.18000E−6	2.24888E−7
0.90625	0.000508471	0.000499963	0.000500188	8.28300E−6	2.24888E−7
0.96875	0.000509661	0.000499971	0.000500196	9.46500E−6	2.24888E−7

TABLE 2.9

Absolute Errors for the Solutions of the Burgers–Huxley Equation Using the Haar Wavelet Method and One Iteration of VIM at Various Collocation Points for x with $t=1$ and $\gamma=0.001$

x	Approximate Solutions Using Haar Wavelet Method (u_{approx})	Approximate Solutions Using VIM (u_{approx})	Exact Solutions (u_{exact})	Absolute Errors Using Haar Wavelet Method	Absolute Errors Using VIM
0.03125	0.000500145	0.000499754	0.000500129	1.6414E−8	3.74813E−7
0.09375	0.000500284	0.000499762	0.000500137	1.47726E−7	3.74813E−7
0.15625	0.000500555	0.00049977	0.000500144	4.10351E−7	3.74813E−7
0.21875	0.000500957	0.000499777	0.000500152	8.04288E−7	3.74813E−7
0.28125	0.00050149	0.000499785	0.00050016	1.32954E−6	3.74813E−7
0.34375	0.000502154	0.000499793	0.000500168	1.9861E−6	3.74813E−7
0.40625	0.00050295	0.000499801	0.000500176	2.7740E−6	3.74813E−7
0.46875	0.000503877	0.000499809	0.000500183	3.6940E−6	3.74813E−7
0.53125	0.000504935	0.000499816	0.000500191	4.7440E−6	3.74813E−7
0.59375	0.000506125	0.000499824	0.000500199	5.9260E−6	3.74813E−7
0.65625	0.000507445	0.000499832	0.000500207	7.2380E−6	3.74813E−7
0.71875	0.000508898	0.00049984	0.000500215	8.6830E−6	3.74813E−7
0.78125	0.000510481	0.000499848	0.000500223	1.02580E−5	3.74813E−7
0.84375	0.000512196	0.000499856	0.00050023	1.19660E−5	3.74813E−7
0.90625	0.000514042	0.000499863	0.000500238	1.38040E−5	3.74813E−7
0.96875	0.00051602	0.000499871	0.000500246	1.57740E−5	3.74813E−7

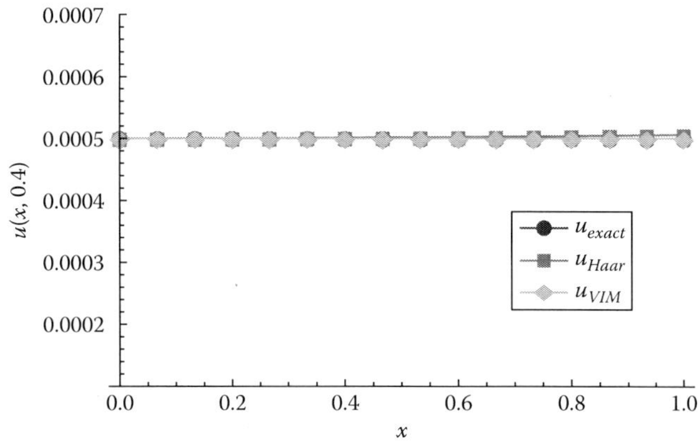

FIGURE 2.11
Comparison of Haar wavelet solutions and VIM solutions with the exact solution of the Burgers–Huxley equation when $t = 0.4$ and $\gamma = 0.001$.

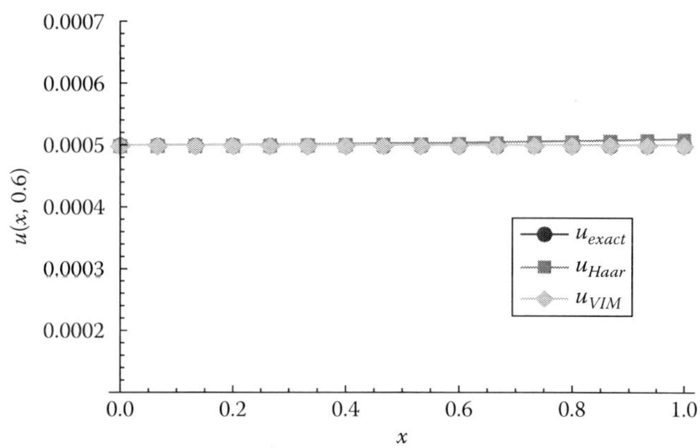

FIGURE 2.12
Comparison of Haar wavelet solutions and VIM solutions with the exact solution of the Burgers–Huxley equation when $t = 0.6$ and $\gamma = 0.001$.

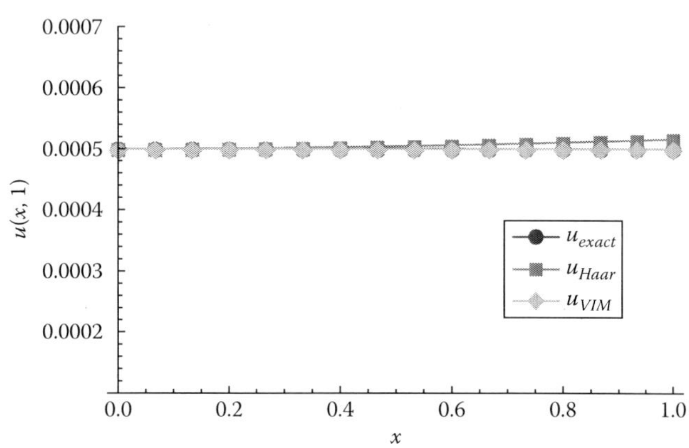

FIGURE 2.13
Comparison of Haar wavelet solutions and VIM solutions with the exact solution of the Burgers–Huxley equation when $t = 1$ and $\gamma = 0.001$.

2.6 Application of Analytical and Numerical Methods for Solving the Huxley Equation

The Huxley equation is a nonlinear partial differential equation of second order of the form

$$u_t = u_{xx} + u(k-u)(u-1), \quad k \neq 0, \tag{2.65}$$

with the initial condition

$$u(x,0) = \frac{1}{2}\left[1 + \tanh\left(\frac{x}{2\sqrt{2}}\right)\right]. \tag{2.66}$$

The exact solution of Equation 2.65 is given by [68]

$$u(x,t) = \frac{1}{2}\left[1 + \tanh\left\{\frac{1}{2\sqrt{2}}\left(x - \frac{2k-1}{\sqrt{2}}t\right)\right\}\right], \quad k \neq 0. \tag{2.67}$$

Taking $k=1$, the boundary conditions are

$$\begin{aligned} u(0,t) &= \frac{1}{2}\left[1 - \tanh\left(\frac{t}{4}\right)\right], \\ u(1,t) &= \frac{1}{2}\left[1 + \tanh\left\{\frac{1}{2\sqrt{2}}\left(1 - \frac{t}{\sqrt{2}}\right)\right\}\right]. \end{aligned} \tag{2.68}$$

2.6.1 Application of Variational Iteration Method for Solving the Huxley Equation

Construct a correctional functional as follows:

$$u_{n+1}(x,t) = u_n(x,t) + \int_0^t \lambda\left[(u_n)_\tau - \widetilde{(u_n)}_{xx} - \widetilde{u_n(k-u_n)(u_n-1)}\right]d\tau, \tag{2.69}$$

where

λ is a general Lagrangian multiplier whose optimal value can be found using variational theory

$\tilde{u}_n(x,t)$ is the restricted variation, that is, $\delta\tilde{u}_n(x,t) = 0$:

$$\delta u_{n+1}(x,t) = \delta u_n(x,t) + \delta\int_0^t \lambda\left[(u_n)_\tau - \widetilde{(u_n)}_{xx} - \widetilde{u_n(k-u_n)(u_n-1)}\right]d\tau \tag{2.70}$$

From Equation 2.70, we have

$$\delta u_{n+1}(x,t) = \delta u_n(x,t) + \delta\int_0^t \lambda\left(\frac{\partial u_n}{\partial\tau}\right)d\tau, \tag{2.71}$$

since $\delta\left(\widetilde{u_n}\right)_{xx} = 0$ and $\delta\left(\widetilde{u_n(k-u_n)(u_n-1)}\right) = 0$.

Integrating the right-hand side of Equation 2.71 yields

$$\delta u_{n+1}(x,t) = \delta u_n(x,t)(1+\lambda) - \int_0^t \delta u_n(x,t)\lambda' d\tau. \tag{2.72}$$

From stationary condition, we know

$$\delta u_{n+1}(x,t) = 0,$$

which yields that $\delta u_n(x,t)(1+\lambda) - \int_0^t \delta u_n(x,\tau)\lambda' d\tau = 0.$

This implies $1+\lambda = 0$ and $\lambda'(t) = 0$.

Hence

$$\lambda = -1. \tag{2.73}$$

Therefore, λ can be identified as -1, and the following variational iteration formula can be obtained as

$$u_{n+1}(x,t) = u_n(x,t) - \int_0^t \left[(u_n)_\tau - \widetilde{(u_n)_{xx}} - u_n\widetilde{(k - u_n)(u_n - 1)} \right] d\tau. \tag{2.74}$$

From the preceding iteration formula, Equation 2.74, we can obtain

$$u_1(x,t) = 0.5 + 0.5\tanh\left(\frac{x}{2\sqrt{2}}\right) - 0.125t\,\text{sech}^2\left(\frac{x}{2\sqrt{2}}\right), \tag{2.75}$$

with the initial condition

$$u_0(x,t) = \frac{1}{2}\left[1 + \tanh\left(\frac{x}{2\sqrt{2}}\right)\right].$$

2.6.2 Application of the Haar Wavelet Method for Solving the Huxley Equation

It is assumed that $\dot{u}''(x,t)$ can be expanded in terms of Haar wavelets as

$$\dot{u}''(x,t) = \sum_{i=1}^{2M} a_s(i)h_i(x) \quad \text{for } t \in [t_s, t_{s+1}] \tag{2.76}$$

where "·" and "′" stand for differentiation with respect to t and x, respectively.

Now, integrating Equation 2.76 with respect to t from t_s to t and twice with respect to x from 0 to x, the following equations are obtained:

$$u''(x,t) = (t - t_s)\sum_{i=1}^{2M} a_s(i)h_i(x) + u''(x,t_s), \tag{2.77}$$

$$u'(x,t)=(t-t_s)\sum_{i=1}^{2M}a_s(i)p_i(x)+u'(x,t_s)-u'(0,t_s)+u'(0,t), \tag{2.78}$$

$$u(x,t)=(t-t_s)\sum_{i=1}^{2M}a_s(i)q_i(x)+u(x,t_s)-u(0,t_s)+x\left[u'(0,t)-u'(0,t_s)\right]+u(0,t), \tag{2.79}$$

$$\dot{u}(x,t)=\sum_{i=1}^{2M}a_s(i)q_i(x)+x\dot{u}'(0,t)+\dot{u}(0,t). \tag{2.80}$$

By using the boundary conditions, at $x=1$, we have

$$\dot{u}'(0,t)=-\sum_{i=1}^{2M}a_s(i)q_i(1)+\dot{u}(1,t)-\dot{u}(0,t). \tag{2.81}$$

Discretizing the results by assuming $x\to x_l$, $t\to t_{s+1}$, we obtain

$$u''(x_l,t_{s+1})=(t_{s+1}-t_s)\sum_{i=1}^{2M}a_s(i)h_i(x_l)+u''(x_l,t_s), \tag{2.82}$$

$$u'(x_l,t_{s+1})=(t_{s+1}-t_s)\sum_{i=1}^{2M}a_s(i)p_i(x_l)+u'(x_l,t_s)-u'(0,t_s)+u'(0,t_{s+1}), \tag{2.83}$$

$$u(x_l,t_{s+1})=(t_{s+1}-t_s)\sum_{i=1}^{2M}a_s(i)q_i(x_l)+u(x_l,t_s)-u(0,t_s)+x_l\left[u'(0,t_{s+1})-u'(0,t_s)\right]+u(0,t_{s+1}), \tag{2.84}$$

$$\dot{u}(x_l,t_{s+1})=\sum_{i=1}^{2M}a_s(i)q_i(x_l)+x_l\dot{u}'(0,t_{s+1})+\dot{u}(0,t_{s+1}). \tag{2.85}$$

Substituting Equations 2.81 and 2.85 in Equation 2.65, we have

$$\sum_{i=1}^{2M}a_s(i)\left[q_i(x_l)-x_lq_i(1)\right]=u''(x_l,t_s)-u(x_l,t_s)\left[1-u(x_l,t_s)\right]^2 -\dot{u}(0,t_{s+1})-x_l\left[\dot{u}(1,t_{s+1})-\dot{u}(0,t_{s+1})\right]. \tag{2.86}$$

From Equation 2.86, the wavelet coefficients $a_s(i)$ can be successively calculated. This process started with

$$u(x_l,t_0)=\frac{1}{2}\left[1+\tanh\left(\frac{x_l}{2\sqrt{2}}\right)\right],$$

$$u'(x_l,t_0)=\frac{1}{4\sqrt{2}}\operatorname{sech}^2\left(\frac{x_l}{2\sqrt{2}}\right),$$

$$u''(x_l,t_0)=-\frac{1}{8}\operatorname{sech}^2\left(\frac{x_l}{2\sqrt{2}}\right)\tanh\left(\frac{x_l}{2\sqrt{2}}\right).$$

2.6.3 Numerical Results for the Huxley Equation

In Tables 2.10 through 2.12, J has been taken as 3, that is, $M=8$ and Δt is taken as 0.0001. Again, the *R.M.S. error* can be calculated for different values of t. For $t=0.4, 0.6$, and 1, the *R.M.S. error* between the Haar wavelet solutions and the exact solutions of Huxley equation are 0.0209303, 0.0354936, and 0.060677, respectively. For $t=0.4, 0.6$, and 1, the *R.M.S. error* between the VIM solutions and the exact solutions of Huxley equation are 8.10868E−4, 1.696E−3, and 4.03601E−3, respectively.

In the case of the Huxley equation, Figure 2.14 presents the comparison graphically between the numerical results obtained by the Haar wavelet method, VIM, and exact solutions for different values of t and $k=1$.

TABLE 2.10

Absolute Errors for the Solutions of the Huxley Equation Using the Haar Wavelet Method and One Iteration of VIM at Various Collocation Points for x with $k=1$ and $t=0.4$

x	Approximate Solutions Using Haar Wavelet Method (u_{approx})	Approximate Solutions Using VIM (u_{approx})	Exact Solutions (u_{exact})	Absolute Errors Using Haar Wavelet Method	Absolute Errors Using VIM
0.03125	0.455737	0.45553	0.455641	9.55529E−5	1.11054E−4
0.09375	0.467153	0.466622	0.466623	5.30592E−4	8.87183E−7
0.15625	0.478927	0.477746	0.477636	1.29070E−3	1.09283 E−4
0.21875	0.491048	0.488891	0.488672	2.37582E−3	2.19028 E−4
0.28125	0.503505	0.500046	0.499718	3.78609E−3	3.27923E−4
0.34375	0.516287	0.511201	0.510765	5.52185E−3	4.35548E−4
0.40625	0.529385	0.522343	0.521802	7.58360E−3	5.41497E−4
0.46875	0.542789	0.533462	0.532817	9.97205E−3	6.45376E−4
0.53125	0.556488	0.544547	0.5438	1.26880E−2	7.46806E−4
0.59375	0.570474	0.555586	0.554741	1.57326E−2	8.45430E−4
0.65625	0.584736	0.56657	0.565629	1.91070E−2	9.40912E−4
0.71875	0.599266	0.577487	0.576454	2.28126E−2	1.03294E−3
0.78125	0.614057	0.588327	0.587206	2.68510E−2	1.12123 E−3
0.84375	0.629099	0.599081	0.597876	3.12230E−2	1.20552E−3
0.90625	0.645576	0.609739	0.608453	3.71230E−2	1.28559E−3
0.96875	0.668244	0.620291	0.61893	4.93140E−2	1.36124E−3

TABLE 2.11

Absolute Errors for the Solutions of the Huxley Equation Using the Haar Wavelet Method and One Iteration of VIM at Various Collocation Points for x with $k=1$ and $t=0.6$

x	Approximate Solutions Using Haar Wavelet Method (u_{approx})	Approximate Solutions Using VIM (u_{approx})	Exact Solutions (u_{exact})	Absolute Errors Using Haar Wavelet Method	Absolute Errors Using VIM
0.03125	0.431155	0.430533	0.430968	1.87446E−4	4.34778E−4
0.09375	0.442789	0.441649	0.441837	9.51983E−4	1.88225E−4
0.15625	0.454998	0.452822	0.452763	2.23552E−3	5.90574E−5
0.21875	0.467771	0.46404	0.463734	4.03766E−3	3.06109E−4
0.28125	0.481098	0.475292	0.47474	6.35830E−3	5.51971E−4
0.34375	0.494968	0.486567	0.485771	9.19760E−3	7.95698E−4
0.40625	0.509372	0.497852	0.496816	1.25560E−2	1.03636E−3
0.46875	0.524298	0.509136	0.507863	1.64343E−2	1.27304E−3
0.53125	0.539737	0.520408	0.518904	2.08334E−2	1.50489E−3
0.59375	0.55568	0.531656	0.529925	2.57546E−2	1.73105E−3
0.65625	0.572117	0.542869	0.540918	3.11994E−2	1.95074E−3
0.71875	0.589041	0.554034	0.551871	3.71696E−2	2.16323E−3
0.78125	0.606441	0.565142	0.562774	4.36670E−2	2.36782E−3
0.84375	0.62431	0.57618	0.573616	5.06940E−2	2.56389E−3
0.90625	0.645842	0.58714	0.584389	6.14530E−2	2.75089E−3
0.96875	0.683829	0.598009	0.595081	8.87480E−2	2.92832E−3

TABLE 2.12

Absolute Errors for the Solutions of the Huxley Equation Using the Haar Wavelet Method and One Iteration of VIM at Various Collocation Points for x with $k=1$ and $t=1$

x	Approximate Solutions Using Haar Wavelet Method (u_{approx})	Approximate Solutions Using VIM (u_{approx})	Exact Solutions (u_{exact})	Absolute Errors Using Haar Wavelet Method	Absolute Errors Using VIM
0.03125	0.383171	0.380539	0.382747	4.23760E−4	2.20814E−3
0.09375	0.395066	0.391704	0.393241	1.82475E−3	1.53709E−3
0.15625	0.407796	0.402974	0.403834	3.96201E−3	8.60175E−4
0.21875	0.421352	0.414338	0.414518	6.83392E−3	1.79977E−4
0.28125	0.435721	0.425783	0.425282	1.04393E−2	5.00891E−4
0.34375	0.450895	0.437298	0.436118	1.47773E−2	1.17981E−3
0.40625	0.466863	0.448869	0.447015	1.98477E−2	1.85419E−3
0.46875	0.483614	0.460485	0.457964	2.56507E−2	2.52146E−3
0.53125	0.501139	0.472132	0.468953	3.21868E−2	3.17912E−3
0.59375	0.519429	0.483797	0.479972	3.94572E−2	3.82474E−3
0.65625	0.538474	0.495467	0.491011	4.74633E−2	4.45598E−3
0.71875	0.558265	0.507129	0.502058	5.62070E−2	5.07062E−3
0.78125	0.578795	0.51877	0.513104	6.56910E−2	5.66659E−3
0.84375	0.600054	0.530378	0.524137	7.59170E−2	6.24192E−3
0.90625	0.632639	0.541941	0.535146	9.74930E−2	6.79482E−3
0.96875	0.718961	0.553445	0.546121	1.72840E−1	7.32368E−3

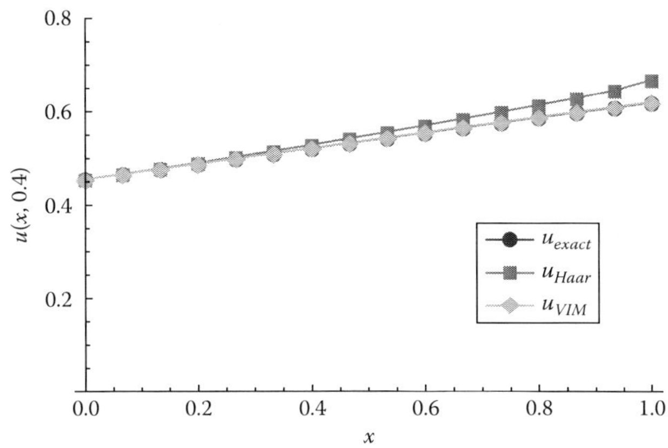

FIGURE 2.14
Comparison of Haar wavelet solutions and VIM solutions with the exact solution of Huxley equations when $t=0.4$ and $k=1$.

2.7 Numerical Solution of the Generalized mKdV Equation

Consider the generalized mKdV equation [80–82]

$$u_t + qu^2u_x + ru_{xxx} = 0, \quad 0 \le x \le 1,\ t \ge 0 \tag{2.87}$$

with the initial condition

$$u(x,0) = \sqrt{\frac{-6r}{q}}\tanh(x). \tag{2.88}$$

The exact solution of Equation 2.87 is given by [80]

$$u(x,t) = \sqrt{\frac{-6r}{q}}\tanh(x+2rt) \tag{2.89}$$

where q and r are parameters.

This exact solution satisfies the following boundary conditions:

$$\begin{aligned} u(0,t) &= \sqrt{\frac{-6r}{q}}\tanh(2rt), \quad t \ge 0 \\ u(1,t) &= \sqrt{\frac{-6r}{q}}\tanh(1+2rt), \quad t \ge 0. \end{aligned} \tag{2.90}$$

The Haar wavelet solution of $u(x,t)$ is sought by assuming that $\dot{u}'''(x,t)$ can be expanded in terms of Haar wavelets as

$$\dot{u}'''(x,t)=\sum_{i=1}^{2M}a_s(i)h_i(x) \quad \text{for } t\in[t_s,t_{s+1}]. \tag{2.91}$$

Integrating Equation 2.91 with respect to t from t_s to t and thrice with respect to x from 0 to x, the following equations are obtained:

$$u'''(x,t)=(t-t_s)\sum_{i=1}^{2M}a_s(i)h_i(x)+u'''(x,t_s), \tag{2.92}$$

$$u''(x,t)=(t-t_s)\sum_{i=1}^{2M}a_s(i)p_i(x)+u''(x,t_s)+u''(0,t)-u''(0,t_s), \tag{2.93}$$

$$u'(x,t)=(t-t_s)\sum_{i=1}^{2M}a_s(i)q_i(x)+u'(x,t_s)+x\left[u''(0,t)-u''(0,t_s)\right]+u'(0,t)-u'(0,t_s), \tag{2.94}$$

$$\begin{aligned}u(x,t)=(t-t_s)\sum_{i=1}^{2M}a_s(i)r_i(x)+u(x,t_s)+\frac{x^2}{2}\left[u''(0,t)-u''(0,t_s)\right]\\+x\left[u'(0,t)-u'(0,t_s)\right]+u(0,t)-u(0,t_s)\end{aligned} \tag{2.95}$$

$$\dot{u}(x,t)=\sum_{i=1}^{2M}a_s(i)r_i(x)+x\dot{u}'(0,t)+\frac{x^2}{2}\dot{u}''(0,t)+\dot{u}(0,t). \tag{2.96}$$

Using finite difference method

$$\dot{u}(0,t)=\frac{u(0,t)-u(0,t_s)}{(t-t_s)},$$

Equation 2.96 becomes

$$\begin{aligned}\dot{u}(x,t)=\sum_{i=1}^{2M}a_s(i)r_i(x)+\frac{x^2}{2}\left[\frac{u''(0,t)-u''(0,t_s)}{t-t_s}\right]\\+x\left[\frac{u'(0,t)-u'(0,t_s)}{t-t_s}\right]+\left[\frac{u(0,t)-u(0,t_s)}{t-t_s}\right].\end{aligned} \tag{2.97}$$

By using the boundary condition at $x=1$, Equation 2.94 becomes

$$u'(1,t)=(t-t_s)\sum_{i=1}^{2M}a_s(i)q_i(1)+u'(1,t_s)-u'(0,t_s)+u'(0,t)+\left[u''(0,t)-u''(0,t_s)\right].$$

This implies

$$u''(0,t)-u''(0,t_s)=-(t-t_s)\sum_{i=1}^{2M}a_s(i)q_i(1)+\left[u'(1,t)-u'(1,t_s)\right]-\left[u'(0,t)-u'(0,t_s)\right]. \quad (2.98)$$

Substituting Equation 2.98 in Equations 2.93 through 2.96, we have

$$\begin{aligned}u''(x,t)&=(t-t_s)\sum_{i=1}^{2M}a_s(i)p_i(x)+u''(x,t_s)\\&\quad+\left[-(t-t_s)\sum_{i=1}^{2M}a_s(i)q_i(1)+\left[u'(1,t)-u'(1,t_s)\right]-\left[u'(0,t)-u'(0,t_s)\right]\right],\end{aligned} \quad (2.99)$$

$$\begin{aligned}u'(x,t)&=(t-t_s)\sum_{i=1}^{2M}a_s(i)q_i(x)+u'(x,t_s)-u'(0,t_s)+u'(0,t)\\&\quad+x\left[-(t-t_s)\sum_{i=1}^{2M}a_s(i)q_i(1)+u'(1,t)-u'(1,t_s)-u'(0,t)+u'(0,t_s)\right],\end{aligned} \quad (2.100)$$

$$\begin{aligned}u(x,t)&=(t-t_s)\sum_{i=1}^{2M}a_s(i)r_i(x)\\&\quad+\frac{x^2}{2}\left[-(t-t_s)\sum_{i=1}^{2M}a_s(i)q_i(1)+u'(1,t)-u'(1,t_s)-u'(0,t)+u'(0,t_s)\right]\\&\quad+u(x,t_s)+x\left[u'(0,t)-u'(0,t_s)\right]+u(0,t)-u(0,t_s),\end{aligned} \quad (2.101)$$

$$\begin{aligned}\dot{u}(x,t)&=\sum_{i=1}^{2M}a_s(i)r_i(x)\\&\quad+\frac{x^2}{2(l-l_s)}\left[-(t-t_s)\sum_{i=1}^{2M}a_s(i)q_i(1)+u'(1,t)-u'(1,t_s)-u'(0,t)+u'(0,t_s)\right]\\&\quad+\frac{x}{t-t_s}\left[u'(0,t)-u'(0,t_s)\right]+\frac{1}{t-t_s}\left[u(0,t)-u(0,t_s)\right].\end{aligned} \quad (2.102)$$

Discretizing the preceding results by assuming $x\to x_l$, $t\to t_{s+1}$, we obtain

$$u'''(x_l,t_{s+1})=(t_{s+1}-t_s)\sum_{i=1}^{2M}a_s(i)h_i(x_l)+u'''(x_l,t_s),$$

$$\begin{aligned}u''(x_l,t_{s+1})&=(t_{s+1}-t_s)\sum_{i=1}^{2M}a_s(i)p_i(x_l)+u''(x_l,t_s)\\&\quad+\left[-(t_{s+1}-t_s)\sum_{i=1}^{2M}a_s(i)q_i(1)+\left[u'(1,t_{s+1})-u'(1,t_s)\right]-\left[u'(0,t_{s+1})-u'(0,t_s)\right]\right],\end{aligned} \quad (2.103)$$

$$u'(x_l,t_{s+1}) = (t_{s+1}-t_s)\sum_{i=1}^{2M} a_s(i)q_i(x_l) + u'(x_l,t_s) + u'(0,t_{s+1}) - u'(0,t_s)$$
$$+ x_l\left[-(t_{s+1}-t_s)\sum_{i=1}^{2M} a_s(i)q_i(1) + \left[u'(1,t_{s+1}) - u'(1,t_s)\right] - \left[u'(0,t_{s+1}) - u'(0,t_s)\right]\right], \tag{2.104}$$

$$u(x_l,t_{s+1}) = (t_{s+1}-t_s)\sum_{i=1}^{2M} a_s(i)r_i(x_l) + u(x_l,t_s) + u(0,t_{s+1}) - u(0,t_s) + x_l\left[u'(0,t_{s+1}) - u'(0,t_s)\right]$$
$$+\frac{x_l^2}{2}\left[-(t_{s+1}-t_s)\sum_{i=1}^{2M} a_s(i)q_i(1) + \left[u'(1,t_{s+1}) - u'(1,t_s)\right] - \left[u'(0,t_{s+1}) - u'(0,t_s)\right]\right], \tag{2.105}$$

$$\dot{u}(x_l,t_{s+1}) = \sum_{i=1}^{2M} a_s(i)r_i(x_l) + \frac{1}{t_{s+1}-t_s}\left[u(0,t_{s+1}) - u(0,t_s)\right] + \frac{x_l}{t_{s+1}-t_s}\left[u'(0,t_{s+1}) - u'(0,t_s)\right]$$
$$+\frac{x_l^2}{2(t_{s+1}-t_s)}\left[-(t_{s+1}-t_s)\sum_{i=1}^{2M} a_s(i)q_i(1) + u'(1,t_{s+1}) - u'(1,t_s) - u'(0,t_{s+1}) + u'(0,t_s)\right]. \tag{2.106}$$

Substituting the preceding equations in Equation 2.87, we have

$$\sum_{i=1}^{2M} a_s(i)r_i(x_l) + \frac{x_l^2}{2(t_{s+1}-t_s)}\left[-(t_{s+1}-t_s)\sum_{i=1}^{2M} a_s(i)q_i(1) + u'(1,t_{s+1}) - u'(1,t_s) - u'(0,t_{s+1}) + u'(0,t_s)\right]$$
$$+\frac{x_l}{t_{s+1}-t_s}\left[u'(0,t_{s+1}) - u'(0,t_s)\right] + \frac{1}{t_{s+1}-t_s}\left[u(0,t_{s+1}) - u(0,t_s)\right]$$
$$= 0.001u'''(x_l,t_s) - 6\left[u(x_l,t_s)\right]^2\left[u'(x_l,t_s)\right]. \tag{2.107}$$

Therefore,

$$\sum_{i=1}^{2M} a_s(i)\left[r_i(x_l) - \frac{x_l^2}{2}q_i(1)\right] = 0.001u'''(x_l,t_s) - 6\left[u(x_l,t_s)\right]^2\left[u'(x_l,t_s)\right]$$
$$-\frac{x_l^2}{2(t_{s+1}-t_s)}\left[u'(1,t_{s+1}) - u'(1,t_s)\right]$$
$$+\frac{x_l^2}{2(t_{s+1}-t_s)}\left[u'(0,t_{s+1}) - u'(0,t_s)\right]$$
$$-\frac{x_l}{t_{s+1}-t_s}\left[u'(0,t_{s+1}) - u'(0,t_s)\right]$$
$$-\frac{1}{t_{s+1}-t_s}\left[u(0,t_{s+1}) - u(0,t_s)\right]. \tag{2.108}$$

From the preceding equation, the wavelet coefficients $a_s(i)$ can be successively calculated. This process starts with

$$u(x_l,t_0)=\sqrt{\frac{-6r}{q}}\tanh(x_l),$$

$$u'(x_l,t_0)=\sqrt{\frac{-6r}{q}}\operatorname{sech}^2(x_l).$$

2.7.1 Numerical Results of the mKdV Equation

Tables 2.13 through 2.16 show the comparisons of the exact solutions with the approximate solutions of the mKdV equation taking $q=6$, $r=-0.001$ and different values of t. In Tables 2.13 through 2.16, J is taken as 3, that is, $M=8$ and Δt is taken as 0.0001.

In the case of $r=-0.001$, the *R.M.S. error* between the numerical solutions and the exact solutions of the mKdV equations for $t=0.2, 0.5, 0.8$, and 1 are 1.7137E−5, 4.33416E−5, 6.95581E−5, and 8.70423E−5, respectively, and for $r=-0.1$ and $t=0.2$, 0.5, 0.8, and 1, the *R.M.S. error* is found to be 0.00209359, 0.00624177, 0.011631, and 0.0159099, respectively. In Tables 2.17 through 2.20, also J has been taken as 3, that is, $M=8$ and Δt is taken as 0.0001.

Similarly, in the case of the mKdV equation, Figures 2.15 through 2.19 demonstrate the comparison graphically between the numerical and exact solutions for different values of t and r.

TABLE 2.13

Absolute Errors for the mKdV Equation at Various Collocation Points of x with $t=0.2$ and $r=-0.001$

x	Approximate Solution (u_{approx})	Exact Solution (u_{exact})	Absolute Error
0.03125	0.000975289	0.000975254	3.45313E−8
0.09375	0.00294375	0.00294344	3.10577E−7
0.15625	0.00488976	0.00488889	8.63705E−7
0.21875	0.00679886	0.00679716	1.6977E−6
0.28125	0.00865771	0.00865489	2.8188E−6
0.34375	0.0104545	0.0104502	4.23521E−6
0.40625	0.0121789	0.012173	5.95649E−6
0.46875	0.0138229	0.0138149	7.99297E−6
0.53125	0.01538	0.0153696	1.03551E−5
0.59375	0.0168459	0.0168328	1.30530E−5
0.65625	0.018218	0.0182019	1.60958E−5
0.71875	0.0194955	0.019476	1.94918E−5
0.78125	0.0206791	0.0206558	2.32477E−5
0.84375	0.0217706	0.0217432	2.73688E−5
0.90625	0.0227729	0.022741	3.18592E−5
0.96875	0.0236899	0.0236532	3.67252E−5

TABLE 2.14

Absolute Errors for the mKdV Equation at Various Collocation Points of x with $t=0.5$ and $r=-0.001$

x	Approximate Solution (u_{approx})	Exact Solution (u_{exact})	Absolute Error
0.03125	0.000956384	0.000956298	8.64056E−8
0.09375	0.00292541	0.00292463	7.77845E−7
0.15625	0.00487254	0.00487037	2.16512E−6
0.21875	0.00678332	0.00677906	4.25957E−6
0.28125	0.00864442	0.00863734	7.07870E−6
0.34375	0.010444	0.0104333	1.06449E−5
0.40625	0.0121718	0.0121568	1.49840E−5
0.46875	0.0138197	0.0137995	2.01238E−5
0.53125	0.0153812	0.0153552	2.60922E−5
0.59375	0.0168522	0.0168192	3.29164E−5
0.65625	0.0182299	0.0181892	4.06215E−5
0.71875	0.0195135	0.0194643	4.92298E−5
0.78125	0.0207037	0.020645	5.87601E−5
0.84375	0.0218024	0.0217332	6.92277E−5
0.90625	0.0228125	0.0227319	8.06444E−5
0.96875	0.0237378	0.0236448	9.30207E−5

TABLE 2.15

Absolute Errors for the mKdV Equation at Various Collocation Points of x with $t=0.8$ and $r=-0.001$

x	Approximate Solution (u_{approx})	Exact Solution (u_{exact})	Absolute Error
0.03125	0.00093748	0.000937341	1.38295E−7
0.09375	0.00290707	0.00290582	1.24525E−6
0.15625	0.00485531	0.00485185	3.46691E−6
0.21875	0.00676778	0.00676095	6.82219E−6
0.28125	0.00863112	0.00861978	1.13399E−5
0.34375	0.0104335	0.0104164	1.70566E−5
0.40625	0.0121647	0.0121406	2.40144E−5
0.46875	0.0138164	0.0137842	3.22585E−5
0.53125	0.0153825	0.0153406	4.18346E−5
0.59375	0.0168584	0.0168056	5.27869E−5
0.65625	0.0182417	0.0181765	6.51565E−5
0.71875	0.0195315	0.0194525	7.89797E−5
0.78125	0.0207283	0.0206341	9.42876E−5
0.84375	0.0218343	0.0217232	1.11105E−4
0.90625	0.0228522	0.0227227	1.29453E−4
0.96875	0.0237858	0.0236364	1.49345E−4

TABLE 2.16

Absolute Errors for the mKdV Equation at Various Collocation Points of x with $t=1$ and $r=-0.001$

x	Approximate Solution (u_{approx})	Exact Solution (u_{exact})	Absolute Error
0.03125	0.000924876	0.000924703	1.72895E−7
0.09375	0.00289483	0.00289328	1.55692E−6
0.15625	0.00484383	0.0048395	4.33497E−6
0.21875	0.00675741	0.00674888	8.53102E−6
0.28125	0.00862225	0.00860807	1.41814E−5
0.34375	0.0104265	0.0104051	2.13321E−5
0.40625	0.0121599	0.0121299	3.00363E−5
0.46875	0.0138143	0.0137739	4.03506E−5
0.53125	0.0153833	0.015331	5.23326E−5
0.59375	0.0168626	0.0167965	6.60379E−5
0.65625	0.0182496	0.0181681	8.15183E−5
0.71875	0.0195434	0.0194446	9.88196E−5
0.78125	0.0207448	0.0206268	1.17981E−4
0.84375	0.0218555	0.0217165	1.39034E−4
0.90625	0.0228786	0.0227166	1.62005E−4
0.96875	0.0238178	0.0236309	1.86910E−4

TABLE 2.17

Absolute Errors for the mKdV Equation at Various Collocation Points of x with $t=0.2$ and $r=-0.1$

x	Approximate Solution (u_{approx})	Exact Solution (u_{exact})	Absolute Error
0.03125	−0.00276621	−0.00276692	7.13784E−7
0.09375	0.0169876	0.0169809	6.65888E−6
0.15625	0.0366175	0.0365968	2.07298E−5
0.21875	0.0559793	0.0559313	4.79820E−5
0.28125	0.0749397	0.0748436	9.60655E−5
0.34375	0.09338	0.0932052	1.74756E−4
0.40625	0.111199	0.110903	2.95389E−4
0.46875	0.128314	0.127843	4.70239E−4
0.53125	0.144661	0.14395	7.11916E−4
0.59375	0.160198	0.159166	1.03281E−3
0.65625	0.174899	0.173455	1.44463E−3
0.71875	0.188756	0.186798	1.95802E−3
0.78125	0.201774	0.199192	2.58235E−3
0.84375	0.213974	0.210648	3.32555E−3
0.90625	0.225384	0.22119	4.19409E−3
0.96875	0.236043	0.23085	5.19293E−3

TABLE 2.18

Absolute Errors for the mKdV Equation at Various Collocation Points of x with $t=0.5$ and $r=-0.1$

x	Approximate Solution (u_{approx})	Exact Solution (u_{exact})	Absolute Error
0.03125	−0.0217032	−0.0217065	3.31576E−6
0.09375	−0.00194595	−0.0019764	3.04463E−5
0.15625	0.0178593	0.0177691	9.02230E−5
0.21875	0.0375721	0.0373766	1.95531E−4
0.28125	0.057063	0.0566968	3.66140E−4
0.34375	0.0762172	0.0755894	6.27708E−4
0.40625	0.0949371	0.0939266	1.01047E−3
0.46875	0.113144	0.111596	1.54778E−3
0.53125	0.130779	0.128504	2.27451E−3
0.59375	0.147801	0.144576	3.22569E−3
0.65625	0.16419	0.159755	4.43515E−3
0.71875	0.179941	0.174007	5.93453E−3
0.78125	0.195064	0.187312	7.75255E−3
0.84375	0.209583	0.199668	9.91447E−3
0.90625	0.223529	0.211087	1.24419E−2
0.96875	0.236946	0.221593	1.53524E−2

TABLE 2.19

Absolute Errors for the mKdV Equation at Various Collocation Points of x with $t=0.8$ and $r=-0.1$

x	Approximate Solution (u_{approx})	Exact Solution (u_{exact})	Absolute Error
0.03125	−0.0404827	−0.0404909	8.13685E−6
0.09375	−0.0208453	−0.0209195	7.42051E−5
0.15625	−0.000970764	−0.00118585	2.15085E−4
0.21875	0.0190084	0.018557	4.51366E−4
0.28125	0.0389709	0.0381558	8.15061E−4
0.34375	0.05881	0.0574616	1.34835E−3
0.40625	0.0784361	0.0763344	2.10172E−3
0.46875	0.0977786	0.0946469	3.13176E−3
0.53125	0.116786	0.112288	4.49874E−3
0.59375	0.135428	0.129163	6.26422E−3
0.65625	0.153689	0.1452	8.48888E−3
0.71875	0.171574	0.160343	1.12307E−2
0.78125	0.189101	0.174557	1.45436E−2
0.84375	0.206301	0.187824	1.84763E−2
0.90625	0.223215	0.200143	2.30719E−2
0.96875	0.239891	0.211525	2.83664E−2

TABLE 2.20

Absolute Errors for the mKdV Equation at Various Collocation Points of x with $t=1$ and $r=-0.1$

x	Approximate Solution (u_{approx})	Exact Solution (u_{exact})	Absolute Error
0.03125	−0.0528499	−0.0528626	1.27575E−5
0.09375	−0.0333573	−0.0334734	1.16031E−4
0.15625	−0.0134928	−0.0138262	3.33334E−4
0.21875	0.00661869	0.00592858	6.90107E−4
0.28125	0.0268634	0.0256371	1.22625E−3
0.34375	0.0471421	0.0451472	1.99484E−3
0.40625	0.0673727	0.0643127	3.06007E−3
0.46875	0.0874922	0.0829976	4.49462E−3
0.53125	0.107457	0.10108	6.37673E−3
0.59375	0.127243	0.118455	8.78725E−3
0.65625	0.146843	0.135037	1.18068E−2
0.71875	0.16627	0.150757	1.55134E−2
0.78125	0.185549	0.165569	1.99805E−2
0.84375	0.204718	0.179443	2.52756E−2
0.90625	0.223827	0.192368	3.14593E−2
0.96875	0.24293	0.204347	3.85831E−2

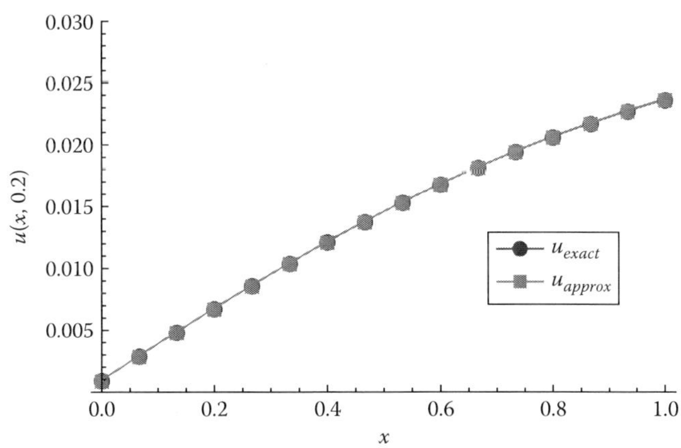

FIGURE 2.15
Comparison of the numerical solution and exact solution of the mKdV equation when $t=0.2$ and $r=-0.001$.

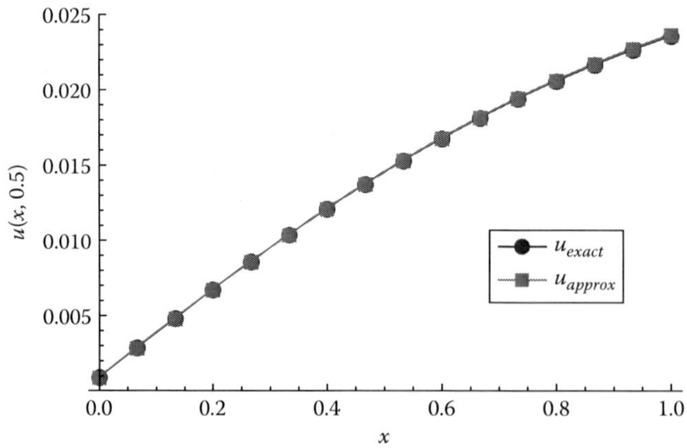

FIGURE 2.16
Comparison of the numerical solution and exact solution of the mKdV equation when $t=0.5$ and $r=-0.001$.

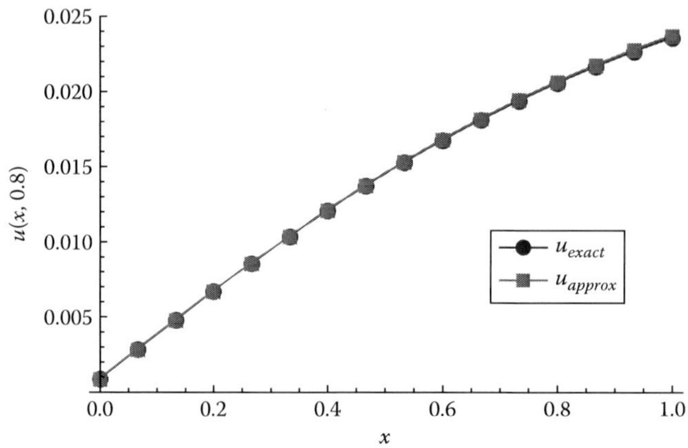

FIGURE 2.17
Comparison of the numerical solution and exact solution of the mKdV equation when $t=0.8$ and $r=-0.001$.

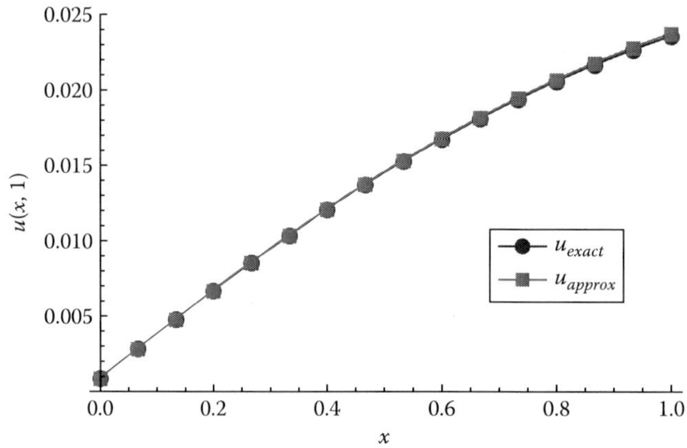

FIGURE 2.18
Comparison of the numerical solution and exact solution of the mKdV equation when $t=1.0$ and $r=-0.001$.

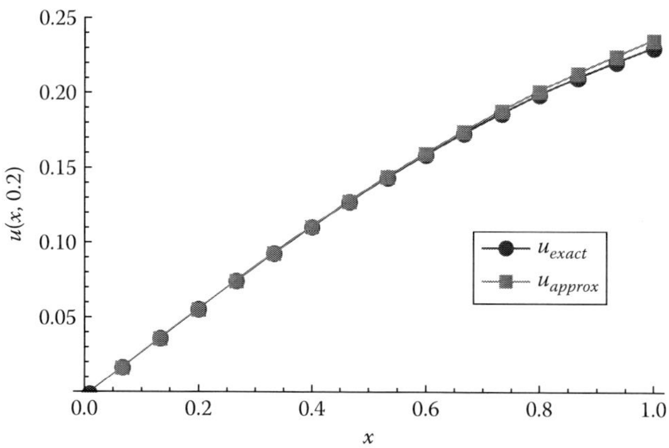

FIGURE 2.19
Comparison of the numerical solution and exact solution of the mKdV equation when $t=0.2$ and $r=-0.1$.

2.8 Error of Collocation Method

The Haar wavelet family for $x \in [0,1)$ is defined as follows:

$$h_i(x) = \begin{cases} 1 & x \in [\xi_1, \xi_2) \\ -1 & x \in [\xi_2, \xi_3) \\ 0 & \text{elsewhere} \end{cases} \tag{2.109}$$

where

$$\xi_1 = \frac{k}{m}, \quad \xi_2 = \frac{k+0.5}{m}, \quad \xi_3 = \frac{k+1}{m}.$$

Consider

$$u(x,\tilde{t}) = (\tilde{t} - t_s)\sum_{i=1}^{2M} u(i)Q_i(x) + \psi(x,\tilde{t}), \quad \tilde{t} \in [t_s, t_{s+1}]. \tag{2.110}$$

Define a projection map

$$P_m : L^2(\Omega) \to V_J$$

by the rule

$$P_m u(x,t_{s+1}) = u_m(x,t_{s+1}) = h\sum_{i=1}^{2M} u(i)Q_i(x) \tag{2.111}$$

where $\Omega = [0,1)$ and V_J is a subspace of $L^2(\Omega)$.

Now, we have to estimate $\|u - P_m u\|$ for arbitrary $u \in L^2(\Omega)$.

Lemma 2.1

Let $u(x,t)$ be defined on $L^2(\Omega)$ and P_m be the projection map as defined earlier, then

$$\|u - P_m u\| \le \frac{\max|u|}{4M^2}.$$

Proof

The integral $\int_0^1 u_m(x,t)dx$ is a ramp, $\frac{u_i}{4M^2}\left[\frac{1}{2M}+(x-\xi_3)\right]$, on the interval $[0,1)$ with the average value $\frac{u_i}{8M^2}\left[\frac{1}{2M}+(1-\xi_3)\right]$.

The error in approximating the ramp by this constant value over the interval $[0,1)$ is

$$r(x)=\frac{u_i}{8M^2}\left[\frac{1}{2M}+(1-\xi_3)\right]-\frac{u_i}{4M^2}\left[\frac{1}{2M}+(x-\xi_3)\right].$$

Hence, using $E(x)$ as the least square of the error on Ω, we have

$$\begin{aligned}
E^2(x) &= \int_0^1 \left[r(x)\right]^2 dx \\
&= \int_0^1 \left(\frac{u_i}{8M^2}\left[\frac{1}{2M}+(1-\xi_3)\right]-\frac{u_i}{4M^2}\left[\frac{1}{2M}+(x-\xi_3)\right]\right)^2 dx \\
&= \int_0^1 \left(\frac{-u_i}{16M^3}+\frac{u_i}{8M^2}-\frac{u_i\xi_3}{8M^2}-\frac{u_i x}{4M^2}+\frac{u_i\xi_3}{4M^2}\right)^2 dx \\
&= \int_0^1 \left(\frac{u_i\xi_3}{8M^2}+\frac{u_i}{8M^2}-\frac{u_i}{16M^3}-\frac{u_i x}{4M^2}\right)^2 dx \\
&= \left(\frac{u_i}{4M^2}\right)^2 \int_0^1 \left(\frac{\xi_3}{2}+\frac{1}{2}-\frac{1}{4M}-x\right)^2 dx \\
&\le \left(\frac{u_i}{4M^2}\right)^2.
\end{aligned}$$

This implies that

$$|E(x)| \le \frac{|u_i|}{4M^2}. \tag{2.112}$$

On the interval Ω, we have

$$\|u - P_m u\| = \max_{x\in\Omega} E(x) \le \frac{\max|u|}{4M^2}. \tag{2.113}$$

2.9 Error Analysis

Let $P_m: L^2(\Omega)\rightarrow V_J$ be a projection map and be defined by

$$P_m u(x,t_{s+1}) = u_m(x,t_{s+1}) = h\sum_{i=1}^{2M} u(i)Q_i(x). \tag{2.114}$$

Let us consider the generalized modified Burgers' equation

$$\frac{\partial u}{\partial t} + u^2\frac{\partial u}{\partial x} = \nu\frac{\partial^2 u}{\partial x^2}. \tag{2.115}$$

Suppose that $u_m = P_m u$ is the approximate solution of Equation 2.115 obtained by wavelet collocation method

$$\frac{\partial u_m}{\partial t} + u_m^2\frac{\partial u_m}{\partial x} = \nu\frac{\partial^2 u_m}{\partial x^2} + e \tag{2.116}$$

then

$$\|e\| \le \frac{A}{4M^2},$$

where

$$A = \max\left|\frac{\partial u}{\partial t}\right| + \nu\max\left|\frac{\partial^2 u}{\partial x^2}\right| + \max\left|u^2\frac{\partial u}{\partial x}\right| + P_m u^2 \max\left|\frac{\partial u}{\partial x}\right|.$$

Proof

Subtracting Equation 2.115 from Equation 2.116, we have

$$\begin{aligned}
e &= \frac{\partial u_m}{\partial t} + u_m^2\frac{\partial u_m}{\partial x} - \nu\frac{\partial^2 u_m}{\partial x^2} - \frac{\partial u}{\partial t} - u^2\frac{\partial u}{\partial x} + \nu\frac{\partial^2 u}{\partial x^2} \\
&= \frac{\partial u_m}{\partial t} - \frac{\partial u}{\partial t} - \nu\left(\frac{\partial^2 u_m}{\partial x^2} - \frac{\partial^2 u}{\partial x^2}\right) + u_m^2\frac{\partial u_m}{\partial x} - u^2\frac{\partial u}{\partial x} \\
&= \frac{\partial(u_m - u)}{\partial t} - \nu\frac{\partial^2(u_m - u)}{\partial x^2} + (u_m)^2\frac{\partial(u_m - u)}{\partial x} - \left(u^2 - {u_m}^2\right)\frac{\partial u}{\partial x} \\
&= \frac{\partial(P_m - I)u}{\partial t} - \nu\frac{\partial^2(P_m - I)u}{\partial x^2} + (P_m u)^2\frac{\partial(P_m - I)u}{\partial x} - (I - P_m)u^2\frac{\partial u}{\partial x}.
\end{aligned}$$

This implies

$$\begin{aligned}\|e\| &\le \|P_m - I\|\max\left|\frac{\partial u}{\partial t}\right| + \nu\|P_m - I\|\max\left|\frac{\partial^2 u}{\partial x^2}\right| + \|P_m - I\|\left(P_m u\right)^2\max\left|\frac{\partial u}{\partial x}\right| + \|P_m - I\|\max\left|u^2\frac{\partial u}{\partial x}\right| \\ &\le \|P_m - I\|\left[\max\left|\frac{\partial u}{\partial t}\right| + \nu\max\left|\frac{\partial^2 u}{\partial x^2}\right| + \left(P_m u\right)^2\max\left|\frac{\partial u}{\partial x}\right| + \max\left|u^2\frac{\partial u}{\partial x}\right|\right] \\ &\le \frac{A}{4M^2}\end{aligned}$$

where

$$A = \max\left|\frac{\partial u}{\partial t}\right| + \nu\max\left|\frac{\partial^2 u}{\partial x^2}\right| + \left(P_m u\right)^2\max\left|\frac{\partial u}{\partial x}\right| + \max\left|u^2\frac{\partial u}{\partial x}\right|.$$

2.10 Conclusion

In this chapter, VIM and the Haar wavelet method have been successfully implemented to compute approximate analytical as well as numerical solutions of nonlinear partial differential equations, namely, the generalized Burgers–Huxley and Huxley equations. The obtained results are then compared with the exact solutions. The results thus obtained have been cited in tables and also graphically to demonstrate the comparison of VIM and the Haar wavelet method. Also, the Burgers' equation, modified Burgers' equation, and mKdV equation have been solved by the Haar wavelet method. The acquired results are then compared with the exact solutions as well as solutions available in open literature. These have been reported in tables and have also been shown in the graphs in order to demonstrate the accuracy and efficiency of the proposed scheme based on Haar wavelet method. The present numerical scheme is reliable and convenient for solving nonlinear partial differential equations. The main advantages of the proposed scheme are its simplicity, applicability, and less computational errors. Moreover, the error may be reduced significantly if we increase the level of resolution, which prompts more number of collocation points.

3

Numerical Solution of a System of Partial Differential Equations

3.1 Introduction

Numerical solutions of nonlinear differential equations are of great importance in physical problems since so far, there exists no general technique for finding analytical and numerical solutions of systems of nonlinear partial differential equations. It is well known that many physical, chemical, and biological problems are characterized by the interaction of diffusion and reaction processes. With the development of science and engineering, nonlinear evolution equations have been used as the models to describe physical phenomena in fluid mechanics, plasma waves, solid-state physics, chemical physics, etc. Systems of nonlinear partial differential equations have also been noticed to arise in many chemical and biological applications. So, for the last few decades, a great deal of attention has been directed toward the solution (both exact and numerical) of these problems. Various methods are available in the literature for the exact and numerical solution of these problems. However, nonlinear partial differential equations are not, in general, easy to handle.

In this chapter, we apply the homotopy perturbation method (HPM), optimal homotopy asymptotic method (OHAM), and Haar wavelet method in order to compute the numerical solutions of the nonlinear system of partial differential equations like Boussinesq–Burgers' equations. Our aim in the present work is to implement HPM and OHAM in order to demonstrate the capability of these methods in handling a system of nonlinear equations so that one can apply it to various types of nonlinearity.

In HPM and OHAM, the concept of homotopy from topology and the conventional perturbation technique were merged to propose a general analytic procedure for the solution of nonlinear problems. Thus, these methods are independent of the existence of a small parameter in the problem at hand and thereby overcome the limitations of conventional perturbation technique. OHAM, however, is the most generalized form of HPM as it employs a more general auxiliary function, $H(p)$, in place of HPM's p.

OHAM provides a simple and easy way to control and adjust the convergence region for strong nonlinearity and is applicable to highly nonlinear fluid problems like Boussinesq–Burgers' equations. In the proposed OHAM procedure, the construction of the homotopy is quite different. The way to ensure the convergence in OHAM is quite different and more rigorous. Unlike other homotopy procedures, OHAM ensures a very rapid convergence since it needs only two or three terms for achieving an accurate solution instead of an infinite series. This is in fact the true power of the method. The convergence control parameters $C_1, C_2, C_3, \ldots$ provide us a convenient way to guarantee the convergence of OHAM series solution. Moreover, the optimal values of convergence control parameters guarantee the certain convergence of OHAM series solution.

3.2 Overview of the Problem

Generalized Boussinesq–Burgers' equations [83–85] are nonlinear partial differential equations of the form

$$u_t - \frac{1}{2} v_x + 2uu_x = 0, \tag{3.1}$$

$$v_t - \frac{1}{2} u_{xxx} + 2(uv)_x = 0, \quad 0 \le x \le 1 \tag{3.2}$$

with initial conditions

$$u(x,0) = \frac{ck}{2} + \frac{ck}{2} \tanh\left(\frac{-kx - \ln b}{2}\right), \tag{3.3}$$

$$v(x,0) = \frac{-k^2}{8} \operatorname{sech}^2\left(\frac{kx + \ln b}{2}\right). \tag{3.4}$$

The exact solutions of Equations 3.1 and 3.2 are given by [86]

$$u(x,t) = \frac{ck}{2} + \frac{ck}{2} \tanh\left(\frac{ck^2 t - kx - \ln b}{2}\right), \tag{3.5}$$

$$v(x,t) = \frac{-k^2}{8} \operatorname{sech}^2\left(\frac{kx - ck^2 t + \ln b}{2}\right). \tag{3.6}$$

The Boussinesq–Burgers' equations arise in the study of fluid flow and describe the propagation of shallow water waves. Here x and t, respectively, represent the normalized space and time, $u(x,t)$ is the horizontal velocity field, and $v(x,t)$ denotes the height of the water surface above a horizontal level at the bottom.

Various analytical methods such as the Darboux transformation method [87], Lax pair, and Bäcklund transformation method [86] have been used in attempting to solve Boussinesq–Burgers' equations.

These exact solutions given in Equations 3.5 and 3.6 satisfy the following boundary conditions:

$$u(0,t) = \frac{ck}{2} + \frac{ck}{2} \tanh\left(\frac{ck^2 t - \ln b}{2}\right),$$

$$u(1,t) = \frac{ck}{2} + \frac{ck}{2} \tanh\left(\frac{ck^2 t - k - \ln b}{2}\right),$$

$$v(0,t) = \frac{-k^2}{8}\operatorname{sech}^2\left(\frac{-ck^2t + \ln b}{2}\right),$$

$$v(1,t) = \frac{-k^2}{8}\operatorname{sech}^2\left(\frac{k - ck^2t + \ln b}{2}\right).$$

3.3 Analytical Solution of a System of Nonlinear Partial Differential Equations

3.3.1 Application of HPM to Boussinesq–Burgers' Equations

Using HPM [25,26], the homotopy for Equations 3.1 and 3.2 can be written as

$$(1-p)\left(\frac{\partial u}{\partial t} - 0\right) + p\left(\frac{\partial u}{\partial t} - \frac{1}{2}\frac{\partial v}{\partial x} + 2u\frac{\partial u}{\partial x}\right) = 0, \tag{3.7}$$

$$(1-p)\left(\frac{\partial v}{\partial t} - 0\right) + p\left(\frac{\partial v}{\partial t} - \frac{1}{2}\frac{\partial^3 u}{\partial x^3} + 2\frac{\partial (uv)}{\partial x}\right) = 0. \tag{3.8}$$

This implies

$$\frac{\partial u}{\partial t} = p\left(\frac{1}{2}\frac{\partial v}{\partial x} - 2u\frac{\partial u}{\partial x}\right), \tag{3.9}$$

$$\frac{\partial v}{\partial t} = p\left(\frac{1}{2}\frac{\partial^3 u}{\partial x^3} - 2\frac{\partial (uv)}{\partial x}\right). \tag{3.10}$$

By substituting $u(x,t) = \sum_{n=0}^{\infty} p^n u_n(x,t)$ and $v(x,t) = \sum_{n=0}^{\infty} p^n v_n(x,t)$ in Equations 3.9 and 3.10, we get

$$\frac{\partial}{\partial t}\left(\sum_{n=0}^{\infty} p^n u_n(x,t)\right) = p\left(\frac{1}{2}\frac{\partial}{\partial x}\left(\sum_{n=0}^{\infty} p^n v_n(x,t)\right) - 2\left(\sum_{n=0}^{\infty} p^n u_n(x,t)\right)\left(\frac{\partial}{\partial x}\left(\sum_{n=0}^{\infty} p^n u_n(x,t)\right)\right)\right), \tag{3.11}$$

$$\frac{\partial}{\partial t}\left(\sum_{n=0}^{\infty} p^n v_n(x,t)\right) = p\left(\frac{1}{2}\frac{\partial^3}{\partial x^3}\left(\sum_{n=0}^{\infty} p^n u_n(x,t)\right) - 2\frac{\partial}{\partial x}\left(\left(\sum_{n=0}^{\infty} p^n u_n(x,t)\right)\left(\sum_{n=0}^{\infty} p^n v_n(x,t)\right)\right)\right). \tag{3.12}$$

Comparing the coefficients of different powers in p for Equations 3.11 and 3.12, we have the following system of partial differential equations:

$$\text{Coefficients of } p^0\text{: } \frac{\partial u_0}{\partial t} = 0, \tag{3.13}$$

$$\frac{\partial v_0}{\partial t} = 0 \tag{3.14}$$

$$\text{Coefficients of } p^1\text{: } \frac{\partial u_1}{\partial t} = \frac{1}{2}\frac{\partial v_0}{\partial x} - 2u_0\frac{\partial u_0}{\partial x}, \tag{3.15}$$

$$\frac{\partial v_1}{\partial t} = \frac{1}{2}\frac{\partial^3 u_0}{\partial x^3} - 2\frac{\partial}{\partial x}(u_0 v_0) \tag{3.16}$$

$$\text{Coefficients of } p^2\text{: } \frac{\partial u_2}{\partial t} = \frac{1}{2}\frac{\partial v_1}{\partial x} - 2\left(u_0\frac{\partial u_1}{\partial x} + u_1\frac{\partial u_0}{\partial x}\right), \tag{3.17}$$

$$\frac{\partial v_2}{\partial t} = \frac{1}{2}\frac{\partial^3 u_1}{\partial x^3} - 2\frac{\partial}{\partial x}(u_0 v_1 + u_1 v_0) \tag{3.18}$$

$$\text{Coefficients of } p^3\text{: } \frac{\partial u_3}{\partial t} = \frac{1}{2}\frac{\partial v_2}{\partial x} - 2\left(u_0\frac{\partial u_2}{\partial x} + u_1\frac{\partial u_1}{\partial x} + u_2\frac{\partial u_0}{\partial x}\right), \tag{3.19}$$

$$\frac{\partial v_3}{\partial t} = \frac{1}{2}\frac{\partial^3 u_2}{\partial x^3} - 2\frac{\partial}{\partial x}(u_0 v_2 + u_1 v_1 + u_2 v_0) \tag{3.20}$$

$$\text{Coefficients of } p^4\text{: } \frac{\partial u_4}{\partial t} = \frac{1}{2}\frac{\partial v_3}{\partial x} - 2\left(u_0\frac{\partial u_3}{\partial x} + u_1\frac{\partial u_2}{\partial x} + u_2\frac{\partial u_1}{\partial x} + u_3\frac{\partial u_0}{\partial x}\right), \tag{3.21}$$

$$\frac{\partial v_4}{\partial t} = \frac{1}{2}\frac{\partial^3 u_3}{\partial x^3} - 2\frac{\partial}{\partial x}(u_0 v_3 + u_1 v_2 + u_2 v_1 + u_3 v_0) \tag{3.22}$$

and so on.

By putting $u(x,0) = \tilde{u}_0$ and $v(x,0) = \tilde{v}_0$ in Equations 3.13 through 3.22 and solving them, we obtain

$$u_1 = \frac{t}{2}\frac{\partial \tilde{v}_0}{\partial x} - 2t\tilde{u}_0\frac{\partial \tilde{u}_0}{\partial x},$$

$$v_1 = \frac{t}{2}\frac{\partial^3 \tilde{u}_0}{\partial x^3} - 2t\frac{\partial}{\partial x}(\tilde{u}_0\tilde{v}_0);$$

$$u_2 = \frac{t^2}{2}\left(\frac{1}{2}\frac{\partial v_1}{\partial x} - 2\left(\tilde{u}_0\frac{\partial u_1}{\partial x} + u_1\frac{\partial \tilde{u}_0}{\partial x}\right)\right),$$

$$v_2 = \frac{t^2}{2}\left(\frac{1}{2}\frac{\partial^3 u_1}{\partial x^3} - 2\frac{\partial}{\partial x}\left(\tilde{u}_0 v_1 + u_1 \tilde{v}_0\right)\right);$$

$$u_3 = \frac{t^3}{6}\left(\frac{1}{2}\frac{\partial v_2}{\partial x} - 2\left(\tilde{u}_0\frac{\partial u_2}{\partial x} + u_1\frac{\partial u_1}{\partial x} + u_2\frac{\partial \tilde{u}_0}{\partial x}\right)\right),$$

$$v_3 = \frac{t^3}{6}\left(\frac{1}{2}\frac{\partial^3 u_2}{\partial x^3} - 2\frac{\partial}{\partial x}\left(\tilde{u}_0 v_2 + u_1 v_1 + u_2 \tilde{v}_0\right)\right);$$

$$u_4 = \frac{t^4}{24}\left(\frac{1}{2}\frac{\partial v_3}{\partial x} - 2\left(\tilde{u}_0\frac{\partial u_3}{\partial x} + u_1\frac{\partial u_2}{\partial x} + u_2\frac{\partial u_1}{\partial x} + u_3\frac{\partial \tilde{u}_0}{\partial x}\right)\right),$$

$$v_4 = \frac{t^4}{24}\left(\frac{1}{2}\frac{\partial^3 u_3}{\partial x^3} - 2\frac{\partial}{\partial x}\left(\tilde{u}_0 v_3 + u_1 v_2 + u_2 v_1 + u_3 \tilde{v}_0\right)\right).$$

In general,

$$u_n = \frac{t^n}{n!}\left(\frac{1}{2}\frac{\partial v_{n-1}}{\partial x} - 2\left(\sum_{i=0}^{n-1} u_i \frac{\partial u_{n-i}}{\partial x}\right)\right), \tag{3.23}$$

$$v_n = \frac{t^n}{n!}\left(\frac{1}{2}\frac{\partial^3 u_{n-1}}{\partial x^3} - 2\frac{\partial}{\partial x}\left(\sum_{i=0}^{n} u_i v_{n-i}\right)\right). \tag{3.24}$$

Finally, the approximate solutions for Boussinesq–Burgers' equations are given by

$$u = \tilde{u}_0(x,t) + u_1(x,t) + u_2(x,t) + \cdots, \tag{3.25}$$

$$v = \tilde{v}_0(x,t) + v_1(x,t) + v_2(x,t) + \cdots. \tag{3.26}$$

3.3.2 Application of OHAM to Boussinesq–Burgers' Equations

Using OHAM [88], the homotopy for Equations 3.1 and 3.2 can be written as

$$(1-p)\frac{\partial\varphi(x,t;p)}{\partial t} = H(p)\left[\frac{\partial\varphi(x,t;p)}{\partial t} - \frac{1}{2}\frac{\partial\psi(x,t;p)}{\partial x} + 2\varphi(x,t;p)\frac{\partial\varphi(x,t;p)}{\partial x}\right], \tag{3.27}$$

$$(1-p)\frac{\partial\psi(x,t;p)}{\partial t} = H'(p)\left[\frac{\partial\psi(x,t;p)}{\partial t} - \frac{1}{2}\frac{\partial^3\varphi(x,t;p)}{\partial x^3} + 2\frac{\partial}{\partial x}\left[\varphi(x,t;p)\psi(x,t;p)\right]\right]. \tag{3.28}$$

Here

$$\varphi(x,t;p)=u_0(x,t)+\sum_{i=1}^{\infty}u_i(x,t)p^i, \tag{3.29}$$

$$\psi(x,t;p)=v_0(x,t)+\sum_{j=1}^{\infty}v_j(x,t)p^j, \tag{3.30}$$

$$H(p)=C_1p+C_2p^2+C_3p^3+\cdots, \tag{3.31}$$

$$H'(p)=D_1p+D_2p^2+D_3p^3+\cdots, \tag{3.32}$$

$$N(\varphi(x,t;p))=N_0(u_0(x,t))+\sum_{k=1}^{\infty}N_k(u_0,u_1,\ldots,u_k)p^k, \tag{3.33}$$

$$N(\psi(x,t;p))=N_0(v_0(x,t))+\sum_{k=1}^{\infty}N_k(v_0,v_1,\ldots,v_k)p^k. \tag{3.34}$$

Substituting Equations 3.29 through 3.34 in Equations 3.27 and 3.28 and equating the coefficients of different powers in p, we have the following system of partial differential equations:

$$\text{Coefficients of } p^0\text{: } \frac{\partial u_0(x,t)}{\partial t}=0, \tag{3.35}$$

$$\frac{\partial v_0(x,t)}{\partial t}=0 \tag{3.36}$$

$$\text{Coefficients of } p^1\text{: } \frac{\partial u_1(x,t)}{\partial t}-\frac{\partial u_0(x,t)}{\partial t}=C_1\left[\frac{\partial u_0(x,t)}{\partial t}-\frac{1}{2}\frac{\partial v_0(x,t)}{\partial x}+2u_0(x,t)\frac{\partial u_0(x,t)}{\partial x}\right], \tag{3.37}$$

$$\frac{\partial v_1(x,t)}{\partial t}-\frac{\partial v_0(x,t)}{\partial t}=D_1\left[\frac{\partial v_0(x,t)}{\partial t}-\frac{1}{2}\frac{\partial^3 u_0(x,t)}{\partial x^3}+\frac{\partial}{\partial x}\left(2u_0(x,t)v_0(x,t)\right)\right] \tag{3.38}$$

$$\begin{aligned}\text{Coefficients of } p^2\text{: } &\frac{\partial u_2(x,t)}{\partial t}-\frac{\partial u_1(x,t)}{\partial t}\\ &=C_1\left[\frac{\partial u_1(x,t)}{\partial t}-\frac{1}{2}\frac{\partial v_1(x,t)}{\partial x}+2u_0(x,t)\frac{\partial u_1(x,t)}{\partial x}+2u_1(x,t)\frac{\partial u_0(x,t)}{\partial x}\right]\\ &\quad+C_2\left[\frac{\partial u_0(x,t)}{\partial t}-\frac{1}{2}\frac{\partial v_0(x,t)}{\partial x}+2u_0(x,t)\frac{\partial u_0(x,t)}{\partial x}\right],\end{aligned} \tag{3.39}$$

$$\frac{\partial v_2(x,t)}{\partial t}-\frac{\partial v_1(x,t)}{\partial t}=D_1\left[\frac{\partial v_1(x,t)}{\partial t}-\frac{1}{2}\frac{\partial^3 u_1(x,t)}{\partial x^3}+\frac{\partial}{\partial x}\left(2u_0(x,t)v_1(x,t)+2u_1(x,t)v_0(x,t)\right)\right]$$
$$+D_2\left[\frac{\partial v_0(x,t)}{\partial t}-\frac{1}{2}\frac{\partial^3 u_0(x,t)}{\partial x^3}+\frac{\partial}{\partial x}\left(2u_0(x,t)v_0(x,t)\right)\right] \quad (3.40)$$

and so on.

For solving Boussinesq–Burgers' equations using OHAM, we consider the following initial conditions for Boussinesq–Burgers' equations (3.1) and (3.2):

$$u(x,0)=\frac{-1}{4}-\frac{1}{4}\tanh\left(\frac{x-\log 2}{2}\right), \quad (3.41)$$

$$v(x,0)=\frac{-1}{8}\operatorname{sech}^2\left(\frac{-x+\log 2}{2}\right). \quad (3.42)$$

Using the initial conditions and solving Equations 3.35 through 3.40, we obtain

$$u_0(x,t)=\frac{-1}{4}-\frac{1}{4}\tanh\left(\frac{x-\log 2}{2}\right), \quad (3.43)$$

$$v_0(x,t)=\frac{-1}{8}\operatorname{sech}^2\left(\frac{-x+\log 2}{2}\right); \quad (3.44)$$

$$u_1(x,t)=\frac{1}{16}t\,C_1\operatorname{sech}^2\left(\frac{x-\log 2}{2}\right), \quad (3.45)$$

$$v_1(x,t)=\frac{-1}{2}t\,D_1\operatorname{cosech}^3(x-\log 2)\sinh^4\left(\frac{x-\log 2}{2}\right); \quad (3.46)$$

$$u_2(x,t)=\frac{1}{128}t\ \operatorname{sech}^4\left(\frac{x-\log 2}{2}\right)\Big[4C_1+4C_1^2-2tC_1^2+4C_2+2tC_1D_1$$
$$+\cosh(x-\log 2)\left[(4+t)C_1^2+4C_2+C_1(4-tD_1)\right]+tC_1^2\sinh(x-\log 2)\Big], \quad (3.47)$$

$$v_2(x,t)=\frac{-1}{256}t\ \operatorname{sech}^5\left(\frac{x-\log 2}{2}\right)\left[-3t\ \cosh\left(\frac{x-\log 2}{2}\right)D_1^2+t\ \cosh\left(\frac{3(x-\log 2)}{2}\right)D_1^2\right.$$
$$-2t(-3+\cosh(x-\log 2))(C_1-D_1)D_1\sinh\left(\frac{x-\log 2}{2}\right)$$
$$\left.+4\operatorname{cosech}\left(\frac{x-\log 2}{2}\right)\left(D_1+D_1^2+D_2\right)\sinh^2(x-\log 2)\right]. \quad (3.48)$$

Using Equations 3.43 through 3.48, the second-order approximate solution is obtained as follows:

$$u(x,t)=\frac{-1}{4}-\frac{1}{4}\tanh\left(\frac{x-\log 2}{2}\right)+\frac{1}{16}t\,C_1\text{sech}^2\left(\frac{x-\log 2}{2}\right)$$
$$+\frac{1}{128}t\ \text{sech}^4\left(\frac{x-\log 2}{2}\right)\times\Big[4C_1+4C_1^2-2tC_1^2+4C_2+2tC_1D_1$$
$$+\cosh(x-\log 2)\left[(4+t)C_1^2+4C_2+C_1(4-tD_1)\right]+tC_1^2\sinh(x-\log 2)\Big], \quad (3.49)$$

$$v(x,t)=\frac{-1}{8}\text{sech}^2\left(\frac{-x+\log 2}{2}\right)+\left(\frac{-1}{2}\right)t\,D_1\text{cosech}^3(x-\log 2)\sinh^4\left(\frac{x-\log 2}{2}\right)$$
$$+\left(\frac{-1}{256}\right)t\ \text{sech}^5\left(\frac{x-\log 2}{2}\right)\Bigg[-3t\ \cosh\left(\frac{x-\log 2}{2}\right)D_1^2+t\ \cosh\left(\frac{3(x-\log 2)}{2}\right)D_1^2$$
$$-2t(-3+\cosh(x-\log 2))(C_1-D_1)D_1\sinh\left(\frac{x-\log 2}{2}\right)$$
$$+4\text{cosech}\left(\frac{x-\log 2}{2}\right)\left(D_1+D_1^2+D_2\right)\sinh^2(x-\log 2)\Bigg]. \quad (3.50)$$

The optimal values of the convergence control constants C_1, C_2, D_1, and D_2 can be obtained using the collocation method.

3.4 Convergence of HPM

The series $u=\lim_{p\to 1} v=v_0+v_1+v_2+\cdots$, given in Equation 1.17 of Chapter 1, is convergent for most cases. The following suggestion has been made by He [26] to find the convergence rate on nonlinear operator $A(v)$:

(i) The second derivative of $N(u)$ with respect to u must be small because the parameter may be relatively large, that is, $p\to 1$.

(ii) The norm of $L^{-1}\dfrac{\partial N}{\partial u}$ must be smaller than 1 so that the series converges.

Let us write Equation 1.13 of Chapter 1 in the following form:

$$L(v)=L(u_0)+p\left[f(r)-N(v)-L(u_0)\right]. \quad (3.51)$$

Applying the inverse operator L^{-1} to both sides of Equation 3.51, we obtain

$$v=u_0+p\left[L^{-1}f(r)-L^{-1}N(v)-u_0\right]. \quad (3.52)$$

Suppose that

$$v = \sum_{i=0}^{\infty} p^i v_i. \tag{3.53}$$

Substituting (3.53) into the right-hand side of Equation 3.52, we have

$$v = u_0 + p\left[L^{-1}f(r) - \left(L^{-1}N\right)\left[\sum_{i=0}^{\infty} p^i v_i\right] - u_0\right]. \tag{3.54}$$

If $p \to 1$, the exact solution may be obtained by using

$$\begin{aligned} u = \lim_{p\to 1} v &= L^{-1}f(r) - \left(L^{-1}N\right)\left[\sum_{i=0}^{\infty} v_i\right] \\ &= L^{-1}f(r) - \sum_{i=0}^{\infty}\left(L^{-1}N\right)v_i. \end{aligned} \tag{3.55}$$

To study the convergence of the method, let us state the following theorem.

Theorem 3.1 [89]

Suppose that V and W are Banach spaces and $N: V \to W$ is a contraction mapping such that for all $v, v^* \in V$

$$\left\|N(v) - N\left(v^*\right)\right\| \le \varepsilon \left\|v - v^*\right\|, \quad \varepsilon \in (0,1). \tag{3.56}$$

Then according to Banach's fixed point theorem, N has a unique fixed point u (say) such that $N(u) = u$.

The sequence generated by the HPM will be assumed in the following form as

$$V_n = N\left(V_{n-1}\right), \quad V_{n-1} = \sum_{i=0}^{n-1} v_i, \quad n = 1, 2, 3, \ldots$$

and suppose that $V_0 = v_0 \in B_r(v)$ where $B_r(v) = \{v^* \in V \mid \|v^* - v\| < r\}$; then we have

(i) $\|V_n - v\| \le \varepsilon^n \|v_0 - v\|$
(ii) $V_n \in B_r(v)$
(iii) $\lim_{n\to\infty} V_n = v$

Proof

(i) The preceding theorem is proved by using the method of induction.

For $n = 1$, $\|V_1 - v\| = \|N(V_0) - N(v)\| \le \varepsilon \|v_0 - v\|$. (from Equation 3.56)

Assume that $\|V_{n-1}-v\| \le \varepsilon^{n-1}\|v_0-v\|$ as the induction hypothesis, then

$$\begin{aligned}\|V_n - v\| &\le \|N(V_{n-1}) - N(v)\| \\ &\le \varepsilon\|V_{n-1} - v\| \\ &\le \varepsilon\,\varepsilon^{n-1}\|v_0 - v\| \quad \text{(from induction hypothesis)} \\ &\le \varepsilon^n\|v_0 - v\|.\end{aligned}$$

Hence, $\|V_n - v\| \le \varepsilon^n\|v_0 - v\|$.

(ii) Using (i), we have

$$\|V_n - v\| \le \varepsilon^n \|v_0 - v\| \le \varepsilon^n r < r.$$

This implies $V_n \in B_r(v)$.

(iii) From (i), we know $\|V_n - v\| \le \varepsilon^n\|v_0 - v\|$.
Again since $0 < \varepsilon < 1$, $\lim_{n\to\infty} \varepsilon^n = 0$.
So $\lim_{n\to\infty}\|V_n - v\| = 0$, which yields $\lim_{n\to\infty} V_n = v$. □

3.5 Convergence of OHAM

Theorem 3.2

Let the solution components u_0, u_1, u_2,... be defined as given in Equations 1.27 through 1.29. The series solution $\sum_{k=0}^{m-1} u_k(x,t)$ defined in Equation 1.31 converges if there exists δ, $0<\delta<1$ such that $\|u_{k+1}\| \le \delta\|u_k\|$ for all $k \ge k_0$ for some $k_0 \in N$.

Proof

Define the sequence $\{S_n\}_{n=0}^{\infty}$ as follows:

$$\begin{aligned}S_0 &= u_0, \\ S_1 &= u_0 + u_1, \\ S_2 &= u_0 + u_1 + u_2, \\ &\cdots \\ S_n &= u_0 + u_1 + u_2 + \cdots + u_n.\end{aligned}$$

We have to show $\{S_n\}_{n=0}^{\infty}$ is a Cauchy sequence in the Hilbert space $\Re$.

Consider

$$\begin{aligned}\|S_{n+1} - S_n\| &= \|u_{n+1}\| \\ &\le \|\delta u_n\| \\ &\le \delta^2\|u_{n-1}\| \\ &\cdots \\ &\le \delta^{n-k_0+1}\|u_{k_0}\|.\end{aligned}$$

Now for every $n,m \in N,\ n \geq m > k_0$

$$\begin{aligned}\|S_n - S_m\| &= \|(S_n - S_{n-1}) + (S_{n-1} - S_{n-2}) + \cdots + (S_{m+1} - S_m)\| \\ &\leq \|S_n - S_{n-1}\| + \|S_{n-1} - S_{n-2}\| + \cdots + \|S_{m+1} - S_m\| \quad \text{(Triangle inequality)} \\ &\leq \delta^{n-k_0}\|u_{k_0}\| + \delta^{n-k_0-1}\|u_{k_0}\| + \cdots + \delta^{m-k_0+1}\|u_{k_0}\| \\ &= \left(\frac{1-\delta^{n-m}}{1-\delta}\right)\delta^{m-k_0+1}\|u_{k_0}\|\end{aligned}$$

This implies $\lim_{n,m\to\infty}\|S_n - S_m\| = 0$ (since $0 < \delta < 1$).

Therefore, $\{S_n\}_{n=0}^{\infty}$ is a Cauchy sequence in the Hilbert space $\mathfrak{R}$, and hence the series solution $\sum_{k=1}^{\infty} u_k(x,t)$ converges. □

3.6 Numerical Results and Discussions

Tables 3.1 and 3.2 show the comparisons of the absolute errors of Boussinesq–Burgers' equations obtained by using two terms and three terms approximations of HPM and OHAM at different values of c, k, b, x, and t. To show the effectiveness and accuracy of proposed schemes, L_2 and L_∞ error norms have been presented in Tables 3.3 and 3.4, respectively.

Graphical representation of results is very useful to demonstrate the efficiency and accuracy of the proposed methods for the discussed problem. Figures 3.1 and 3.2 cite the comparison graphically between the approximate solutions obtained by five terms HPM, three terms OHAM, and exact solutions for different values of x and $t=0.5$. Figures 3.3 and 3.4, respectively, show one soliton approximate solution of $u(x,t)$ and $v(x,t)$, obtained by OHAM for Boussinesq–Burgers' equations.

From Figures 3.1 through 3.4, one can see a very good agreement between the exact solutions and the solutions obtained by HPM and OHAM. Tables 3.1 through 3.4 depict the performance of OHAM in comparison with HPM and clearly witness the reliability and efficiency of OHAM for the solutions of Boussinesq–Burgers' equations.

3.7 Numerical Approach to Boussinesq–Burgers' Equations

The Haar wavelet solutions of $u(x,t)$ and $v(x,t)$ are sought by assuming that $\dot{u}'''(x,t)$ and $\dot{v}'(x,t)$ can be expanded in terms of Haar wavelets as

$$\dot{u}'''(x,t) = \sum_{i=1}^{2M} a_s(i)h_i(x), \tag{3.57}$$

$$\dot{v}'(x,t) = \sum_{i=1}^{2M} b_s(i)h_i(x), \quad \text{for } t \in [t_s, t_{s+1}], \tag{3.58}$$

where "." and " ′" stand for differentiation with respect to t and x, respectively.

TABLE 3.1

Absolute Errors in the Solutions of Boussinesq–Burgers' Equations Using Two Terms Approximation for HPM and OHAM at Various Points with $c = \frac{1}{2}$, $k = -1$, $b = 2$, and $C_1 = -1.01653$, $D_1 = 0.934599$ Obtained by Equation 1.33

(x, t)	$\lvert u_{exact} - u_{HPM} \rvert$	$\lvert v_{exact} - v_{HPM} \rvert$	$\lvert u_{exact} - u_{OHAM} \rvert$	$\lvert v_{exact} - v_{OHAM} \rvert$
(0.1, 0.1)	4.03765E-5	5.49815E-5	5.43565E-5	5.30293E-5
(0.1, 0.2)	1.57743E-4	2.24577E-4	3.17229E-5	8.55529E-6
(0.1, 0.3)	3.46200E-4	5.15476E-4	6.20012E-5	1.91443E-4
(0.1, 0.4)	5.99524E-4	9.33935E-4	2.20592E-4	5.01892E-4
(0.1, 0.5)	9.11191E-4	1.48572E-3	4.37526E-4	9.45668E-4
(0.2, 0.1)	3.45343E-5	6.17326E-5	6.27431E-5	3.12906E-5
(0.2, 0.2)	1.33936E-4	2.51026E-4	6.06192E-5	6.49797E-5
(0.2, 0.3)	2.91679E-4	5.73647E-4	1.52649E-7	2.94577E-4
(0.2, 0.4)	5.00967E-4	1.03482E-3	1.11857E-4	6.62723E-4
(0.2, 0.5)	7.54752E-4	1.63915E-3	2.68365E-5	1.17404E-3
(0.3, 0.1)	2.80601E-5	6.75811E-5	7.13609E-5	8.76219E-6
(0.3, 0.2)	1.07664E-4	2.73710E-4	9.11785E-5	1.21023E-4
(0.3, 0.3)	2.31765E-4	6.23007E-4	6.64983E-5	3.93977E-4
(0.3, 0.4)	3.93103E-4	1.11945E-3	4.58083E-6	8.14078E-4
(0.3, 0.5)	5.84236E-4	1.76633E-3	8.71313E-5	1.38462E-3
(0.4, 0.1)	2.10557E-5	7.23025E-5	8.00687E-5	1.40750E-5
(0.4, 0.2)	7.93485E-5	2.91751E-4	1.22900E-5	1.75296E-4
(0.4, 0.3)	1.67436E-4	6.61637E-4	1.35937E-4	4.86954E-4
(0.4, 0.4)	2.77729E-4	1.18453E-3	1.26768E-4	9.51621E-4
(0.4, 0.5)	4.02528E-4	1.86226E-3	1.03094E-4	1.57112E-3
(0.5, 0.1)	1.36435E-5	7.57082E-5	8.87114E-5	3.67202E-5
(0.5, 0.2)	4.94924E-5	3.04426E-4	1.55217E-4	2.26450E-4
(0.5, 0.3)	9.98476E-5	6.87978E-4	2.07217E-4	5.71013E-4
(0.5, 0.4)	1.56938E-4	1.22742E-3	2.52482E-4	1.07147E-3
(0.5, 0.5)	2.12957E-4	1.92305E-3	2.98817E-4	1.72811E-3

Integrating Equation 3.57 with respect to t from t_s to t and thrice with respect to x from 0 to x, the following equations are obtained:

$$u'''(x,t) = (t - t_s)\sum_{i=1}^{2M} a_s(i)h_i(x) + u'''(x,t_s), \tag{3.59}$$

$$u''(x,t) = (t - t_s)\sum_{i=1}^{2M} a_s(i)p_i(x) + u''(x,t_s) - u''(0,t_s) + u''(0,t), \tag{3.60}$$

$$u'(x,t) = (t - t_s)\sum_{i=1}^{2M} a_s(i)q_i(x) + u'(x,t_s) - u'(0,t_s) + x\left[u''(0,t) - u''(0,t_s)\right] + u'(0,t), \tag{3.61}$$

TABLE 3.2

Absolute Errors in the Solutions of Boussinesq–Burgers' Equations Using Three Terms Approximation for HPM and OHAM at Various Points with $c=\frac{1}{2}$, $k=-1$, $b=2$, and $C_1=0.9786175$, $C_2=-3.9162929$, $D_1=1.0514603$, $D_2=-4.209964$ Obtained by Equation 1.33

(x,t)	$\lvert u_{exact}-u_{HPM}\rvert$	$\lvert v_{exact}-v_{HPM}\rvert$	$\lvert u_{exact}-u_{OHAM}\rvert$	$\lvert v_{exact}-v_{OHAM}\rvert$
(0.1, 0.1)	9.11428E-7	1.19150E-6	3.15534E-6	5.85344E-7
(0.1, 0.2)	7.40859E-6	9.41690E-6	7.33961E-7	2.12165E-6
(0.1, 0.3)	2.53911E-5	3.13655E-5	1.36454E-6	1.12982E-5
(0.1, 0.4)	6.10825E-5	7.32950E-5	3.08338E-6	3.43727E-5
(0.1, 0.5)	1.21007E-4	1.40972E-4	2.06021E-5	7.71116E-5
(0.2, 0.1)	1.02449E-6	1.06292E-6	3.23055E-6	8.39207E-7
(0.2, 0.2)	8.29954E-6	8.34741E-6	1.05314E-6	3.45590E-6
(0.2, 0.3)	2.83495E-5	2.76201E-5	7.84907E-9	2.08340E-6
(0.2, 0.4)	6.79737E-5	6.41010E-5	6.84685E-6	8.49823E-6
(0.2, 0.5)	1.34218E-4	1.22412E-4	2.86627E-5	3.29106E-5
(0.3, 0.1)	1.12268E-6	8.94544E-7	3.34664E-6	2.35740E-6
(0.3, 0.2)	9.06757E-6	6.96343E-6	1.53865E-6	9.48698E-6
(0.3, 0.3)	3.08802E-5	2.28283E-5	1.62153E-6	1.67671E-5
(0.3, 0.4)	7.38211E-5	5.24659E-5	1.08559E-5	2.02210E-5
(0.3, 0.5)	1.45333E-4	9.91676E-5	3.66844E-5	1.65572E-5
(0.4, 0.1)	1.20223E-6	6.91094E-7	3.50880E-6	3.92344E-6
(0.4, 0.2)	9.68307E-6	5.30519E-6	2.20317E-6	1.57786E-5
(0.4, 0.3)	3.28850E-5	1.71337E-5	3.52562E-6	3.22743E 5
(0.4, 0.4)	7.83973E-5	3.87483E-5	1.50654E-5	5.08384E-5
(0.4, 0.5)	1.53919E-4	7.19747E-5	4.45217E-5	6.96457E-5
(0.5, 0.1)	1.25997E-6	4.59601E-7	3.72005E-6	5.48493E-6
(0.5, 0.2)	1.01214E-5	3.43102E-6	3.05227E-6	2.21052E-5
(0.5, 0.3)	3.42835E-5	1.07397E-5	5.69591E-6	4.80354E-5
(0.5, 0.4)	8.15178E-5	2.34456E-5	1.94224E-5	8.22156E-5
(0.5, 0.5)	1.59629E-4	4.18318E-5	5.20368E-5	1.24363E-4

$$u(x,t)=(t-t_s)\sum_{i=1}^{2M}a_s(i)r_i(x)+u(x,t_s)-u(0,t_s)$$
$$+\frac{x^2}{2}\left[u''(0,t)-u''(0,t_s)\right]+x\left[u'(0,t)-u'(0,t_s)\right]+u(0,t), \tag{3.62}$$

$$\dot{u}(x,t)=\sum_{i=1}^{2M}a_s(i)r_i(x)+\frac{x^2}{2}\dot{u}''(0,t)+x\dot{u}'(0,t)+\dot{u}(0,t). \tag{3.63}$$

Integrating Equation 3.58 with respect to t from t_s to t and once with respect to x from 0 to x, the following equations are obtained:

$$v'(x,t)=(t-t_s)\sum_{i=1}^{2M}b_s(i)h_i(x)+v'(x,t_s), \tag{3.64}$$

TABLE 3.3

L_2 and L_∞ Error Norm for Boussinesq–Burgers' Equations Using Two Terms Approximation for HPM and OHAM at Various Points of x

	HPM				OHAM			
	Error in the Case of Two Terms Approximation for $u(x,t)$		Error in the Case of Two Terms Approximation for $v(x,t)$		Error in the Case of Two Terms Approximation for $u(x,t)$		Error in the Case of Two Terms Approximation for $v(x,t)$	
x	L_2	L_∞	L_2	L_∞	L_2	L_∞	L_2	L_∞
0.1	5.16927E-4	9.11191E-4	1.48572E-3	1.48572E-3	2.22663E-4	4.37526E-4	4.86974E-4	9.45668E-4
0.2	4.30076E-4	7.54752E-4	9.11434E-4	1.63915E-3	1.35752E-4	1.11857E-4	6.17988E-4	1.17404E-3
0.3	3.35248E-4	5.84236E-4	9.83942E-4	1.76633E-3	7.13313E-5	9.11785E-5	7.41598E-4	1.38462E-3
0.4	2.34067E-4	4.02528E-4	1.03916E-3	1.86226E-3	1.15493E-4	1.35937E-4	8.53470E-4	1.57112E-3
0.5	1.28519E-4	2.12957E-4	1.07484E-3	1.92305E-3	2.13513E-4	2.98817E-4	9.50063E-4	1.72811E-3

TABLE 3.4

L_2 and L_∞ Error Norm for Boussinesq–Burgers' Equations Using Three Terms Approximation for HPM and OHAM at Various Points of x

	HPM				OHAM			
	Error in the Case of Three Terms Approximation for $u(x,t)$		Error in the Case of Three Terms Approximation for $v(x,t)$		Error in the Case of Three Terms Approximation for $u(x,t)$		Error in the Case of Three Terms Approximation for $v(x,t)$	
x	L_2	L_∞	L_2	L_∞	L_2	L_∞	L_2	L_∞
0.1	6.17644E-5	1.21007E-4	7.25523E-5	1.40972E-4	9.44786E-6	2.06021E-5	3.81056E-5	7.71116E-5
0.2	6.85690E-5	1.34218E-4	6.31305E-5	1.22412E-4	1.32663E-5	2.86627E-5	1.53122E-5	3.29106E-5
0.3	7.43079E-5	1.45333E-4	5.12978E-5	9.91676E-5	1.72034E-5	3.66844E-5	1.45583E-5	2.02210E-5
0.4	7.87576E-5	1.53919E-4	3.74272E-5	4.18318E-5	2.11601E-5	4.45217E-5	4.18116E-5	6.96457E-5
0.5	8.17386E-5	1.59629E-4	2.20314E-5	7.19747E-5	2.50625E-5	5.20368E-5	7.07837E-5	1.24363E-4

$$v(x,t)=(t-t_s)\sum_{i=1}^{2M} b_s(i)p_i(x)+v(x,t_s)-v(0,t_s)+v(0,t), \tag{3.65}$$

$$\dot{v}(x,t)=\sum_{i=1}^{2M} b_s(i)p_i(x)+\dot{v}(0,t). \tag{3.66}$$

Discretizing the above results by assuming $x \to x_l$, $t \to t_{s+1}$, from Equations 3.64 through 3.66, we obtain

$$v'(x_l,t_{s+1})=(t_{s+1}-t_s)\sum_{i=1}^{2M} b_s(i)h_i(x_l)+v'(x_l,t_s), \tag{3.67}$$

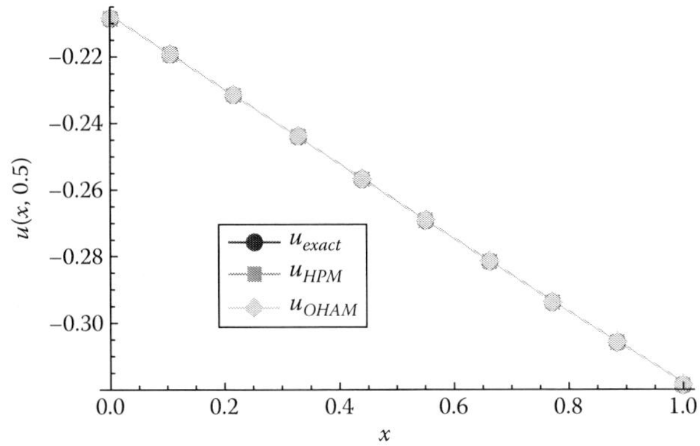

FIGURE 3.1
Comparison of a five terms HPM solution and a three terms OHAM solution with the exact solution of $u(x,t)$ for Boussinesq–Burgers' equations when $c=0.5$, $k=-1$, $b=2$, and $t=0.5$.

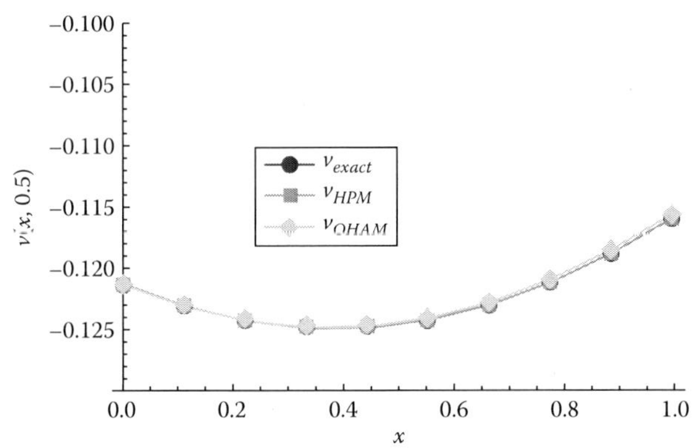

FIGURE 3.2
Comparison of a five terms HPM solution and a three terms OHAM solution with the exact solution of $v(x,t)$ for Boussinesq–Burgers' equations when $c=0.5$, $k=-1$, $b=2$, and $t=0.5$.

$$v\left(x_l,t_{s+1}\right)=\left(t_{s+1}-t_s\right)\sum_{i=1}^{2M}b_s\left(i\right)p_i\left(x_l\right)+v\left(x_l,t_s\right)-v\left(0,t_s\right)+v\left(0,t_{s+1}\right), \tag{3.68}$$

$$\dot{v}\left(x_l,t_{s+1}\right)=\sum_{i=1}^{2M}b_s\left(i\right)p_i\left(x_l\right)+\dot{v}\left(0,t_{s+1}\right). \tag{3.69}$$

Using the finite difference method

$$\dot{u}\left(0,t\right)=\frac{u\left(0,t\right)-u\left(0,t_s\right)}{\left(t-t_s\right)}.$$

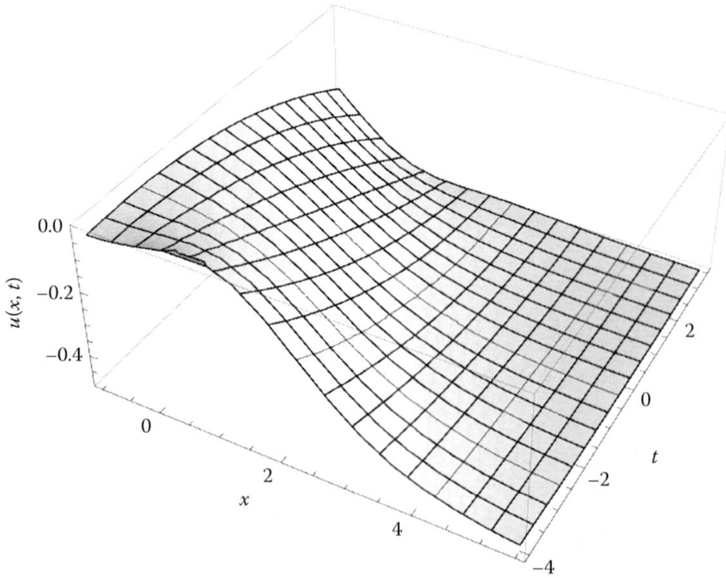

FIGURE 3.3
One soliton approximate solution of $u(x,t)$, obtained by OHAM for Boussinesq–Burgers' equations with parameters $c=0.5$, $k=-1$, and $b=2$.

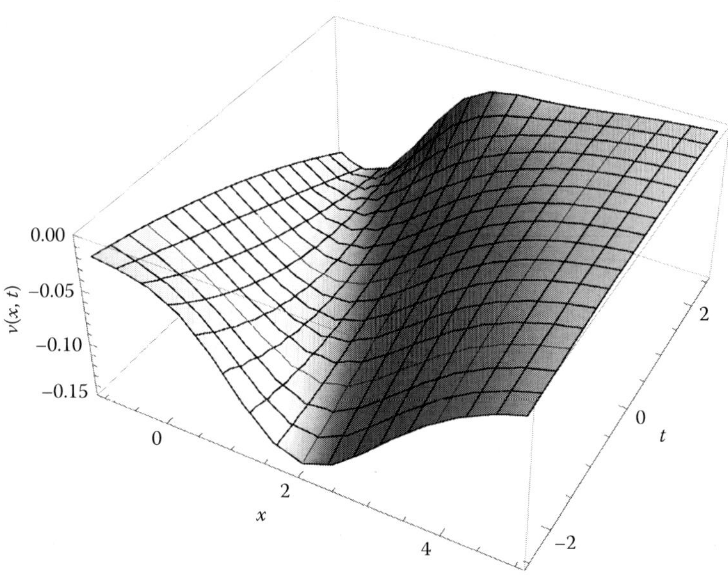

FIGURE 3.4
One soliton approximate solution of $v(x,t)$, obtained by OHAM for Boussinesq–Burgers' equations with parameters $c=\frac{1}{2}$, $k=-1$, and $b=2$.

Equation 3.63 becomes

$$\dot{u}(x,t)=\sum_{i=1}^{2M}a_s(i)r_i(x)+\frac{x^2}{2}\left(\frac{u''(0,t)-u''(0,t_s)}{t-t_s}\right)$$
$$+x\left(\frac{u'(0,t)-u'(0,t_s)}{t-t_s}\right)+\left(\frac{u(0,t)-u(0,t_s)}{t-t_s}\right). \quad (3.70)$$

By using the boundary condition at $x=1$, Equation 3.61 becomes

$$u'(1,t)=(t-t_s)\sum_{i=1}^{2M}a_s(i)q_i(1)+u'(1,t_s)-u'(0,t_s)+\left[u''(0,t)-u''(0,t_s)\right]+u'(0,t).$$

This implies

$$u''(0,t)-u''(0,t_s)=-(t-t_s)\sum_{i=1}^{2M}a_s(i)q_i(1)+u'(1,t)-u'(1,t_s)-u'(0,t)+u'(0,t_s). \quad (3.71)$$

Substituting Equation 3.71 in Equations 3.60, 3.61, 3.62, and 3.70 and discretizing the resultant equations by assuming $x\to x_l$, $t\to t_{s+1}$, we obtain

$$u'''(x_l,t_{s+1})=(t_{s+1}-t_s)\sum_{i=1}^{2M}a_s(i)h_i(x_l)+u'''(x_l,t_s), \quad (3.72)$$

$$u''(x_l,t_{s+1})=(t_{s+1}-t_s)\sum_{i=1}^{2M}a_s(i)p_i(x_l)+u''(x_l,t_s)$$
$$-(t_{s+1}-t_s)\sum_{i=1}^{2M}a_s(i)q_i(1)+\left[u'(1,t_{s+1})-u'(1,t_s)\right]$$
$$-\left[u'(0,t_{s+1})-u'(0,t_s)\right], \quad (3.73)$$

$$u'(x_l,t_{s+1})=(t_{s+1}-t_s)\sum_{i=1}^{2M}a_s(i)q_i(x_l)+u'(x_l,t_s)-u'(0,t_s)+u'(0,t_{s+1})$$
$$+x_l\left[-(t_{s+1}-t_s)\sum_{i=1}^{2M}a_s(i)q_i(1)+\left[u'(1,t_{s+1})-u'(1,t_s)\right]-\left[u'(0,t_{s+1})-u'(0,t_s)\right]\right], \quad (3.74)$$

$$u(x_l,t_{s+1}) = (t_{s+1}-t_s)\sum_{i=1}^{2M} a_s(i) r_i(x_l) + u(x_l,t_s) - u(0,t_s) + u(0,t_{s+1}) + x_l\left[u'(0,t_{s+1}) - u'(0,t_s)\right]$$
$$+\frac{x_l^2}{2}\left[-(t_{s+1}-t_s)\sum_{i=1}^{2M} a_s(i) q_i(1) + \left[u'(1,t_{s+1}) - u'(1,t_s)\right] - \left[u'(0,t_{s+1}) - u'(0,t_s)\right]\right], \tag{3.75}$$

$$\dot{u}(x_l,t_{s+1}) = \sum_{i=1}^{2M} a_s(i) r_i(x_l) + \frac{x_l}{t_{s+1}-t_s}\left[u'(0,t_{s+1}) - u'(0,t_s)\right] + \frac{1}{t_{s+1}-t_s}\left[u(0,t_{s+1}) - u(0,t_s)\right]$$
$$+\frac{x_l^2}{2(t_{s+1}-t_s)}\left[-(t_{s+1}-t_s)\sum_{i=1}^{2M} a_s(i) q_i(1) + \left[u'(1,t_{s+1}) - u'(1,t_s)\right] - \left[u'(0,t_{s+1}) - u'(0,t_s)\right]\right]. \tag{3.76}$$

Substituting the preceding equations in Equations 3.1 and 3.2, we have

$$\sum_{i=1}^{2M} a_s(i)\left[r_i(x_l) - \frac{x_l^2}{2}q_i(1)\right] + \frac{x_l^2}{2(t_{s+1}-t_s)}\left[u'(1,t_{s+1}) - u'(1,t_s) - u'(0,t_{s+1}) + u'(0,t_s)\right]$$
$$+\frac{x_l}{t_{s+1}-t_s}[u'(0,t_{s+1}) - u'(0,t_s)] + \frac{1}{t_{s+1}-t_s}[u(0,t_{s+1}) - u(0,t_s)]$$
$$= \frac{1}{2}\left((t_{s+1}-t_s)\sum_{i=1}^{2M} b_s(i) h_i(x_l) + v'(x_l,t_s)\right)$$
$$-2\left[(t_{s+1}-t_s)\sum_{i=1}^{2M} a_s(i)\left[r_i(x_l) - \frac{x_l^2}{2}q_i(1)\right] + u(x_l,t_s) - u(0,t_s) + u(0,t_{s+1})\right.$$
$$\left. + x_l\left[u'(0,t_{s+1}) - u'(0,t_s)\right] + \frac{x_l^2}{2}\left[u'(1,t_{s+1}) - u'(1,t_s) - u'(0,t_{s+1}) + u'(0,t_s)\right]\right]$$
$$\times\left[(t_{s+1}-t_s)\sum_{i=1}^{2M} a_s(i)\left[q_i(x_l) - x_l q_i(1)\right] + u'(x_l,t_s) - u'(0,t_s) + u'(0,t_{s+1})\right.$$
$$\left. + x_l\left[u'(1,t_{s+1}) - u'(1,t_s) - u'(0,t_{s+1}) + u'(0,t_s)\right]\right]; \tag{3.77}$$

$$\sum_{i=1}^{2M} b_s(i)p_i(x_l)+\dot{v}(0,t_{s+1})$$
$$=\frac{1}{2}\left((t_{s+1}-t_s)\sum_{i=1}^{2M} a_s(i)h_i(x_l)+u'''(x_l,t_s)\right)$$
$$-2\left[(t_{s+1}-t_s)\sum_{i=1}^{2M} a_s(i)\left[r_i(x_l)-\frac{x_l^2}{2}q_i(1)\right]+u(x_l,t_s)-u(0,t_s)+u(0,t_{s+1})\right.$$
$$\left.+x_l\left[u'(0,t_{s+1})-u'(0,t_s)\right]+\frac{x_l^2}{2}\left[\left[u'(1,t_{s+1})-u'(1,t_s)\right]-\left[u'(0,t_{s+1})-u'(0,t_s)\right]\right]\right]$$
$$\times\left((t_{s+1}-t_s)\sum_{i=1}^{2M} b_s(i)h_i(x_l)+v'(x_l,t_s)\right)$$
$$-2\left((t_{s+1}-t_s)\sum_{i=1}^{2M} b_s(i)p_i(x_l)+v(x_l,t_s)-v(0,t_s)+v(0,t_{s+1})\right)$$
$$\times\left[(t_{s+1}-t_s)\sum_{i=1}^{2M} a_s(i)\left[q_i(x_l)-x_lq_i(1)\right]+u'(x_l,t_s)-u'(0,t_s)+u'(0,t_{s+1})\right.$$
$$\left.+x_l\left[u'(1,t_{s+1})-u'(1,t_s)-u'(0,\ t_{s+1})+u'(0,t_s)\right]\right]. \tag{3.78}$$

From Equations 3.77 and 3.78, the wavelet coefficients $a_s(i)$ and $b_s(i)$ can be successively calculated using mathematical software. This process starts with

$$u(x_l,t_0)=\frac{-1}{4}-\frac{1}{4}\tanh\left(\frac{x_l-\log 2}{2}\right),$$

$$u'(x_l,t_0)=\frac{-1}{8}\operatorname{sech}^2\left(\frac{x_l-\log 2}{2}\right),$$

$$u''(x_l,t_0)=\frac{1}{8}\operatorname{sech}^2\left(\frac{x_l-\log 2}{2}\right)\tanh\left(\frac{x_l-\log 2}{2}\right),$$

$$v(x_l,t_0)=\frac{-1}{8}\operatorname{sech}^2\left(\frac{-x_l+\log 2}{2}\right),$$

$$v'(x_l,t_0)=\frac{-1}{8}\operatorname{sech}^2\left(\frac{-x_l+\log 2}{2}\right)\tanh\left(\frac{-x_l+\log 2}{2}\right).$$

3.8 Convergence of Haar Wavelet Approximation

The convergence of the method may be discussed on the same lines as given by learned researcher Saha Ray, 2012 [38].

Theorem 3.3

Let $f(x) \in L^2(R)$ be a continuous function defined on $[0,1)$. Then the error at the Jth level may be defined as $E_J(x) = |f(x) - f_J(x)| = \left| f(x) - \sum_{i=1}^{2M} a_i h_i(x) \right| = \left| \sum_{i=2M}^{\infty} a_i h_i(x) \right|$.

The error norm for $E_J(x)$ is obtained as

$$\|E_J(x)\|_2 \le \frac{K^2}{12} 2^{-2J} \tag{3.79}$$

where $|f'(x)| \le K$ for all $x \in [0,1)$ and $K > 0$ and M is a positive number related to the Jth level of resolution of the wavelet given by $M = 2^J$.

Proof

The error at Jth level resolution is defined as

$$E_J(x) = |f(x) - f_J(x)| = \left| f(x) - \sum_{i=1}^{2M} a_i h_i(x) \right| = \left| \sum_{i=2M}^{\infty} a_i h_i(x) \right|;$$

$$\|E_J(x)\|^2 = \int_{-\infty}^{\infty} \left(\sum_{i=2M}^{\infty} a_i h_i(x), \sum_{s=2M}^{\infty} a_s h_s(x) \right) dx = \sum_{i=2M}^{\infty} \sum_{s=2M}^{\infty} a_i a_s \int_{-\infty}^{\infty} h_i(x) h_s(x) dx \le \sum_{i=2M}^{\infty} |a_i|^2 .$$

Now,

$$a_i = \int_0^1 2^{j/2} f(x) h(2^j x - k) dx,$$

where $h_i(x) = 2^{j/2} h(2^j x - k)$; $k = 0,1,2,\ldots,2^j - 1$; $j = 0,1,2,\ldots,J$; and

$$h(2^j x - k) = \begin{cases} 1, & \frac{k}{2^j} \le x < \frac{k+0.5}{2^j} \\ -1, & \frac{k+0.5}{2^j} \le x < \frac{k+1}{2^j} \\ 0, & \text{elsewhere} \end{cases}.$$

Hence

$$a_i = 2^{\frac{j}{2}} \left[\int_{\frac{k}{2^j}}^{\frac{k+0.5}{2^j}} f(x) dx - \int_{\frac{k+0.5}{2^j}}^{\frac{k+1}{2^j}} f(x) dx \right]$$

$$= 2^{\frac{j}{2}} \left[\left(\frac{k+0.5}{2^j} - \frac{k}{2^j} \right) f(\xi_1) - \left(\frac{k+1}{2^j} - \frac{k+0.5}{2^j} \right) f(\xi_2) \right] \quad \text{(applying mean value theorem)}$$

where $\xi_1 \in \left(\frac{k}{2^j}, \frac{k+0.5}{2^j} \right)$ and $\xi_2 \in \left(\frac{k+0.5}{2^j}, \frac{k+1}{2^j} \right)$.

Consequently,

$$a_i = 2^{\frac{j}{2}}\left[\left(\frac{0.5}{2^j}\right)f(\xi_1)-\left(\frac{0.5}{2^j}\right)f(\xi_2)\right]$$

$$= 2^{\frac{j}{2}-j-1}\left[f(\xi_1)-f(\xi_2)\right]$$

$$= 2^{-\frac{j}{2}-1}(\xi_1-\xi_2)f'(\xi),$$

where $\xi \in (\xi_1, \xi_2)$ (applying mean value theorem).

This implies $a_i^2 = 2^{-j-2}(\xi_2-\xi_1)^2 f'(\xi)^2 \le 2^{-j-2}\,2^{-2j}K^2 = 2^{-3j-2}K^2$.

Therefore,

$$\left\|E_J(x)\right\|^2 \le \sum_{i=2M}^{\infty} a_i^2 \le \sum_{i=2M}^{\infty} 2^{-3j-2}K^2 = K^2\sum_{j=J+1}^{\infty}\sum_{i=2^j}^{2^{j+1}-1} 2^{-3j-2}$$

$$= K^2\sum_{j=J+1}^{\infty} 2^{-3j-2}\left(2^{j+1}-1-2^j+1\right)$$

$$= K^2\sum_{j=J+1}^{\infty}\left(2^{-2j-1}-2^{-2j-2}\right)$$

$$= K^2\sum_{j=J+1}^{\infty} 2^{-2j}\left(2^{-1}-2^{-2}\right)$$

$$= \frac{K^2}{4}\sum_{j=J+1}^{\infty} 2^{-2j} = \frac{K^2}{4}\frac{2^{-2(j+1)}}{\left(1-\frac{1}{4}\right)} = \frac{K^2}{12}2^{-2j}.$$

From Equation 3.79, it can be observed that the error bound is inversely proportional to the level of resolution J. Hence, a more accurate result can be obtained by increasing the level of resolution in the Haar wavelet method. □

3.9 Numerical Results

Tables 3.5 through 3.7 show the comparisons of the exact solutions with the approximate solutions of Boussinesq–Burgers' equations at different collocation points. In Tables 3.5 through 3.7, J has been taken as 4, that is, $M=16$, and Δt is taken as 0.0001.

The *R.M.S. error* between the numerical solutions and the exact solutions of $u(x,t)$ for Boussinesq–Burgers' equations at $t=0.5$, 1.0, and 1.5 is 0.000142255, 0.000216937, and 0.000935793, respectively, and for $v(x,t)$, the *R.M.S. error* is found to be 0.0118472, 0.0236667, and 0.0346156, respectively.

Figures 3.5 through 3.8 cite the comparison graphically between the numerical and exact solutions for different values of t.

TABLE 3.5

Absolute Errors in the Solutions of Boussinesq–Burgers' Equations at Various Collocation Points of x with $t=0.5$

x	Approximate Solution (u_{approx})	Approximate Solution (v_{approx})	Exact Solution (u_{exact})	Exact Solution (v_{exact})	Absolute Error $\lvert u_{exact}-u_{approx}\rvert$	Absolute Error $\lvert v_{exact}-v_{approx}\rvert$
0.015625	−0.197367	−0.118653	−0.197359	−0.119458	8.35980E-6	8.05119E-4
0.046875	−0.201157	−0.119271	−0.201104	−0.120218	5.29589E-5	9.47342E-4
0.078125	−0.204944	−0.122098	−0.204872	−0.120927	7.19897E-5	1.17066E-3
0.109375	−0.20869	−0.12456	−0.208662	−0.121582	2.82735E-5	2.97813E-3
0.140625	−0.212446	−0.124475	−0.212471	−0.122183	2.42459E-5	2.29235E-3
0.171875	−0.216279	−0.123392	−0.216297	−0.122728	1.87152E-5	6.63754E-4
0.203125	−0.220164	−0.124375	−0.22014	−0.123217	2.37371E-5	1.15814E-3
0.234375	−0.224021	−0.127389	−0.223998	−0.123648	2.32677E-5	3.74124E-3
0.265625	−0.227829	−0.129227	−0.227868	−0.12402	3.87521E-5	5.2064E-3
0.296875	−0.23166	−0.128271	−0.231749	−0.124334	8.83945E-5	3.93697E-3
0.328125	−0.23557	−0.127024	−0.235638	−0.124587	6.77895E-5	2.4362E-3
0.359375	−0.239511	−0.128457	−0.239535	−0.124781	2.37165E-5	3.67564E-3
0.390625	−0.243396	−0.131492	−0.243436	−0.124914	4.01933E-5	6.57804E-3
0.421875	−0.24723	−0.132477	−0.247341	−0.124986	1.11453E-4	7.49073E-3
0.453125	−0.251101	−0.130649	−0.251247	−0.124997	1.46677E-4	5.65231E-3
0.484375	−0.255048	−0.129425	−0.255153	−0.124947	1.04846E-4	4.47776E-3
0.515625	−0.258996	−0.131357	−0.259056	−0.124836	5.93702E-5	6.52131E-3
0.546875	−0.262864	−0.134199	−0.262954	−0.124664	9.07736E-5	9.53507E-3
0.578125	−0.266683	−0.134123	−0.266847	−0.124432	1.63839E-4	9.69100E-3
0.609375	−0.270555	−0.1315	−0.270731	−0.12414	1.76238E-4	7.35999E-3
0.640625	−0.274493	−0.130521	−0.274605	−0.123789	1.11967E-4	6.73224E-3
0.671875	−0.278397	−0.132961	−0.278467	−0.123379	6.97724E-5	9.58204E-3
0.703125	−0.282198	−0.135353	−0.282315	−0.122911	1.17448E-4	1.2442E-2
0.734375	−0.285962	−0.13405	−0.286148	−0.122387	1.86803E-4	1.16634E-2
0.765625	−0.289792	−0.130795	−0.289964	−0.121806	1.71833E-4	8.98884E-3
0.796875	−0.293671	−0.130309	−0.293761	−0.12117	8.93953E-5	9.13864E-3
0.828125	−0.297477	−0.133207	−0.297537	−0.120481	5.94947E-5	1.2726E-2
0.859375	−0.301163	−0.13486	−0.30129	−0.119739	1.26730E-4	1.51218E-2
0.890625	−0.304831	−0.132242	−0.30502	−0.118946	1.88492E-4	1.32964E-2
0.921875	−0.308579	−0.128677	−0.308724	−0.118103	1.44573E-4	1.05737E-2
0.953125	−0.312353	−0.129219	−0.312401	−0.117212	4.80031E-5	1.20066E-2
0.984375	−0.316031	−0.13366	−0.316049	−0.116275	1.80034E-5	1.73849E-2

TABLE 3.6

Absolute Errors in the Solutions of Boussinesq–Burgers' Equations at Various Collocation Points of x with $t = 1.0$

x	Approximate Solution (u_{approx})	Approximate Solution (v_{approx})	Exact Solution (u_{exact})	Exact Solution (v_{exact})	Absolute Error $\|u_{exact} - u_{approx}\|$	Absolute Error $\|v_{exact} - v_{approx}\|$
0.015625	−0.227876	−0.123436	−0.227868	−0.12402	8.38233E-6	5.84734E-4
0.046875	−0.231802	−0.124054	−0.231749	−0.124334	5.32620E-5	2.79865E-4
0.078125	−0.235711	−0.12688	−0.235638	−0.124587	7.30976E-5	2.29300E-3
0.109375	−0.239566	−0.129343	−0.239535	−0.124781	3.09406E-5	4.56229E-3
0.140625	−0.243417	−0.129258	−0.243436	−0.124914	1.90610E-5	4.34445E-3
0.171875	−0.247331	−0.128175	−0.247341	−0.124986	9.87667E-6	3.18901E-3
0.203125	−0.251285	−0.129158	−0.251247	−0.124997	3.75151E-5	4.16092E-3
0.234375	−0.255196	−0.132172	−0.255153	−0.124947	4.33927E-5	7.22497E-3
0.265625	−0.259045	−0.13401	−0.259056	−0.124836	1.07795E-5	9.1736E-3
0.296875	−0.262903	−0.133054	−0.262954	−0.124664	5.10099E-5	8.38921E-3
0.328125	−0.266827	−0.131807	−0.266847	−0.124432	1.93935E-5	7.37414E-3
0.359375	−0.270768	−0.133239	−0.270731	−0.12414	3.72955E-5	9.09897E-3
0.390625	−0.27464	−0.136275	−0.274605	−0.123789	3.50153E-5	1.24855E-2
0.421875	−0.278446	−0.137259	−0.278467	−0.123379	2.05208E-5	1.38802E-2
0.453125	−0.282277	−0.135432	−0.282315	−0.122911	3.85771E-5	1.25206E-2
0.484375	−0.28617	−0.134208	−0.286148	−0.122387	2.17555E-5	1.18209E-2
0.515625	−0.290051	−0.13614	−0.289964	−0.121806	8.69255E-5	1.43344E-2
0.546875	−0.293837	−0.138982	−0.293761	−0.12117	7.62424E-5	1.78123E-2
0.578125	−0.297561	−0.138906	−0.297537	−0.120481	2.47286E-5	1.84257E-2
0.609375	−0.301325	−0.136283	−0.30129	−0.119739	3.44909E-5	1.65447E-2
0.640625	−0.305141	−0.135304	−0.30502	−0.118946	1.21288E-4	1.63586E-2
0.671875	−0.30891	−0.137744	−0.308724	−0.118103	1.86102E-4	1.96411E-2
0.703125	−0.312561	−0.140136	−0.312401	−0.117212	1.60845E-4	2.2924E-2
0.734375	−0.316162	−0.138833	−0.316049	−0.116275	1.1339E-4	2.25577E-2
0.765625	−0.319817	−0.135577	−0.319667	−0.115293	149407E-4	2.02845E-2
0.796875	−0.323506	−0.135092	−0.323254	−0.114268	2.51678E-4	2.08239E-2
0.828125	−0.327108	−0.137989	−0.326809	−0.113201	2.99826E-4	2.47885E-2
0.859375	−0.330578	−0.139643	−0.330329	−0.112095	2.48857E-4	2.75487E-2
0.890625	−0.334015	−0.137025	−0.333814	−0.11095	2.00973E-4	2.60745E-2
0.921875	−0.337519	−0.13346	−0.337263	−0.10977	2.55958E-4	2.36893E-2
0.953125	−0.341035	−0.134002	−0.340674	−0.108556	3.6035E-4	2.54455E-2
0.984375	−0.344442	−0.138443	−0.344047	−0.10731	394479E-4	3.11327E-2

TABLE 3.7

Absolute Errors in the Solutions of Boussinesq–Burgers' Equations at Various Collocation Points of x with $t = 1.5$

x	Approximate Solution (u_{approx})	Approximate Solution (v_{approx})	Exact Solution (u_{exact})	Exact Solution (v_{exact})	Absolute Error $\lvert u_{exact} - u_{approx} \rvert$	Absolute Error $\lvert v_{exact} - v_{approx} \rvert$
0.015625	−0.259065	−0.124493	−0.259056	−0.124836	8.80296E-6	3.42744E-4
0.046875	−0.263011	−0.125112	−0.262954	−0.124664	5.70662E-5	4.47176E-4
0.078125	−0.26693	−0.127938	−0.266847	−0.124432	8.3701E-5	3.50573E-3
0.109375	−0.270783	−0.130401	−0.270731	−0.12414	5.17642E-5	6.26042E-3
0.140625	−0.27462	−0.130316	−0.274605	−0.123789	1.53796E-5	6.52672E-3
0.171875	−0.278509	−0.129233	−0.278467	−0.123379	4.15244E-5	5.85325E-3
0.203125	−0.282425	−0.130215	−0.282315	−0.122911	1.09137E-4	7.30402E-3
0.234375	−0.286287	−0.13323	−0.286148	−0.122387	1.38385E-4	1.08429E-2
0.265625	−0.290075	−0.135067	−0.289964	−0.121806	1.10593E-4	1.32615E-2
0.296875	−0.29386	−0.134111	−0.293761	−0.12117	9.95855E-5	1.29412E-2
0.328125	−0.2977	−0.132864	−0.297537	−0.120481	1.63072E-4	1.23836E-2
0.359375	−0.301544	−0.134297	−0.30129	−0.119739	2.54057E-4	1.45584E-2
0.390625	−0.305308	−0.137332	−0.30502	−0.118946	2.88254E-4	1.83867E-2
0.421875	−0.308995	−0.138317	−0.308724	−0.118103	2.71105E-4	2.0214E-2
0.453125	−0.312694	−0.13649	−0.312401	−0.117212	2.9305E-4	1.92774E-2
0.484375	−0.316444	−0.135265	−0.316049	−0.116275	3.94684E-4	1.89901E-2
0.515625	−0.320169	−0.137198	−0.319667	−0.115293	5.02116E-4	2.19049E-2
0.546875	−0.323789	−0.14004	−0.323254	−0.114268	5.34300E-4	2.57723E-2
0.578125	−0.327335	−0.139964	−0.326809	−0.113201	5.25882E-4	2.67630E-2
0.609375	−0.330908	−0.137341	−0.330329	−0.112095	5.78576E-4	2.52464E-2
0.640625	−0.334522	−0.136362	−0.333814	−0.11095	7.07734E-4	2.54116E-2
0.671875	−0.338077	−0.138802	−0.337263	−0.10977	8.13912E-4	2.90315E-2
0.703125	−0.341503	−0.141194	−0.340674	−0.108556	8.28591E-4	3.26377E-2
0.734375	−0.344867	−0.13989	−0.344047	−0.10731	8.19196E-4	3.25803E-2
0.765625	−0.348272	−0.136635	−0.347381	−0.106034	8.9094E-4	3.06012E-2
0.796875	−0.3517	−0.136149	−0.350674	−0.104729	1.02614E-3	3.14198E-2
0.828125	−0.35503	−0.139047	−0.353926	−0.103399	1.10396E-3	3.56484E-2
0.859375	−0.358215	−0.140701	−0.357136	−0.102044	1.07891E-3	3.86573E-2
0.890625	−0.361357	−0.138083	−0.360304	−0.100666	1.05272E-3	3.74164E-2
0.921875	−0.364553	−0.134517	−0.363428	−0.0992682	1.12470E-3	3.52490E-2
0.953125	−0.367749	−0.135059	−0.366508	−0.0978518	1.24090E-3	3.72076E-2
0.984375	−0.370825	−0.1395	−0.369543	−0.0964187	1.28116E-3	4.30817E-2

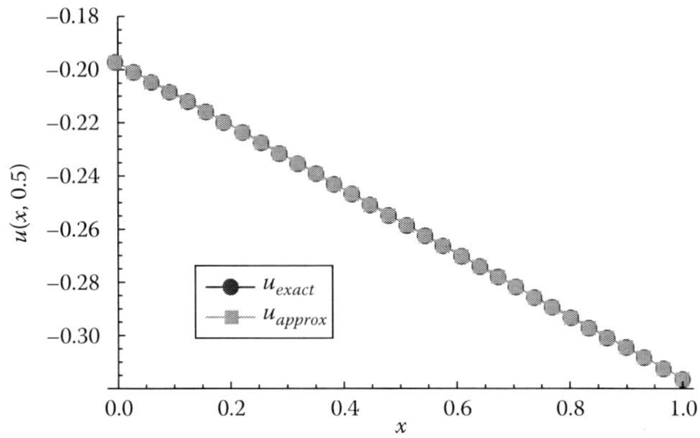

FIGURE 3.5
Comparison of a numerical solution and the exact solution of Boussinesq–Burgers' equations when $t = 0.5$.

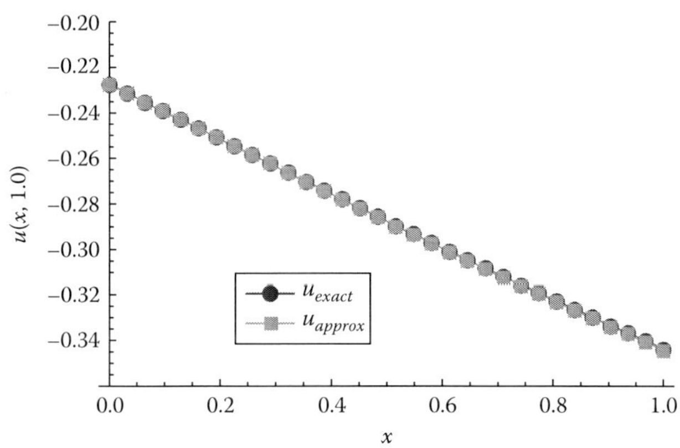

FIGURE 3.6
Comparison of a numerical solution and the exact solution of Boussinesq–Burgers' equations when $t = 1.0$.

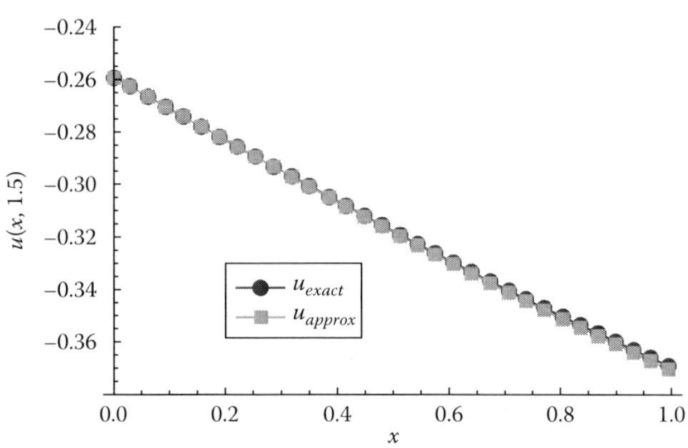

FIGURE 3.7
Comparison of a numerical solution and the exact solution of Boussinesq–Burgers' equations when $t = 1.5$.

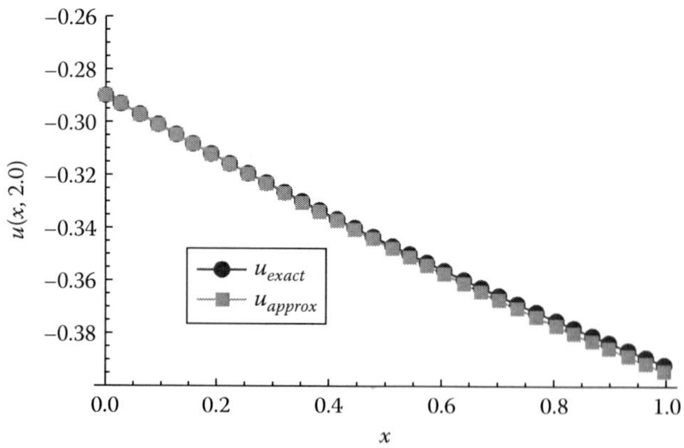

FIGURE 3.8
Comparison of a numerical solution and the exact solution of Boussinesq–Burgers' equations when $t = 2.0$.

3.10 Conclusion

In this chapter, the Boussinesq–Burgers' equations have been solved by using OHAM. The obtained results are then compared with exact solutions as well as HPM. Also, the Haar wavelet method has been implemented to find numerical solutions of Boussinesq–Burgers' equations. These results have been cited in the tables and also graphically demonstrated in order to justify the accuracy and efficiency of the proposed schemes.

This present chapter explores the reliability and powerfulness of OHAM over other perturbation methods for the system of nonlinear partial differential equations like Boussinesq–Burgers' equations. An advantage of OHAM over perturbation methods is that it does not depend on small parameters. Unlike other analytical methods, namely, HPM, OHAM allows fine tuning of the convergence region and rate of convergence by suitably identifying convergence control parameters $C_1, C_2, C_3, \ldots$. Additionally, we conclude that OHAM provides a simple and easy way to control and adjust the convergence region for strong nonlinearity and is applicable to highly nonlinear fluid problems like Boussinesq–Burgers' equations. Consequently, the present success of these methods for the highly nonlinear problem verifies that these methods are a useful tool for nonlinear problems in science and engineering.

4

Numerical Solution of Fractional Differential Equations by the Haar Wavelet Method

4.1 Introduction to Fractional Calculus

Fractional calculus is a 300-year-old topic but not very popular in the science and engineering community. The history of fractional calculus started at the end of the seventeenth century, and the birth of fractional calculus was because of a letter exchange. At that time, scientific journals did not exist and scientists exchanged their information by means of letters. Fractional calculus was first anticipated by Leibnitz, who was one of the founders of standard calculus, in a letter written in 1695. The primary effort to give a logical definition is because of Liouville. Subsequently, many mathematicians such as Abel, Caputo, Euler, Fourier, Grünwald, Hadamard, Hardy, Heaviside, Holmgren, Laplace, Leibniz, Letnikov, Riemann, Riesz, and Weyl made predominant contributions to the theory of fractional calculus.

Fractional calculus is a branch of calculus that generalizes the derivative of a function to arbitrary order. It is often referred to as generalized integral and differential calculus of arbitrary order [41]. In contemporary years, fractional calculus has become the focus of curiosity for many researchers in exclusive disciplines of applied science and engineering because of the fact that a realistic modeling of a physical phenomenon can be efficiently executed by way of utilizing fractional calculus. Indeed, fractional derivatives [90,91] provide an excellent instrument for the description of memory and hereditary properties of various materials and processes. This is the key advantage of fractional derivatives as compared to classical integer order derivatives in which such effects are abandoned. Many authors pointed out that fractional differential equations (FDEs) are inherently multidisciplinary with their application across diverse disciplines of applied science and engineering for the description of properties of various real physical phenomena. However, a reliable and effective general technique for solving them cannot be found even in the most useful works on fractional derivatives and integrals. So, in this chapter, our aim is to implement a reliable, appropriate, and efficient wavelet-based technique for the numerical solution of fractional differential equations.

In this chapter, the brief description for fractional calculus and the numerical solutions for nonlinear fractional differential equations are discussed. This present work is intended to make this subject available as a popular subject to the science and engineering community.

4.2 Fractional Derivative and Integration

Fractional calculus is a field of applied mathematics that deals with derivatives and integrals of arbitrary orders. Several approaches were used to define the derivatives of fractional order. At the outset, Liouville attempts to provide the logical definition of fractional order derivatives. Since then, several definitions of fractional integrals and derivatives have been proposed. This fractional calculus involves different definitions of fractional operators such as the Riemann–Liouville, the Caputo, the Weyl, the Hadamard, the Marchaud, the Riesz, the Grünwald–Letnikov, and the Erdelyi–Kober [90–92]. The Riemann–Liouville fractional derivative is mostly used by mathematicians, but this is not suitable for real-world physical problems as it requires the definition of fractional order initial conditions, which have no physically significant explanation yet. An alternative definition introduced by Caputo has the advantage of defining integer order initial conditions for fractional order differential equations. As a result, on this work, we will use the Caputo fractional derivative by means of Caputo in his work on the theory of viscoelasticity [93].

4.2.1 Riemann–Liouville Integral and Derivative Operator

The definition of first fractional integral was initially given by Liouville after a rigorous investigation in a series of papers from 1832 to 1837. Later on, research and further improvements have been made by many others directed to the development of the integral-based Riemann–Liouville fractional integral operator, which has been an essential foundation in fractional calculus ever since. The primal effort of Liouville was later purified in 1865 by the Swedish mathematician Holmgren [94], who made substantial contributions to the growing study of fractional calculus. But it was Riemann [95] who reconstructed it to fit Abel's integral equation and hence made it enormously more useful. Even now, there exist numerous forms of fractional integral operators, starting from divided-difference types to infinite-sum types; however, the Riemann–Liouville operator remains to be probably the most frequently used when fractional integration is carried out. Probably, the most conventionally encountered definition of an integral of fractional order is the Riemann–Liouville integral [41], in which the fractional integral operator $J^{\alpha}(\alpha > 0)$, of a function $f(t)$, is defined as [41,90]

$$J^{\alpha} f(t) = \frac{1}{\Gamma(\alpha)} \int_{0}^{t} (t-\tau)^{\alpha-1} f(\tau) d\tau, \quad \alpha > 0,\ \alpha \in \Re^{+} \tag{4.1}$$

where

$\Gamma(\cdot)$ is the well-known gamma function

$\Re^{+}$ is the set of positive real numbers

The gamma function Γ is outlined by

$$\Gamma(n) = \int_{0}^{\infty} e^{-t} t^{n-1} dt \quad \text{and for real number } n, \Gamma(n) = (n-1)!$$

Some properties of the operator J^α are as follows:

$$J^\alpha J^\beta f(t) = J^{\alpha+\beta} f(t), \quad (\alpha > 0, \beta > 0)$$

$$J^\alpha t^\gamma = \frac{\Gamma(1+\gamma)}{\Gamma(1+\gamma+\alpha)} t^{\alpha+\gamma}, \quad (\gamma > -1).$$

The Riemann–Liouville fractional derivative is defined by

$$D^\alpha f(t) = D^m J^{m-\alpha} f(t)$$
$$= \frac{d^m}{dt^m}\left[\frac{1}{\Gamma(m-\alpha)}\int_0^t \frac{f(\tau)}{(t-\tau)^{\alpha-m+1}} d\tau\right], \quad m-1 < \alpha < m, \; m \in N. \tag{4.2}$$

The left Riemann–Liouville fractional derivative can be defined by

$${}_aD_t^\alpha f(t) = \frac{1}{\Gamma(n-\alpha)}\left(\frac{d}{dt}\right)^n \int_a^t (t-\tau)^{n-\alpha-1} f(\tau) d\tau. \tag{4.3}$$

The right Riemann–Liouville fractional derivative can be defined by

$${}_tD_b^\alpha f(t) = \frac{1}{\Gamma(n-\alpha)}\left(-\frac{d}{dt}\right)^n \int_t^b (\tau-t)^{n-\alpha-1} f(\tau) d\tau. \tag{4.4}$$

Fractional Riemann–Liouville derivatives have various interesting properties. For example, the fractional derivative of a constant is not zero, that is,

$${}_aD_t^\alpha C = \frac{C(t-\alpha)^{-\alpha}}{\Gamma(1-\alpha)}. \tag{4.5}$$

Lemma 4.1

The integral formula of Riemann–Liouville fractional derivative

$$\int_\Gamma \left({}_0D_t^\alpha f(t)\right) g(t) dt = \int_\Gamma f(t) \left({}_tD_T^\alpha g(t)\right) dt$$

is valid under the assumption that $f, g \in C(\Gamma)$ and ${}_0D_t^\alpha f(t)$, ${}_tD_T^\alpha g(t)$ exist for all $t \in \Gamma$ and are continuous in t.

4.2.2 Caputo Fractional Derivative

An alternative definition of fractional derivative introduced by M. Caputo in 1967 [93] is called Caputo fractional derivative. The Caputo fractional derivative ${}_0D_t^{\alpha}$ of a function $f(t)$ is defined as [41,93]

$$D_t^{\alpha} f(t) = J^{m-\alpha} D^m f(t)$$
$$= \begin{cases} \dfrac{1}{\Gamma(m-\alpha)} \displaystyle\int_0^t (t-\tau)^{(m-\alpha-1)} \frac{d^m f(\tau)}{d\tau^m} d\tau, & \text{if } m-1<\alpha<m, m \in N \\ \dfrac{d^m f(t)}{dt^m}, & \text{if } \alpha = m, m \in N \end{cases} \tag{4.6}$$

where the parameter α is the order of the derivative and is allowed to be real or even complex. For Caputo's derivative, we have

$$D^{\alpha} C = 0 \quad (C \text{ is a constant}). \tag{4.7}$$

Similar to integer order differentiation, the Caputo derivative is linear:

$$D^{\alpha}(\gamma f(t) + \delta g(t)) = \gamma D^{\alpha} f(t) + \delta D^{\alpha} g(t), \tag{4.8}$$

where γ and δ are constants, and satisfies the so-called Leibnitz rule

$$D^{\alpha}(g(t) f(t)) = \sum_{k=0}^{\infty} \binom{\alpha}{k} g^{(k)}(t) D^{\alpha-k} f(t) \tag{4.9}$$

if $f(\tau)$ is continuous in $[0, t]$ and $g(\tau)$ has continuous derivatives sufficient number of times in $[0, t]$.

The following are two basic properties of the Caputo fractional derivative:

$${}_0D_t^{\alpha} t^{\beta} = \frac{\Gamma(1+\beta)}{\Gamma(1+\beta-\alpha)} t^{\beta-\alpha}, \quad 0<\alpha<\beta+1,\ \beta>-1 \tag{4.10}$$

$$J^{\alpha} D^{\alpha} f(t) = f(t) - \sum_{k=0}^{n-1} f^k(0^+) \frac{t^k}{k!}, \quad n-1<\alpha \le n \quad \text{and} \quad n \in N. \tag{4.11}$$

4.2.3 Grünwald–Letnikov Fractional Derivative

The Grünwald–Letnikov fractional derivative was first introduced by Anton Karl Grünwald (1838–1920) from Prague in 1867 and by Aleksey Vasilievich Letnikov (1837–1888) from Moscow in 1868. The Grünwald–Letnikov fractional derivative of order p (>0) is defined as [41]

$${}_aD_t^{\alpha} f(t) = \lim_{\substack{h \to 0 \\ nh = t-a}} h^{-p} \sum_{r=0}^{n} \omega_r^p f(t-rh) \tag{4.12}$$

where $\omega_r^p = (-1)^r \binom{p}{r}$.

$$\omega_0^p = 1 \quad \text{and} \quad \omega_r^p = \left(1 - \frac{p+1}{r}\right)\omega_{r-1}^p \quad r = 1, 2, \ldots$$

4.2.4 Riesz Fractional Derivative

Here, we present some significant definitions, namely, the right Riemann–Liouville derivative, left Riemann–Liouville derivative, Riesz fractional derivative, and Riesz fractional integral, which are to be used subsequently in consequent chapters.

The left and right Riemann–Liouville fractional derivatives of a function *f*(*x*) of order $\alpha(n-1<\alpha<n)$ are defined as [41,90]

$$_{-\infty}D_x^{\alpha} f(x) = \frac{1}{\Gamma(n-\alpha)} \frac{\partial^n}{\partial x^n} \int_{-\infty}^{x} (x-\zeta)^{n-1-\alpha} f(\zeta) d\zeta, \tag{4.13}$$

$$_{x}D_{\infty}^{\alpha} f(x) = \frac{(-1)^n}{\Gamma(n-\alpha)} \frac{\partial^n}{\partial x^n} \int_{x}^{\infty} (\zeta-x)^{n-1-\alpha} f(\zeta) d\zeta. \tag{4.14}$$

The Riesz fractional derivative of a function *f*(*x*) is defined as [41,90]

$$\frac{\partial^{\alpha}}{\partial |x|^{\alpha}} f(x) = \frac{-1}{2\cos\left(\frac{\alpha\pi}{2}\right)} \left[_{-\infty}D_x^{\alpha} f(x) + {}_{x}D_{+\infty}^{\alpha} f(x) \right]. \tag{4.15}$$

The fractional Riesz integral of a function *f*(*x*) is defined as [41,90]

$$\begin{aligned} _{RZ}I^{\alpha} f(x) &= \frac{1}{2\cos\left(\frac{\alpha\pi}{2}\right)} \left[I_+^{\alpha} + I_-^{\alpha} \right] f(x) \\ &= \frac{1}{2\Gamma(\alpha)\cos\left(\frac{\alpha\pi}{2}\right)} \int_{-\infty}^{+\infty} |x-\zeta|^{\alpha-1} f(\zeta) d\zeta, \quad \alpha > 0,\ \alpha \neq 1, 3, 5, \ldots \end{aligned} \tag{4.16}$$

where

$$I_+^{\alpha} f(x) = \frac{1}{\Gamma(n-\alpha)} \int_{x}^{\infty} (\zeta - x)^{n-\alpha-1} f(\zeta) d\zeta$$

$$I_-^{\alpha} f(x) = \frac{1}{\Gamma(n-\alpha)} \int_{-\infty}^{x} (x - \zeta)^{n-\alpha-1} f(\zeta) d\zeta.$$

4.3 Outline of the Present Study

In this chapter, we have considered both the analytical and the numerical approach for solving some particular nonlinear fractional differential equations like the fractional Burgers–Fisher equation, fractional Fisher's type equation, and time- and space-fractional Fokker–Planck equation (FPE), which have a wide variety of applications in various physical phenomena. These fractional differential equations have proved particularly beneficial in the context of an anomalous diffusion model, a fluid dynamics model, heat conduction, elasticity, and capillary-gravity waves.

Consider the generalized one-dimensional Burgers–Fisher equation of fractional order

$$\frac{\partial^\alpha u}{\partial t^\alpha} + \xi u^\eta \frac{\partial u}{\partial x} = \mu \frac{\partial^2 u}{\partial x^2} + \beta u\left(1 - u^\eta\right), \tag{4.17}$$

where ξ, μ, and β are parameters and $0 < \alpha \leq 1$. This equation has a wide range of applications in the fluid dynamics model, heat conduction, elasticity, and capillary-gravity waves. When $\xi = 0$ and $\eta = 1$, Equation 4.17 reduces to a Fisher-type equation.

The generalized time-fractional Fisher's biological population diffusion equation is given by

$$\frac{\partial^\alpha u}{\partial t^\alpha} = \frac{\partial^2 u}{\partial x^2} + F(u), \quad u(x,0) = \varphi(x), \tag{4.18}$$

where

$u(x,t)$ denotes the population density and $t > 0$, $x \in \Re$

$F(u)$ is a continuous nonlinear function satisfying the following conditions $F(0) = F(1) = 0$, $F'(0) > 0 > F'(1)$

The derivatives in Equations 4.17 and 4.18 are the Caputo derivative of order α.

Next, we consider the time- and space-fractional FPE. The classical FPE was introduced by Adriaan Fokker and Max Planck, commonly used to describe the Brownian motion of particles [96]. An FPE describes the change of probability of a random function in space and time; hence, it is naturally used to describe solute transport. The general FPE for the motion of a concentration field, $u(x,t)$, of one space variable, x, at time t has the form [97–99]

$$\frac{\partial u}{\partial t} = \left[-\frac{\partial}{\partial x} A(x) + \frac{\partial^2}{\partial x^2} B(x) \right] u(x,t), \tag{4.19}$$

with the initial condition

$$u(x,0) = f(x), \quad x \in \Re \tag{4.20}$$

where $A(x)$ and $B(x) > 0$ are referred to as the drift and diffusion coefficients. The drift and diffusion coefficients may also depend on time.

There is a more general form of FPE called the nonlinear FPE, which is of the form [97–99]

$$\frac{\partial u}{\partial t}=\left[-\frac{\partial}{\partial x}A(x,t,u)+\frac{\partial^2}{\partial x^2}B(x,t,u)\right]u(x,t). \tag{4.21}$$

The nonlinear FPE has important applications in various fields such as plasma physics, surface physics, population dynamics, biophysics, engineering, neuroscience, nonlinear hydrodynamics, polymer physics, laser physics, pattern formation, psychology, and marketing [100].

In recent years, there has been a great deal of interest in fractional diffusion equations. These equations arise in continuous-time random walks, modeling of anomalous diffusive and subdiffusive systems, unification of diffusion, and wave propagation phenomenon [101].

Consider the generalized nonlinear time- and space-fractional FPE [102]

$$\frac{\partial^\alpha u}{\partial t^\alpha}=\left[-\frac{\partial^\beta}{\partial x^\beta}A(x,t,u)+\frac{\partial^{2\beta}}{\partial x^{2\beta}}B(x,t,u)\right]u(x,t),\quad t>0,\ x>0 \tag{4.22}$$

where α and β are parameters describing the order of the fractional time and space derivatives, respectively. The function $u(x,t)$ is assumed to be a casual function of time and space, that is, vanishing for $t<0$ and $x<0$. The fractional derivatives are considered in the Caputo sense.

Various mathematical methods such as the Adomian decomposition method (ADM) [103], variational iteration method (VIM) [103], operational Tau method (OTM) [104], and homotopy perturbation method (HPM) [105] have been used in attempting to solve fractional FPEs.

4.4 Application of Analytical and Numerical Techniques to the Fractional Burgers–Fisher Equation

4.4.1 Haar Wavelet-Based Scheme for the Fractional Burgers–Fisher Equation

Consider the generalized fractional order Burgers–Fisher equation given in Equation 4.17 with the following initial and boundary conditions:

$$u(x,0)=\left[\frac{1}{2}+\frac{1}{2}\tanh\left(\frac{-\xi\eta}{2(\eta+1)}x\right)\right]^{1/\eta}, \tag{4.23}$$

$$u(0,t)=\left(\frac{1}{2}+\frac{1}{2}\tanh\left[\frac{-\xi\eta}{2(\eta+1)}\left(-\left(\frac{\xi}{\eta+1}+\frac{\beta(\eta+1)}{\xi}\right)t\right)\right]\right)^{1/\eta},\quad t\ge 0 \tag{4.24}$$

$$u(1,t)=\left(\frac{1}{2}+\frac{1}{2}\tanh\left[\frac{-\xi\eta}{2(\eta+1)}\left(1-\left(\frac{\xi}{\eta+1}+\frac{\beta(\eta+1)}{\xi}\right)t\right)\right]\right)^{1/\eta},\quad t\ge 0 \tag{4.25}$$

and when $\alpha = 1$, the exact solution of Equation 4.17 is given by [106]

$$u(x,t) = \left(\frac{1}{2} + \frac{1}{2}\tanh\left[\frac{-\xi\eta}{2(\eta+1)}\left(x - \left(\frac{\xi}{\eta+1} + \frac{\beta(\eta+1)}{\xi}\right)t\right)\right]\right)^{1/\eta}. \tag{4.26}$$

Let us divide both space and time intervals [0, 1] into m equal subintervals; each of width $\Delta = 1/m$. Here, we have taken $\eta = 1$, $\mu = 1$, and $\xi = \beta = 0.01$. Therefore, Equation 4.17 reduces to

$$\frac{\partial^\alpha u}{\partial t^\alpha} - \frac{\partial^2 u}{\partial x^2} + 0.01\left(u\frac{\partial u}{\partial x} + u(u-1)\right) = 0. \tag{4.27}$$

The Haar wavelet solution of $u(x,t)$ is sought by assuming that $\partial^2 u(x,t)/\partial x^2$ can be expanded in terms of Haar wavelets as

$$\frac{\partial^2 u(x,t)}{\partial x^2} = \sum_{i=1}^{m}\sum_{j=1}^{m} c_{ij} h_i(x) h_j(t). \tag{4.28}$$

Integrating Equation 4.28 with respect to x from 0 to x, we get

$$\frac{\partial u(x,t)}{\partial x} - p(t) = \sum_{i=1}^{m}\sum_{j=1}^{m} c_{ij} Q h_i(x) h_j(t). \tag{4.29}$$

Again, integrating Equation 4.29 with respect to x from 0 to x, we get

$$u(x,t) = \sum_{i=1}^{m}\sum_{j=1}^{m} c_{ij} Q^2 h_i(x) h_j(t) + q(t) + xp(t). \tag{4.30}$$

Putting $x = 0$ in Equation 4.30, we get

$$q(t) = u(0,t). \tag{4.31}$$

Putting $x = 1$ in Equation 4.30, we get

$$p(t) = u(1,t) - u(0,t) - \sum_{i=1}^{m}\sum_{j=1}^{m} c_{ij}\left[Q^2 h_i(x)\right]_{x=1} h_j(t). \tag{4.32}$$

Again, $q(t) + xp(t)$ can be approximated using the Haar wavelet function as

$$q(t) + xp(t) = \sum_{i=1}^{m}\sum_{j=1}^{m} r_{ij} h_i(x) h_j(t). \tag{4.33}$$

This implies

$$u(0,t)+x\left[u(1,t)-u(0,t)-\sum_{i=1}^{m}\sum_{j=1}^{m}c_{ij}\left[Q^2h_i(x)\right]_{x=1}h_j(t)\right]=\sum_{i=1}^{m}\sum_{j=1}^{m}r_{ij}h_i(x)h_j(t). \quad (4.34)$$

Substituting Equation 4.33 in Equation 4.30, we get

$$u(x,t)=\sum_{i=1}^{m}\sum_{j=1}^{m}c_{ij}Q^2h_i(x)h_j(t)+\sum_{i=1}^{m}\sum_{j=1}^{m}r_{ij}h_i(x)h_j(t). \quad (4.35)$$

The nonlinear term presented in Equation 4.27 can be approximated using the Haar wavelet function as

$$u\frac{\partial u}{\partial x}+u(u-1)=\sum_{i=1}^{m}\sum_{j=1}^{m}d_{ij}h_i(x)h_j(t). \quad (4.36)$$

Therefore, from Equations 4.29, 4.32, and 4.35, we have

$$\begin{aligned}&\left(\sum_{i=1}^{m}\sum_{j=1}^{m}c_{ij}Q^2h_i(x)h_j(t)+\sum_{i=1}^{m}\sum_{j=1}^{m}r_{ij}h_i(x)h_j(t)\right)\\&\times\left[\sum_{i=1}^{m}\sum_{j=1}^{m}c_{ij}Qh_i(x)h_j(t)+u(1,t)-u(0,t)-\sum_{i=1}^{m}\sum_{j=1}^{m}c_{ij}\left[Q^2h_i(x)\right]_{x=1}h_j(t)\right]\\&+\left(\sum_{i=1}^{m}\sum_{j=1}^{m}c_{ij}Q^2h_i(x)h_j(t)+\sum_{i=1}^{m}\sum_{j=1}^{m}r_{ij}h_i(x)h_j(t)\right)\\&\times\left(\sum_{i=1}^{m}\sum_{j=1}^{m}c_{ij}Q^2h_i(x)h_j(t)+\sum_{i=1}^{m}\sum_{j=1}^{m}r_{ij}h_i(x)h_j(t)-1\right)=\sum_{i=1}^{m}\sum_{j=1}^{m}d_{ij}h_i(x)h_j(t).\end{aligned} \quad (4.37)$$

Substituting Equations 4.28 and 4.36 in Equation 4.27, we will have

$$\frac{\partial^\alpha u}{\partial t^\alpha}=\sum_{i=1}^{m}\sum_{j=1}^{m}c_{ij}h_i(x)h_j(t)-0.01\sum_{i=1}^{m}\sum_{j=1}^{m}d_{ij}h_i(x)h_j(t). \quad (4.38)$$

Now, applying J^α to both sides of Equation 4.38 yields

$$u(x,t)-u(x,0)=J^\alpha\left(\sum_{i=1}^{m}\sum_{j=1}^{m}c_{ij}h_i(x)h_j(t)\right)-0.01J^\alpha\left(\sum_{i=1}^{m}\sum_{j=1}^{m}d_{ij}h_i(x)h_j(t)\right). \quad (4.39)$$

Substituting Equations 4.23 and 4.35 in Equation 4.39, we get

$$\sum_{i=1}^{m}\sum_{j=1}^{m} c_{ij}Q^2h_i(x)h_j(t)+\sum_{i=1}^{m}\sum_{j=1}^{m} r_{ij}h_i(x)h_j(t)-\left[\frac{1}{2}+\frac{1}{2}\tanh\left(\frac{-\xi\eta}{2(\eta+1)}x\right)\right]$$

$$=\sum_{i=1}^{m}\sum_{j=1}^{m} c_{ij}h_i(x)Q_t^{\alpha}h_j(t)-0.01\left(\sum_{i=1}^{m}\sum_{j=1}^{m} d_{ij}h_i(x)Q_t^{\alpha}h_j(t)\right). \tag{4.40}$$

Now, substituting the collocation points $x_l=(l-0.5)/m$ and $t_k=(k-0.5)/m$ for $l,k=1,2,\ldots,m$ in Equations 4.34, 4.37, and 4.40, we have $3m^2$ equations in $3m^2$ unknowns in c_{ij}, r_{ij}, and d_{ij}. By solving this system of equations using mathematical software, the Haar wavelet coefficients c_{ij}, r_{ij}, and d_{ij} can be obtained.

4.4.2 Application of Optimal Homotopy Asymptotic Method to the Time-Fractional Burgers–Fisher Equation

Using optimal homotopy asymptotic method (OHAM), the homotopy for Equation 4.27 can be written as

$$(1-p)\frac{\partial^{\alpha}\varphi(x,t;p)}{\partial t^{\alpha}}=H(p)\left[\frac{\partial^{\alpha}\varphi(x,t;p)}{\partial t^{\alpha}}-\frac{\partial^{\alpha}\varphi(x,t;p)}{\partial x^2}\right.$$

$$\left.+\varphi(x,t;p)\frac{\partial\varphi(x,t;p)}{\partial x}-\varphi(x,t;p)\big(1-\varphi(x,t;p)\big)\right]. \tag{4.41}$$

Here

$$\varphi(x,t;p)=u_0(x,t)+\sum_{i=1}^{\infty}u_i(x,t)p^i, \tag{4.42}$$

$$H(p)=C_1p+C_2p^2+C_3p^3+\cdots, \tag{4.43}$$

$$N\big(\varphi(x,t;p)\big)=N_0\big(u_0(x,t)\big)+\sum_{k=1}^{\infty}N_k(u_0,u_1,\ldots,u_k)p^k. \tag{4.44}$$

Substituting Equations 4.42 through 4.44 in Equation 4.41 and equating the coefficients of like powers of p, we have the following system of partial differential equations:

$$\text{Coefficients of } p^0:\quad \frac{\partial^{\alpha}u_0(x,t)}{\partial t^{\alpha}}=0 \tag{4.45}$$

$$\text{Coefficients of } p^1:\quad \frac{\partial^{\alpha}u_1(x,t)}{\partial t^{\alpha}}-\frac{\partial^{\alpha}u_0(x,t)}{\partial t^{\alpha}}=C_1\left[\frac{\partial^{\alpha}u_0(x,t)}{\partial t^{\alpha}}-\frac{\partial^2u_0(x,t)}{\partial x^2}\right.$$

$$\left.+0.01\left[u_0(x,t)\frac{\partial u_0(x,t)}{\partial x}+\big(u_0(x,t)\big)^2-u_0(x,t)\right]\right] \tag{4.46}$$

Coefficients of p^2: $\frac{\partial^\alpha u_2(x,t)}{\partial t^\alpha}-\frac{\partial^\alpha u_1(x,t)}{\partial t^\alpha}$

$$=C_1\left[\frac{\partial^\alpha u_1(x,t)}{\partial t^\alpha}-\frac{\partial^2 u_1(x,t)}{\partial x^2}+0.01\left[u_0(x,t)\frac{\partial u_1(x,t)}{\partial x}+2u_0(x,t)u_1(x,t)+u_1(x,t)\frac{\partial u_0(x,t)}{\partial x}-u_1(x,t)\right]\right]$$

$$+C_2\left[\frac{\partial^\alpha u_0(x,t)}{\partial t^\alpha}-\frac{\partial^2 u_0(x,t)}{\partial x^2}+0.01\left[u_0(x,t)\frac{\partial u_0(x,t)}{\partial x}+\left(u_0(x,t)\right)^2-u_0(x,t)\right]\right] \tag{4.47}$$

and so on.

For solving a fractional order Burgers–Fisher equation using OHAM, we consider the initial condition Equation 4.23, and solving Equation 4.45 to Equation 4.47, we obtain

$$u_0(x,t)=\left[\frac{1}{2}+\frac{1}{2}\tanh\left(\frac{-0.01}{4}x\right)\right], \tag{4.48}$$

$$u_1(x,t)=\frac{-0.00250625C_1\text{sech}^2(0.0025x)t^\alpha}{\Gamma(1+\alpha)}, \tag{4.49}$$

$$\begin{aligned}u_2(x,t)&=u_1(x,t)+C_1\left[u_1(x,t)-\frac{C_1\left(\text{sech}^4(0.0025x)-2\text{sech}^2(0.0025x)\tanh^2(0.0025x)\right)t^{2\alpha}}{\Gamma(1+2\alpha)}\right.\\&+0.01\left[\frac{C_1\text{sech}^2(0.0025x)\left(-1+\tanh(0.0025x)\right)\tanh(0.0025x)t^{2\alpha}}{\Gamma(1+2\alpha)}\right.\\&+\frac{C_1\text{sech}^4(0.0025x)t^{2\alpha}}{\Gamma(1+2\alpha)}+\frac{0.00250625C_1\text{sech}^2(0.0025x)\left(-1+\tanh(0.0025x)\right)t^{2\alpha}}{\Gamma(1+2\alpha)}\\&\left.\left.-\frac{0.00250625C_1\text{sech}^2(0.0025x)t^{2\alpha}}{\Gamma(1+2\alpha)}\right]\right]\\&+C_2\left[-\frac{\partial^2 u_0(x,t)}{\partial x^2}+0.01\left[u_0(x,t)\frac{\partial u_0(x,t)}{\partial x}+\left(u_0(x,t)\right)^2-u_0(x,t)\right]\right]\frac{t^\alpha}{\Gamma(1+\alpha)},\end{aligned} \tag{4.50}$$

and so on.

Using Equations 4.48 through 4.50, the second-order approximate solution is obtained as follows:

$$
\begin{aligned}
u(x,t) &= u_0(x,t)+u_1(x,t)+u_2(x,t) \\
&= \frac{1}{2}+\frac{1}{2}\tanh\left(\frac{-0.01}{4}x\right)-\frac{0.00250625C_1\operatorname{sech}^2(0.0025x)t^{\alpha}}{\Gamma(1+\alpha)}+u_1(x,t) \\
&+C_1\left[u_1(x,t)-\frac{C_1\left(\operatorname{sech}^4(0.0025x)-2\operatorname{sech}^2(0.0025x)\tanh^2(0.0025x)\right)t^{2\alpha}}{\Gamma(1+2\alpha)}\right. \\
&+0.01\left[\frac{C_1\operatorname{sech}^2(0.0025x)\left(-1+\tanh(0.0025x)\right)\tanh(0.0025x)t^{2\alpha}}{\Gamma(1+2\alpha)}\right. \\
&+\frac{C_1\operatorname{sech}^4(0.0025x)t^{2\alpha}}{\Gamma(1+2\alpha)}+\frac{0.00250625C_1\operatorname{sech}^2(0.0025x)\left(-1+\tanh(0.0025x)\right)t^{2\alpha}}{\Gamma(1+2\alpha)} \\
&\left.\left.-\frac{0.00250625C_1\operatorname{sech}^2(0.0025x)t^{2\alpha}}{\Gamma(1+2\alpha)}\right]\right] \\
&+C_2\left[-\frac{\partial^2 u_0(x,t)}{\partial x^2}+0.01\left\{u_0(x,t)\frac{\partial u_0(x,t)}{\partial x}+\left(u_0(x,t)\right)^2-u_0(x,t)\right\}\right]\frac{t^{\alpha}}{\Gamma(1+\alpha)}.
\end{aligned}
\tag{4.51}
$$

The optimal values of the convergence control parameters C_1 and C_2 can be obtained by using the collocation method from Equation 1.33.

4.5 Numerical Results for a Fractional Burgers–Fisher Equation

Table 4.1 shows the comparison of the absolute errors of the Burgers–Fisher equation obtained by using the Haar wavelet method and OHAM at different values of x and t, taking $\alpha=1$. Similarly, Tables 4.2 through 4.4 exhibit the comparison of approximate solutions obtained by the Haar wavelet method and OHAM for a fractional order Burgers–Fisher equation, taking $\alpha=0.75$, 0.5, and 0.25, respectively. In Tables 4.1 through 4.4, m has been taken as 16. The obtained results in Tables 4.1 through 4.4 demonstrate that these methods are well suited for solving the fractional Burgers–Fisher equation. Both methods are quite efficient and effective.

Figures 4.1 and 4.2 cite the comparison graphically between the numerical solutions obtained by the Haar wavelet method, OHAM, and exact solutions for different values of t and x.

TABLE 4.1

Absolute Errors in the Solution of the Fractional Order Burgers–Fisher Equation Given in Equation 4.27 Using the Haar Wavelet Method and Three Terms for Second-Order OHAM with Convergence Control Parameters $C_1=0$, $C_2=-0.99999$ at Various Points of x and t for $\alpha=1$

	$\lvert u_{Exact}-u_{Haar}\rvert$				$\lvert u_{Exact}-u_{OHAM}\rvert$			
x	$t=0.2$	$t=0.4$	$t=0.6$	$t=0.8$	$t=0.2$	$t=0.4$	$t=0.6$	$t=0.8$
0.1	5.4804E−5	2.3476E−5	7.85260E−6	3.9181E−5	4.229E−11	8.408E−10	3.403E−9	8.7368E−9
0.2	2.3553E−5	7.7785E−6	3.91080E−5	7.0440E−5	8.333E−11	3.384E−10	2.273E−9	6.7268E−9
0.3	7.0426E−5	3.9091E−5	7.75940E−6	2.3578E−5	2.089E−10	1.642E−10	1.142E−9	4.7168E−9
0.4	3.9169E−5	7.8222E−6	2.35157E−5	5.4870E−5	3.346E−10	6.667E−10	1.13E−11	2.7068E−9
0.5	7.9054E−6	2.3463E−5	5.48121E−5	8.6199E−6	4.602E−10	1.1692E−9	1.119E−9	6.968E−10
0.6	5.4768E−5	2.3384E−5	7.97308E−6	3.9384E−5	5.858E−10	1.6717E−9	2.249E−9	1.3132E−9
0.7	2.3489E−5	7.9370E−6	3.93167E−5	7.0791E−5	7.115E−10	2.1742E−9	3.381E−9	3.3232E−9
0.8	7.0337E−5	3.8884E−5	7.48940E−6	2.4026E−5	8.371E−10	2.6767E−9	4.511E−9	5.3332E−9
0.9	3.9031E−5	7.5074E−6	2.39232E−5	5.5543E−5	9.627E−10	3.1792E−9	5.642E−9	7.3432E−9
1.0	8.5852E−5	5.4286E−5	2.28326E−5	8.8514E−6	1.0883E−9	3.6817E−9	6.772E−9	9.3532E−9

TABLE 4.2

Comparison between the Approximate Solutions of Fractional Order Burgers–Fisher Equation Given in Equation 4.27 Using the Haar Wavelet Method and Three Terms for Second-Order OHAM with Convergence Control Parameters $C_1=-0.000104528$, $C_2=-0.99979$ at Various Points of x and t for $\alpha=0.75$

	$t=0.2$		$t=0.4$		$t=0.6$		$t=0.8$	
x	u_{Haar}	u_{OHAM}	u_{Haar}	u_{OHAM}	u_{Haar}	u_{OHAM}	u_{Haar}	u_{OHAM}
0.1	0.50043	0.500691	0.5009	0.501247	0.501369	0.501734	0.501839	0.502182
0.2	0.500271	0.500566	0.50074	0.501122	0.501207	0.501609	0.501674	0.502057
0.3	0.500187	0.500441	0.500656	0.500997	0.501119	0.501484	0.501584	0.501932
0.4	0.500028	0.500316	0.500495	0.500872	0.500944	0.501359	0.5014	0.501807
0.5	0.499873	0.500191	0.500335	0.500747	0.500758	0.501234	0.501197	0.501682
0.6	0.499789	0.500066	0.500248	0.500622	0.500653	0.501109	0.50108	0.501557
0.7	0.499641	0.499941	0.500092	0.500497	0.500446	0.500984	0.500839	0.501432
0.8	0.499559	0.499816	0.500005	0.500372	0.500326	0.500859	0.500697	0.501307
0.9	0.499423	0.499691	0.499855	0.500247	0.500092	0.500734	0.500409	0.501182
1.0	0.499345	0.499566	0.499767	0.500122	0.499953	0.500609	0.500236	0.501057

4.6 Application of Analytical and Numerical Methods to a Fractional Fisher's Type Equation

4.6.1 Haar Wavelet-Based Scheme for the Generalized Fisher's Equation

Consider the generalized Fisher's equation [107] of fractional order

$$\frac{\partial^\alpha u}{\partial t^\alpha}=\frac{\partial^2 u}{\partial x^2}+u\left(1-u^6\right),\quad 0<\alpha\le 1,\ 0\le x\le 1, \tag{4.52}$$

TABLE 4.3

Comparison between the Approximate Solutions of the Fractional Order Burgers–Fisher Equation Given in Equation 4.27 Using the Haar Wavelet Method and Three Terms for Second-Order OHAM with Convergence Control Parameters $C_1 = 0.000163239$, $C_2 = -1.00032796$ at Various Points of x and t for $\alpha = 0.5$

	$t=0.2$		$t=0.4$		$t=0.6$		$t=0.8$	
x	u_{Haar}	u_{OHAM}	u_{Haar}	u_{OHAM}	u_{Haar}	u_{OHAM}	u_{Haar}	u_{OHAM}
0.1	0.500429	0.50114	0.500898	0.501664	0.501368	0.502066	0.501837	0.502404
0.2	0.500273	0.501015	0.500736	0.501539	0.501201	0.501941	0.501666	0.502279
0.3	0.50019	0.50089	0.500646	0.501414	0.501105	0.501816	0.501566	0.502154
0.4	0.500056	0.500765	0.500484	0.501289	0.500921	0.501691	0.501363	0.502029
0.5	0.499953	0.50064	0.500329	0.501164	0.500725	0.501566	0.501132	0.501904
0.6	0.499898	0.500515	0.500238	0.501039	0.500605	0.501441	0.500986	0.501779
0.7	0.49986	0.50039	0.500098	0.500914	0.500386	0.501316	0.500698	0.501654
0.8	0.499838	0.500265	0.50001	0.500789	0.500246	0.501191	0.500513	0.501529
0.9	0.499889	0.50014	0.499894	0.500664	0.499999	0.501066	0.500153	0.501404
1.0	0.499912	0.500015	0.499814	0.500539	0.499838	0.500941	0.499922	0.501279

TABLE 4.4

Comparison between the Approximate Solutions of the Fractional Order Burgers–Fisher Equation Given in Equation 4.27 Using the Haar Wavelet Method and Three Terms for Second-Order OHAM with Convergence Control Parameters $C_1 = -0.00019986$, $C_2 = -0.999602$ at Various Points of x and t for $\alpha = 0.25$

	$t=0.2$		$t=0.4$		$t=0.6$		$t=0.8$	
x	u_{Haar}	u_{OHAM}	u_{Haar}	u_{OHAM}	u_{Haar}	u_{OHAM}	u_{Haar}	u_{OHAM}
0.1	0.500427	0.501724	0.500897	0.502074	0.501366	0.502309	0.501836	0.50249
0.2	0.50027	0.501599	0.500733	0.501949	0.501196	0.502184	0.50166	0.502365
0.3	0.500184	0.501474	0.500639	0.501824	0.501095	0.502059	0.501552	0.50224
0.4	0.500056	0.501349	0.50048	0.501699	0.500909	0.501934	0.50134	0.502115
0.5	0.49997	0.501224	0.500339	0.501574	0.500717	0.501809	0.501099	0.50199
0.6	0.49992	0.501099	0.500248	0.501449	0.500591	0.501684	0.500937	0.501865
0.7	0.499923	0.500974	0.500142	0.501324	0.500385	0.501559	0.500637	0.50174
0.8	0.499916	0.500849	0.500064	0.501199	0.500243	0.501434	0.500432	0.501615
0.9	0.500043	0.500724	0.500011	0.501074	0.500027	0.501309	0.500059	0.50149
1.0	0.500097	0.500599	0.499954	0.500949	0.499869	0.501184	0.499805	0.501365

with the initial condition

$$u(x,0) = \frac{1}{\left(1+e^{3x/2}\right)^{1/3}}. \tag{4.53}$$

When $\alpha = 1$, the exact solution of Equation 4.52 is given by [108]

$$u(x,t) = \left(\frac{1}{2} + \frac{1}{2}\tanh\left[\frac{-3}{4}\left(x - \frac{5}{2}t\right)\right]\right)^{1/3}. \tag{4.54}$$

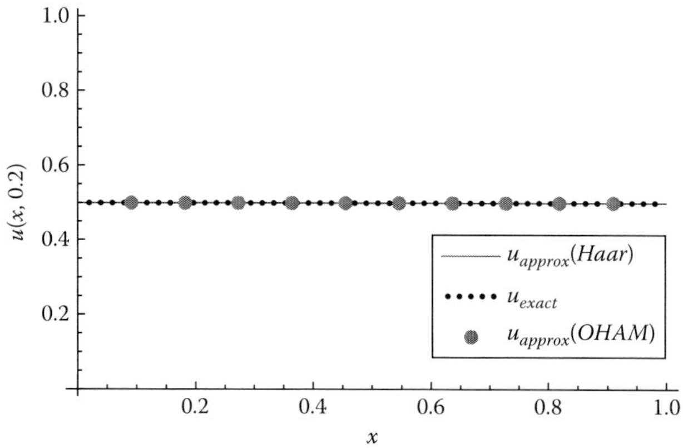

FIGURE 4.1
Comparison of a Haar wavelet solution and an OHAM solution with the exact solution of Burgers–Fisher equation when $t=0.2$.

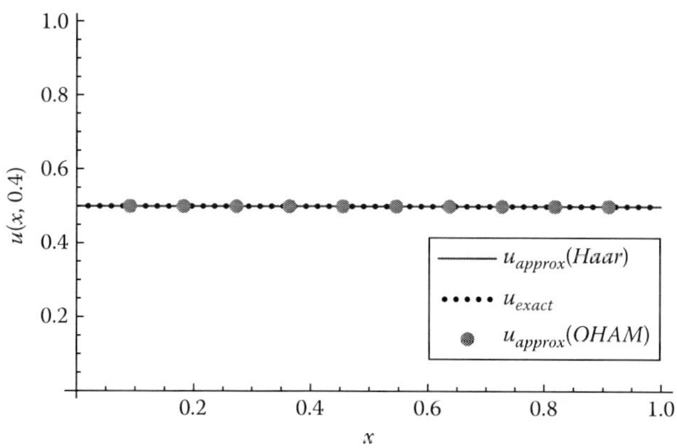

FIGURE 4.2
Comparison of a Haar wavelet solution and an OHAM solution with the exact solution of Burgers–Fisher equation when $t=0.4$.

Let us divide both the space and time intervals [0, 1] into m equal subintervals, each of width $\Delta=1/m$.

The Haar wavelet solution of $u(x,t)$ is sought by assuming that $\partial^2 u(x,t)/\partial x^2$ can be expanded in terms of Haar wavelets as

$$\frac{\partial^2 u(x,t)}{\partial x^2}=\sum_{i=1}^{m}\sum_{j=1}^{m} c_{ij}h_i(x)h_j(t). \tag{4.55}$$

Integrating Equation 4.55 twice with respect to x from 0 to x, we get

$$u(x,t)=\sum_{i=1}^{m}\sum_{j=1}^{m} c_{ij}Q^2h_i(x)h_j(t)+q(t)+xp(t). \tag{4.56}$$

Putting $x=0$ in Equation 4.56, we get

$$q(t)=u(0,t). \tag{4.57}$$

Putting $x=1$ in Equation 4.56, we get

$$p(t)=u(1,t)-u(0,t)-\sum_{i=1}^{m}\sum_{j=1}^{m}c_{ij}\left[Q^2h_i(x)\right]_{x=1}h_j(t). \tag{4.58}$$

Substituting Equations 4.57 and 4.58 in Equation 4.56, we have

$$\begin{aligned}u(x,t)=&\sum_{i=1}^{m}\sum_{j=1}^{m}c_{ij}Q^2h_i(x)h_j(t)+u(0,t)\\&+x\left[u(1,t)-u(0,t)-\sum_{i=1}^{m}\sum_{j=1}^{m}c_{ij}\left[Q^2h_i(x)\right]_{x=1}h_j(t)\right].\end{aligned} \tag{4.59}$$

The nonlinear term presented in Equation 4.52 can be approximated using the Haar wavelet function as

$$u\left(1-u^6\right)=\sum_{i=1}^{m}\sum_{j=1}^{m}d_{ij}h_i(x)h_j(t), \tag{4.60}$$

which implies

$$\begin{aligned}&\left(\sum_{i=1}^{m}\sum_{j=1}^{m}c_{ij}Q^2h_i(x)h_j(t)+u(0,t)+x\left[u(1,t)-u(0,t)-\sum_{i=1}^{m}\sum_{j=1}^{m}c_{ij}\left[Q^2h_i(x)\right]_{x=1}h_j(t)\right]\right)\\&\times\left[1-\left(\sum_{i=1}^{m}\sum_{j=1}^{m}c_{ij}Q^2h_i(x)h_j(t)+u(0,t)+x\left[u(1,t)-u(0,t)-\sum_{i=1}^{m}\sum_{j=1}^{m}c_{ij}\left[Q^2h_i(x)\right]_{x=1}h_j(t)\right]\right)^6\right]\\&=\sum_{i=1}^{m}\sum_{j=1}^{m}d_{ij}h_i(x)h_j(t).\end{aligned} \tag{4.61}$$

Substituting Equations 4.55 and 4.60 in Equation 4.52, we have

$$\frac{\partial^\alpha u}{\partial t^\alpha}=\sum_{i=1}^{m}\sum_{j=1}^{m}c_{ij}h_i(x)h_j(t)+\sum_{i=1}^{m}\sum_{j=1}^{m}d_{ij}h_i(x)h_j(t). \tag{4.62}$$

Now, applying J^α to both sides of Equation 4.62 yields

$$u(x,t)-u(x,0)=J^\alpha\left(\sum_{i=1}^{m}\sum_{j=1}^{m}c_{ij}h_i(x)h_j(t)\right)+J^\alpha\left(\sum_{i=1}^{m}\sum_{j=1}^{m}d_{ij}h_i(x)h_j(t)\right). \tag{4.63}$$

Substituting Equations 4.53 and 4.59 in Equation 4.63, we get

$$\sum_{i=1}^{m}\sum_{j=1}^{m} c_{ij}Q^2h_i(x)h_j(t)+u(0,t)+x\left[u(1,t)-u(0,t)-\sum_{i=1}^{m}\sum_{j=1}^{m} c_{ij}\left[Q^2h_i(x)\right]_{x=1}h_j(t)\right]$$

$$-\frac{1}{\left(1+e^{(3/2)x}\right)^{1/2}}=\sum_{i=1}^{m}\sum_{j=1}^{m} c_{ij}h_i(x)Q_t^{\alpha}h_j(t)+\sum_{i=1}^{m}\sum_{j=1}^{m} d_{ij}h_i(x)Q_t^{\alpha}h_j(t). \tag{4.64}$$

Now, substituting the collocation points $x_l=(l-0.5)/m$ and $t_k=(k-0.5)/m$ for $l,k=1,2,\ldots,m$ in Equations 4.64 and 4.61, we have $2m^2$ equations in $2m^2$ unknowns in c_{ij} and d_{ij}. By solving these systems of equations using mathematical software, the Haar wavelet coefficients c_{ij} and d_{ij} can be obtained.

4.6.2 Application of OHAM to the Generalized Fisher's Equation

Using OHAM, the homotopy for Equation 4.52 can be written as

$$(1-p)\frac{\partial^{\alpha}\varphi(x,t;p)}{\partial t^{\alpha}}=H(p)\left[\frac{\partial^{\alpha}\varphi(x,t;p)}{\partial t^{\alpha}}-\frac{\partial^{2}\varphi(x,t;p)}{\partial x^{2}}-\varphi(x,t;p)\left[1-\left(\varphi(x,t;p)\right)^{6}\right]\right]. \tag{4.65}$$

Here

$$\varphi(x,t;p)=u_0(x,t)+\sum_{i=1}^{\infty}u_i(x,t)p^{i}, \tag{4.66}$$

$$H(p)=C_1p+C_2p^2+C_3p^3+\cdots, \tag{4.67}$$

$$N\left(\varphi(x,t;p)\right)=N_0\left(u_0(x,t)\right)+\sum_{k=1}^{\infty}N_k\left(u_0,u_1,\ldots,u_k\right)p^{k}. \tag{4.68}$$

Substituting Equations 4.66 through 4.68 in Equation 4.65 and equating the coefficients of like powers of p, we have the following system of partial differential equations:

$$\text{Coefficients of } p^0:\quad \frac{\partial^{\alpha}u_0(x,t)}{\partial t^{\alpha}}=0 \tag{4.69}$$

$$\text{Coefficients of } p^1:\quad \frac{\partial^{\alpha}u_1(x,t)}{\partial t^{\alpha}}-\frac{\partial^{\alpha}u_0(x,t)}{\partial t^{\alpha}}$$

$$=C_1\left[\frac{\partial^{\alpha}u_0(x,t)}{\partial t^{\alpha}}-\frac{\partial^{2}u_0(x,t)}{\partial x^{2}}+\left(u_0(x,t)\right)^{7}-u_0(x,t)\right] \tag{4.70}$$

Coefficients of p^2: $$\frac{\partial^\alpha u_2(x,t)}{\partial t^\alpha}-\frac{\partial^\alpha u_1(x,t)}{\partial t^\alpha}$$

$$=C_1\left[\frac{\partial^\alpha u_1(x,t)}{\partial t^\alpha}-\frac{\partial^2 u_1(x,t)}{\partial x^2}+7\left(u_0(x,t)\right)^6 u_1(x,t)-u_1(x,t)\right]$$

$$+C_2\left[\frac{\partial^\alpha u_0(x,t)}{\partial t^\alpha}-\frac{\partial^2 u_0(x,t)}{\partial x^2}+\left(u_0(x,t)\right)^7-u_0(x,t)\right] \tag{4.71}$$

and so on.

We consider the initial condition given in Equation 4.53, and solving the preceding equations, we obtain

$$u_0(x,t)=\frac{1}{\left(1+e^{3x/2}\right)^{1/3}}, \tag{4.72}$$

$$u_1(x,t)=\frac{-5C_1e^{3x/2}t^\alpha}{4\left(1+e^{3x/2}\right)^{4/3}\Gamma(1+\alpha)}, \tag{4.73}$$

and so on.

The fourth order approximate solution can be obtained by using the formula

$$u(x,t)=u_0(x,t)+u_1(x,t)+u_2(x,t)+u_3(x,t)+u_4(x,t).$$

The optimal values of the convergence control constants C_1, C_2, C_3, and C_4 can be obtained using the collocation method from Equation 1.33.

4.7 Numerical Results for a Fractional Fisher's Equation

Tables 4.5 through 4.8 present the comparison of approximate solutions obtained by the Haar wavelet method and OHAM for the fractional order generalized Fisher's equation given in Equation 4.52. The results obtained in the tables demonstrate that these methods are well suited for solving the fractional order generalized Fisher's equation. Both methods are quite efficient and effective.

In the case of the generalized Fisher's equation, Figures 4.3 and 4.4 present the comparison graphically between the numerical results obtained by the Haar wavelet method, OHAM, and exact solutions for different values of t and x.

TABLE 4.5

Absolute Errors in the Solution of the Generalized Fisher's Equation (4.52) Using the Haar Wavelet Method and Five Terms for Fourth-Order OHAM with Convergence Control Parameters $C_1 = -0.637012$, $C_2 = -0.151156$, $C_3 = 0.023432$, $C_4 = -0.0012788$ at Various Points of x and t for $\alpha = 1$

	$\lvert u_{Exact} - u_{Haar} \rvert$				$\lvert u_{Exact} - u_{OHAM} \rvert$			
x	$t=0.2$	$t=0.4$	$t=0.6$	$t=0.8$	$t=0.2$	$t=0.4$	$t=0.6$	$t=0.8$
0.1	0.0051104	0.0054849	0.0042629	0.0027556	6.67439E−5	0.0040747	0.023752	0.0740709
0.2	0.0096957	0.0106181	0.0085377	0.0058225	1.99503E−4	0.0023417	0.019345	0.0678136
0.3	0.013553	0.0151142	0.0125286	0.0089272	4.65337E−4	4.3112E−4	0.013810	0.0581267
0.4	0.0163976	0.018593	0.0158324	0.0116856	7.18406E−4	0.0015578	0.007399	0.0453319
0.5	0.0179265	0.0206477	0.0180051	0.0136684	9.47663E−4	0.0035178	4.521E−4	0.030035
0.6	0.017924	0.0209528	0.0186691	0.0145062	0.0011444	0.0053436	0.006645	0.013062
0.7	0.0163414	0.01935	0.0175984	0.013973	0.0013029	0.0069445	0.013494	0.0046328
0.8	0.0133823	0.015939	0.0148102	0.0120755	0.0014207	0.0082540	0.019736	0.0220718
0.9	0.0095361	0.0111158	0.0106057	0.0090918	0.0014982	0.009231	0.025078	0.0383517
1.0	0.0055837	0.0055837	0.0055837	0.0055837	0.0015382	0.0098632	0.029315	0.0527218

TABLE 4.6

Approximate Solutions of the Generalized Fisher's Equation (4.52) Using the Haar Wavelet Method and Five Terms for Fourth-Order OHAM with Convergence Control Parameters $C_1 = -0.649458$, $C_2 = 0.053658$, $C_3 = -0.1822726$, $C_4 = 0.0894301$ at Various Points of x and t for $\alpha = 0.75$

	$t=0.2$		$t=0.4$		$t=0.6$		$t=0.8$	
x	u_{Haar}	u_{OHAM}	u_{Haar}	u_{OHAM}	u_{Haar}	u_{OHAM}	u_{Haar}	u_{OHAM}
0.1	0.859185	0.899389	0.920547	0.931947	0.957952	0.927394	0.978665	0.902524
0.2	0.838644	0.888856	0.905313	0.930017	0.948042	0.930519	0.972681	0.905646
0.3	0.817612	0.877253	0.889588	0.927859	0.93764	0.934718	0.966206	0.91116
0.4	0.796431	0.864492	0.873714	0.92523	0.927088	0.939825	0.959581	0.919341
0.5	0.775484	0.850503	0.858074	0.921857	0.916772	0.945568	0.953191	0.930278
0.6	0.755093	0.835244	0.84299	0.917456	0.90701	0.951569	0.947357	0.943809
0.7	0.735424	0.818704	0.828628	0.911753	0.897971	0.957356	0.942244	0.959482
0.8	0.716399	0.800908	0.81491	0.904497	0.889575	0.962388	0.937776	0.976567
0.9	0.697656	0.781918	0.801474	0.895481	0.881463	0.966095	0.933591	0.994112
1.0	0.678526	0.761827	0.787651	0.884559	0.872963	0.967927	0.929017	1.01104

4.8 Solution of a Fractional FPE

4.8.1 Application of Haar Wavelets to Time-Fractional FPE

Consider the nonlinear time-fractional FPE [103,105]

$$\frac{\partial^\alpha u}{\partial t^\alpha} = \left[-\frac{\partial}{\partial x}\left(\frac{4u}{x} - \frac{x}{3} \right) + \frac{\partial^2 u}{\partial x^2} \right] u(x,t), \quad t > 0,\ x > 0,\ 0 < \alpha \le 1, \tag{4.74}$$

TABLE 4.7

Approximate Solutions of the Generalized Fisher's Equation (4.52) Using the Haar Wavelet Method and Five Terms for Fourth-Order OHAM with Convergence Control Parameters $C_1 = -0.5059152$, $C_2 = -0.0211535$, $C_3 = -0.05081612$, and $C_4 = 0.0574318$ at Various Points of x and t for $\alpha = 0.5$

	$t=0.2$		$t=0.4$		$t=0.6$		$t=0.8$	
x	u_{Haar}	u_{OHAM}	u_{Haar}	u_{OHAM}	u_{Haar}	u_{OHAM}	u_{Haar}	u_{OHAM}
0.1	0.859121	0.927731	0.920482	0.943924	0.957888	0.943779	0.9786	0.939948
0.2	0.838402	0.921616	0.905071	0.943123	0.947799	0.94491	0.972438	0.940694
0.3	0.817123	0.914826	0.889099	0.942438	0.937151	0.946615	0.965717	0.941956
0.4	0.795694	0.907206	0.872977	0.941713	0.926351	0.948908	0.958844	0.944068
0.5	0.774576	0.898605	0.857166	0.940787	0.915864	0.951841	0.952284	0.947508
0.6	0.754149	0.888882	0.842046	0.93948	0.906066	0.955446	0.946413	0.952761
0.7	0.734584	0.877915	0.827789	0.937586	0.897131	0.959662	0.941405	0.96016
0.8	0.715744	0.865607	0.814255	0.934862	0.888921	0.964283	0.937121	0.969739
0.9	0.697156	0.851889	0.800975	0.931031	0.880963	0.968937	0.933091	0.98117
1.0	0.678052	0.83673	0.787178	0.925803	0.872489	0.973109	0.928543	0.993777

TABLE 4.8

Approximate Solutions of the Generalized Fisher's Equation (4.52) Using the Haar Wavelet Method and Five Terms for Fourth-Order OHAM with Convergence Control Parameters $C_1 = -0.33833012$, $C_2 = -0.04303056$, $C_3 = 0.1230816$, and $C_4 = -0.0545852$ at Various Points of x and t for $\alpha = 0.25$

	$t=0.2$		$t=0.4$		$t=0.6$		$t=0.8$	
x	u_{Haar}	u_{OHAM}	u_{Haar}	u_{OHAM}	u_{Haar}	u_{OHAM}	u_{Haar}	u_{OHAM}
0.1	0.859016	0.949105	0.920377	0.959635	0.957783	0.964066	0.978495	0.966453
0.2	0.837987	0.943681	0.904655	0.955839	0.947384	0.960802	0.972023	0.963262
0.3	0.816234	0.937961	0.88821	0.9522	0.936261	0.957955	0.964827	0.960634
0.4	0.79426	0.93184	0.871543	0.948657	0.924917	0.95553	0.95741	0.958645
0.5	0.772676	0.925203	0.855266	0.945143	0.913963	0.953545	0.950383	0.957415
0.6	0.752001	0.917922	0.839899	0.941569	0.903919	0.951996	0.944265	0.957051
0.7	0.732474	0.909857	0.825678	0.937799	0.895021	0.950809	0.939295	0.957571
0.8	0.713871	0.900851	0.812383	0.933634	0.887048	0.949804	0.935249	0.958845
0.9	0.695534	0.890742	0.799352	0.928815	0.87934	0.948678	0.931468	0.96056
1.0	0.676507	0.87937	0.785633	0.923035	0.870944	0.947027	0.926998	0.96224

subject to the initial condition

$$u(x,0) = x^2. \tag{4.75}$$

When $\alpha = 1$, the exact solution of Equation 4.74 is given by [103,105]

$$u(x,t) = x^2 e^t. \tag{4.76}$$

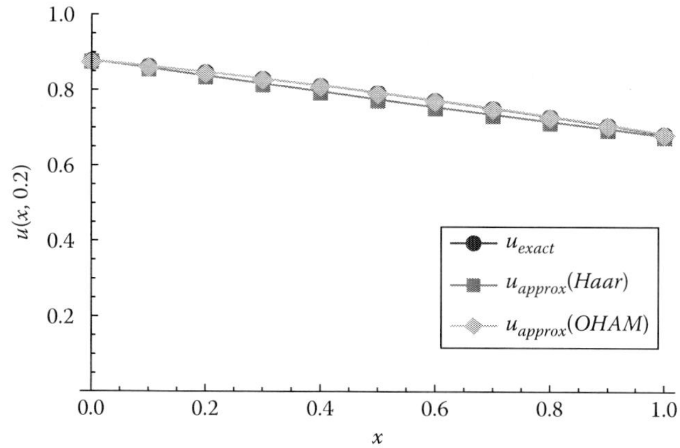

FIGURE 4.3
Comparison of a Haar wavelet solution and an OHAM solution with the exact solution of generalized Fisher's equation when $t = 0.2$.

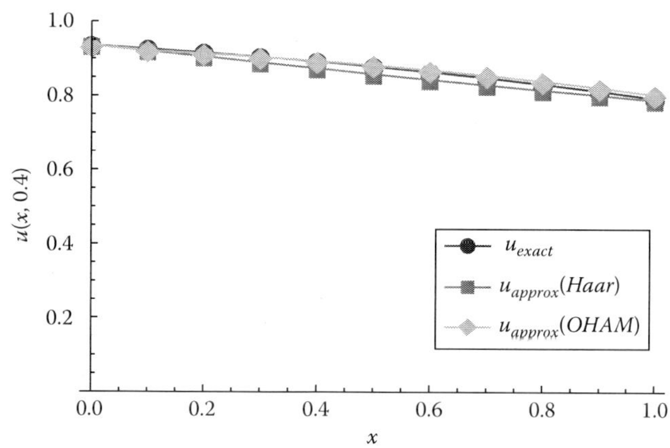

FIGURE 4.4
Comparison of a Haar wavelet solution and an OHAM solution with the exact solution of generalized Fisher's equation when $t = 0.4$.

Let us divide both the space and time intervals [0, 1] into m equal subintervals, each of width $\Delta = 1/m$. The Haar wavelet solution of $u(x,t)$ is sought by assuming that $\partial^2 u(x,t)/\partial x^2$ can be expanded in terms of Haar wavelets as

$$\frac{\partial^2 u(x,t)}{\partial x^2} = \sum_{i=1}^{m} \sum_{j=1}^{m} c_{ij} h_i(x) h_j(t). \tag{4.77}$$

Integrating Equation 4.77 with respect to x from 0 to x, we get

$$\frac{\partial u(x,t)}{\partial x} - p(t) = \sum_{i=1}^{m}\sum_{j=1}^{m} c_{ij} Q h_i(x) h_j(t). \tag{4.78}$$

Again, integrating Equation 4.78 with respect to x from 0 to x, we get

$$u(x,t) = \sum_{i=1}^{m}\sum_{j=1}^{m} c_{ij} Q^2 h_i(x) h_j(t) + q(t) + xp(t). \tag{4.79}$$

Putting $x = 0$ in Equation 4.79, we get

$$q(t) = u(0,t) = 0. \tag{4.80}$$

Putting $x = 1$ in Equation 4.79, we get

$$\begin{aligned} p(t) &= u(1,t) - u(0,t) - \sum_{i=1}^{m}\sum_{j=1}^{m} c_{ij}\left[Q^2 h_i(x)\right]_{x=1} h_j(t) \\ &= u(1,t) - \sum_{i=1}^{m}\sum_{j=1}^{m} c_{ij}\left[Q^2 h_i(x)\right]_{x=1} h_j(t). \end{aligned} \tag{4.81}$$

Putting Equation 4.81 in Equations 4.78 and 4.79, we have

$$\frac{\partial u(x,t)}{\partial x} \sum_{i=1}^{m}\sum_{j=1}^{m} c_{ij} Q h_i(x) h_j(t) + u(1,t) - \sum_{i=1}^{m}\sum_{j=1}^{m} c_{ij}\left[Q^2 h_i(x)\right]_{x=1} h_j(t), \tag{4.82}$$

$$u(x,t) = \sum_{i=1}^{m}\sum_{j=1}^{m} c_{ij} Q^2 h_i(x) h_j(t) + x\left[u(1,t) - \sum_{i=1}^{m}\sum_{j=1}^{m} c_{ij}\left[Q^2 h_i(x)\right]_{x=1} h_j(t)\right]. \tag{4.83}$$

The nonlinear term presented in Equation 4.74 can be approximated using the Haar wavelet function as

$$\frac{\partial^2 u^2}{\partial x^2} - \frac{\partial}{\partial x}\left(\frac{4u^2}{x}\right) = \sum_{i=1}^{m}\sum_{j=1}^{m} d_{ij} h_i(x) h_j(t).$$

This implies

$$2u\left(\frac{\partial^2 u}{\partial x^2} + \frac{2u}{x^2}\right) + 2\left(\frac{\partial u}{\partial x}\right)\left(\frac{\partial u}{\partial x} - \frac{4u}{x}\right) = \sum_{i=1}^{m}\sum_{j=1}^{m} d_{ij} h_i(x) h_j(t). \tag{4.84}$$

Therefore, substituting Equations 4.77, 4.82, and 4.83 in Equation 4.84, we have

$$
\begin{aligned}
&2\left(\sum_{i=1}^{m}\sum_{j=1}^{m} c_{ij}Q^2h_i(x)h_j(t)+x\left[u(1,t)-\sum_{i=1}^{m}\sum_{j=1}^{m} c_{ij}\left[Q^2h_i(x)\right]_{x=1}h_j(t)\right]\right)\\
&\quad\times\left[\sum_{i=1}^{m}\sum_{j=1}^{m} c_{ij}h_i(x)h_j(t)+\frac{2}{x^2}\left(\sum_{i=1}^{m}\sum_{j=1}^{m} c_{ij}Q^2h_i(x)h_j(t)+x\left[u(1,t)-\sum_{i=1}^{m}\sum_{j=1}^{m} c_{ij}\left[Q^2h_i(x)\right]_{x=1}h_j(t)\right]\right)\right]\\
&\quad+2\left(\sum_{i=1}^{m}\sum_{j=1}^{m} c_{ij}Qh_i(x)h_j(t)+u(1,t)-\sum_{i=1}^{m}\sum_{j=1}^{m} c_{ij}\left[Q^2h_i(x)\right]_{x=1}h_j(t)\right)\\
&\quad\times\left[\left(\sum_{i=1}^{m}\sum_{j=1}^{m} c_{ij}Qh_i(x)h_j(t)+u(1,t)-\sum_{i=1}^{m}\sum_{j=1}^{m} c_{ij}\left[Q^2h_i(x)\right]_{x=1}h_j(t)\right)\right.\\
&\quad\left.-\frac{4}{x}\left(\sum_{i=1}^{m}\sum_{j=1}^{m} c_{ij}Q^2h_i(x)h_j(t)+x\left[u(1,t)-\sum_{i=1}^{m}\sum_{j=1}^{m} c_{ij}\left[Q^2h_i(x)\right]_{x=1}h_j(t)\right]\right)\right]\\
&=\sum_{i=1}^{m}\sum_{j=1}^{m} d_{ij}h_i(x)h_j(t). \qquad (4.85)
\end{aligned}
$$

Substituting Equation 4.84 in Equation 4.74, we have

$$\frac{\partial^\alpha u}{\partial t^\alpha}=\sum_{i=1}^{m}\sum_{j=1}^{m} d_{ij}h_i(x)h_j(t)+\frac{x}{3}\frac{\partial u}{\partial x}+\frac{u}{3}. \qquad (4.86)$$

Now, applying J^α to both sides of Equation 4.86 yields

$$u(x,t)-u(x,0)=J^\alpha\left(\sum_{i=1}^{m}\sum_{j=1}^{m} d_{ij}h_i(x)h_j(t)+\frac{x}{3}\frac{\partial u}{\partial x}+\frac{u}{3}\right). \qquad (4.87)$$

Substituting Equations 4.75, 4.82, and 4.83 in Equation 4.87, we get

$$
\begin{aligned}
&\sum_{i=1}^{m}\sum_{j=1}^{m} c_{ij}Q^2h_i(x)h_j(t)+x\left[u(1,t)-\sum_{i=1}^{m}\sum_{j=1}^{m} c_{ij}\left[Q^2h_i(x)\right]_{x=1}h_j(t)\right]-x^2\\
&\quad=\sum_{i=1}^{m}\sum_{j=1}^{m} d_{ij}h_i(x)Q_t^\alpha h_j(t)\\
&\quad+\frac{x}{3}\left(\sum_{i=1}^{m}\sum_{j=1}^{m} c_{ij}Qh_i(x)Q_t^\alpha h_j(t)+J^\alpha u(1,t)-\sum_{i=1}^{m}\sum_{j=1}^{m} c_{ij}\left[Q^2h_i(x)\right]_{x=1}Q_t^\alpha h_j(t)\right)\\
&\quad+\frac{1}{3}\left(\sum_{i=1}^{m}\sum_{j=1}^{m} c_{ij}Q^2h_i(x)Q_t^\alpha h_j(t)+x\left[J^\alpha u(1,t)-\sum_{i=1}^{m}\sum_{j=1}^{m} c_{ij}\left[Q^2h_i(x)\right]_{x=1}Q_t^\alpha h_j(t)\right]\right). \qquad (4.88)
\end{aligned}
$$

Now, substituting the collocation points $x_l = (l-0.5)/m$ and $t_k = (k-0.5)/m$ for $l, k = 1, 2, \ldots, m$ in Equations 4.85 and 4.88, we have $2m^2$ equations in $2m^2$ unknowns in c_{ij} and d_{ij}. By solving this system of equations using mathematical software, the Haar wavelet coefficients c_{ij} and d_{ij} can be obtained.

4.8.2 Application of a Two-Dimensional Haar Wavelet for Solving Time- and Space-Fractional FPE

Consider the time- and space-fractional FPE [103,105]

$$\frac{\partial^{\alpha} u}{\partial t^{\alpha}} = \left[-\frac{\partial^{\beta}}{\partial x^{\beta}}\left(\frac{x}{6}\right) + \frac{\partial^{2\beta}}{\partial x^{2\beta}}\left(\frac{x^2}{12}\right) \right] u(x,t), \quad t > 0, \ x > 0, \tag{4.89}$$

where $0 < \alpha, \beta \le 1$ subject to the initial condition

$$u(x,0) = x^2. \tag{4.90}$$

When $\alpha = 1$ and $\beta = 1$, the exact solution of Equation 4.89 is given by [103,105]

$$u(x,t) = x^2 e^{t/2}. \tag{4.91}$$

Let us divide space interval [0, 1] into m equal subintervals, each of width $\Delta = 1/m$.

Haar wavelet solution of $u(x,t)$ is sought by assuming that $(\partial^{\alpha} u(x,t))/\partial t^{\alpha}$ can be expanded in terms of Haar wavelets as

$$\frac{\partial^{\alpha} u(x,t)}{\partial t^{\alpha}} = \sum_{i=1}^{m}\sum_{j=1}^{m} a_{ij} h_i(x) h_j(t). \tag{4.92}$$

Applying J^{α} to both sides of Equation 4.92, we get

$$u(x,t) = x^2 + \sum_{i=1}^{m}\sum_{j=1}^{m} a_{ij} h_i(x) Q^{\alpha} h_j(t). \tag{4.93}$$

Now

$$\frac{\partial^{\beta}}{\partial x^{\beta}}\left(x u(x,t)\right) = \frac{\partial^{\beta}}{\partial x^{\beta}}\left(x^3\right) + \sum_{i=1}^{m}\sum_{j=1}^{m} a_{ij} \frac{\partial^{\beta}}{\partial x^{\beta}}\left(x h_i(x)\right) Q^{\alpha} h_j(t) \tag{4.94}$$

and

$$\frac{\partial^{2\beta}}{\partial x^{2\beta}}\left(x^2 u(x,t)\right) = \frac{\partial^{2\beta}}{\partial x^{2\beta}}\left(x^4\right) + \sum_{i=1}^{m}\sum_{j=1}^{m} a_{ij} \frac{\partial^{2\beta}}{\partial x^{2\beta}}\left(x^2 h_i(x)\right) Q^{\alpha} h_j(t). \tag{4.95}$$

Substituting Equations 4.92, 4.94, and 4.95 in Equation 4.89, we get

$$\sum_{i=1}^{m}\sum_{j=1}^{m}a_{ij}h_i(x)h_j(t)=\frac{-1}{6}\left(\frac{\partial^{\beta}}{\partial x^{\beta}}\left(x^3\right)+\sum_{i=1}^{m}\sum_{j=1}^{m}a_{ij}\frac{\partial^{\beta}}{\partial x^{\beta}}\left(xh_i(x)\right)Q^{\alpha}h_j(t)\right)$$
$$+\frac{1}{12}\left(\frac{\partial^{2\beta}}{\partial x^{2\beta}}\left(x^4\right)+\sum_{i=1}^{m}\sum_{j=1}^{m}a_{ij}\frac{\partial^{2\beta}}{\partial x^{2\beta}}\left(x^2h_i(x)\right)Q^{\alpha}h_j(t)\right). \tag{4.96}$$

Now substituting the collocation points $x_l=(l-0.5)/m$ for $l=1,2,\ldots,m$ in Equation 4.96, we have m^2 equations in m^2 unknowns a_{ij}. By solving this system of equations using mathematical software, the Haar wavelet coefficients a_{ij} can be obtained.

4.9 Numerical Results for a Fractional FPE

Table 4.9 shows the comparison of exact solutions with the approximate solutions of different numerical methods for time-fractional FPE. Agreement between present numerical results with other approximate solutions and exact solutions appear very satisfactory through illustrations in Tables 4.9 and 4.10. Table 4.10 shows the comparison of approximate solutions of fractional order time-fractional FPE obtained by using the two-dimensional Haar wavelet method with the solutions of ADM and VIM presented in Reference [103]. Similarly, Tables 4.11 and 4.12 show the comparison of approximate solutions obtained by different numerical methods for time- and space-fractional FPE. It is found that the solutions obtained by using the present method are in good agreement with the results presented by Odibat et al. [103] and even better than the results obtained by OTM presented by Vanani et al. [104]. However, the errors may be reduced significantly if we increase the level of resolution that prompts a greater number of collocation points.

TABLE 4.9

Comparison of the Present Method Solution with Other Numerical Methods for Classical Order Time-Fractional FPE (4.74) at Various Points of x and t for $\alpha=1$

t	x	u_{ADM} [103]	u_{VIM} [103]	u_{Exact}	u_{Haar}
0.2	0.25	0.076333	0.076333	0.076338	0.0756165
	0.5	0.305333	0.305333	0.305351	0.304392
	0.75	0.687000	0.687000	0.687039	0.686321
	1.0	1.221333	1.221333	1.221403	1.2214
0.4	0.25	0.093167	0.093167	0.093239	0.0958469
	0.5	0.372667	0.372667	0.372956	0.376435
	0.75	0.838500	0.838500	0.839151	0.841761
	1.0	1.490667	1.490667	1.491825	1.49182
0.6	0.25	0.113500	0.113500	0.113882	0.110663
	0.5	0.454000	0.454000	0.455530	0.451238
	0.75	1.021500	1.021500	1.024942	1.02172
	1.0	1.816000	1.816000	1.822119	1.82212

TABLE 4.10

Comparison of the Present Method Solution with Other Numerical Methods for Time-Fractional FPE (4.74) at Various Points of x and t, Taking $\alpha=0.5$ and $\alpha=0.75$

		$\alpha=0.5$			$\alpha=0.75$		
t	x	u_{ADM} [103]	u_{VIM} [103]	u_{Haar}	u_{ADM} [103]	u_{VIM} [103]	u_{Haar}
0.2	0.25	0.110744	0.091795	0.0900792	0.087699	0.084593	0.0714745
	0.5	0.442978	0.367179	0.421013	0.350796	0.338372	0.329117
	0.75	0.996699	0.826154	0.990531	0.789291	0.761337	0.773339
	1.0	1.771910	1.468717	1.79902	1.403180	1.353488	1.40468
0.4	0.25	0.143997	0.118678	0.13581	0.111718	0.106178	0.0973803
	0.5	0.575909	0.474712	0.587481	0.446872	0.424712	0.431178
	0.75	1.295980	1.068102	1.35217	1.005460	0.955602	0.998822
	1.0	2.303960	1.898849	2.43004	1.787490	1.698849	1.80046
0.6	0.25	0.176478	0.146209	0.167654	0.138479	0.129926	0.116878
	0.5	0.705914	0.584835	0.749162	0.553918	0.519702	0.534521
	0.75	1.588310	1.315878	1.742	1.246320	1.169330	1.24986
	1.0	2.823650	2.339338	3.14621	2.215670	2.078809	2.26291

TABLE 4.11

Comparison of Approximate Solutions Obtained by Using VIM, ADM, and the Haar Wavelet Method for Time- and Space-Fractional FPE (4.89) at Various Points of x and t, Taking $\alpha=1$ and $\beta=1$

t	x	u_{ADM} [103]	u_{VIM} [103]	u_{Exact}	u_{Haar}
0.2	0.25	0.069062	0.069062	0.069073	0.0689468
	0.5	0.276259	0.276250	0.276293	0.274611
	0.75	0.621563	0.621563	0.621659	0.619337
0.4	0.25	0.076250	0.076250	0.076338	0.0753937
	0.5	0.305000	0.305000	0.305351	0.299222
	0.75	0.686250	0.686250	0.687039	0.676175
0.6	0.25	0.084062	0.084063	0.084366	0.0818405
	0.5	0.336250	0.336250	0.337465	0.323833
	0.75	0.756562	0.756562	0.759296	0.733012

4.10 Convergence Analysis of the Two-Dimensional Haar Wavelet Method

In this section, we have introduced the error analysis for the two-dimensional Haar wavelet method.

We assume that $f(x,y)\in C^2([a,b]\times[a,b])$ and there exist $M>0$, for which

$$\left|\frac{\partial^2 f(x,y)}{\partial x \partial y}\right| \le M, \quad \text{for all } x,y\in[a,b]\times[a,b].$$

TABLE 4.12

Comparison of Approximate Solutions of Fractional Order Time- and Space-Fractional FPE (4.89) Obtained by Using VIM, ADM, OTM, and the Haar Wavelet Method at Various Points of x and t, Taking $\alpha=\beta=0.5$ and $\alpha=\beta=0.75$

		$\alpha=0.5$ and $\beta=0.5$				$\alpha=0.75$ and $\beta=0.75$			
t	x	u_{ADM} [103]	u_{VIM} [103]	u_{OTM} [103]	u_{Haar}	u_{ADM} [103]	u_{VIM} [103]	u_{OTM} [103]	u_{Haar}
0.2	0.25	0.060440	0.06111	0.061929	0.0601168	0.063002	0.062922	0.062920	0.0633685
	0.5	0.244329	0.24618	0.248365	0.244247	0.258161	0.256856	0.256782	0.256326
	0.75	0.559866	0.56056	0.562348	0.559936	0.592855	0.587790	0.588104	0.595415
0.4	0.25	0.059620	0.05996	0.061392	0.0591215	0.063371	0.063291	0.063305	0.063968
	0.5	0.242066	0.24303	0.246833	0.241821	0.264157	0.262868	0.262916	0.260722
	0.75	0.558992	0.55902	0.562276	0.558771	0.615589	0.610213	0.611786	0.618446
0.6	0.25	0.059004	0.05898	0.060883	0.0583544	0.063713	0.063642	0.063669	0.0644986
	0.5	0.240363	0.24033	0.245395	0.239941	0.269702	0.268564	0.268707	0.264632
	0.75	0.558407	0.55777	0.562273	0.557834	0.636878	0.631709	0.634637	0.639038

Next, we may proceed as follows. Suppose $f_{nm}(x,y)=\sum_{i=0}^{n-1}\sum_{j=0}^{m-1} c_{ij}h_i(x)h_j(y)$, where $n=2^{\alpha+1}$, $\alpha=0,1,2,\ldots$ and $m=2^{\beta+1}$, $\beta=0,1,2,\ldots$.

Then

$$f(x,y)-f_{nm}(x,y)=\sum_{i=n}^{\infty}\sum_{j=m}^{\infty}c_{ij}h_i(x)h_j(y)+\sum_{i=n}^{\infty}\sum_{j=0}^{m-1}c_{ij}h_i(x)h_j(y)+\sum_{i=0}^{n-1}\sum_{j=m}^{\infty}c_{ij}h_i(x)h_j(y).$$

From Parseval's formula, we have

$$\begin{aligned}\left\|f(x,y)-f_{nm}(x,y)\right\|^2 &= \int_a^b\int_a^b \left(f(x,y)-f_{nm}(x,y)\right)^2 dxdy \\ &= \sum_{p=n}^{\infty}\sum_{s=m}^{\infty}\sum_{i=n}^{\infty}\sum_{j=m}^{\infty} c'_{ij}c'_{ps}\int_a^b h_i(x)h_p(x)dx\int_a^b h_j(y)h_s(y)dy \\ &+ \sum_{p=n}^{\infty}\sum_{s=0}^{m-1}\sum_{i=n}^{\infty}\sum_{j=0}^{m-1} c'_{ij}c'_{ps}\int_a^b h_i(x)h_p(x)dx\int_a^b h_j(y)h_s(y)dy \\ &+ \sum_{p=0}^{n-1}\sum_{s=m}^{\infty}\sum_{i=0}^{n-1}\sum_{j=m}^{\infty} c'_{ij}c'_{ps}\int_a^b h_i(x)h_p(x)dx\int_a^b h_j(y)h_s(y)dy \\ &= \sum_{i=n}^{\infty}\sum_{j=m}^{\infty}c'^2_{ij}+\sum_{i=n}^{\infty}\sum_{j=0}^{m-1}c'^2_{ij}+\sum_{i=0}^{n-1}\sum_{j=m}^{\infty}c'^2_{ij},\end{aligned}$$

where

$$c_{ij}'^2 = \frac{c_{ij}(b-a)^2}{2^{i+j}}$$

$$c_{ij} = \int_a^b \left(\int_a^b f(x,y)h_i(y)dy \right) h_j(x)dx$$

$$= \int_a^b \left(\int_{a+k\left(\frac{b-a}{2^i}\right)}^{a+\left(k+\frac{1}{2}\right)\left(\frac{b-a}{2^i}\right)} f(x,y)dy - \int_{a+\left(k+\frac{1}{2}\right)\left(\frac{b-a}{2^i}\right)}^{a+(k+1)\left(\frac{b-a}{2^i}\right)} f(x,y)dy \right) h_j(x)dx.$$

Using the mean value theorem of integral calculus, we have

$$a+k\frac{(b-a)}{2^i} \le y_1 \le a+\left(k+\frac{1}{2}\right)\frac{(b-a)}{2^i}, \quad a+\left(k+\frac{1}{2}\right)\frac{(b-a)}{2^i} \le y_2 \le a+(k+1)\frac{(b-a)}{2^i}.$$

Hence, we obtain

$$c_{ij} = (b-a)\int_a^b \left(f(x,y_1)2^{-i-1} - f(x,y_2)2^{-k-1} \right) h_j(x)dx.$$

Again by using the mean value theorem

$$c_{ij} = 2^{-i-1}(b-a)\int_a^b \left(f(x,y_1) - f(x,y_2) \right) h_j(x)dx.$$

Using Lagrange's mean value theorem

$$c_{ij} = 2^{-i-1}(b-a)\int_a^b \left((y_1-y_2)\frac{\partial f(x,y^*)}{\partial y} \right) h_j(x)dx \quad \text{where } y_1 \le y^* \le y_2$$

$$= 2^{-i-1}(b-a)(y_1-y_2)\left(\int_{a+k\left(\frac{b-a}{2^j}\right)}^{a+\left(k+\frac{1}{2}\right)\left(\frac{b-a}{2^j}\right)} \frac{\partial f(x,y^*)}{\partial y}dx - \int_{a+\left(k+\frac{1}{2}\right)\left(\frac{b-a}{2^j}\right)}^{a+\left(k+\frac{1}{2}\right)\left(\frac{b-a}{2^j}\right)} \frac{\partial f(x,y^*)}{\partial y}dx \right)$$

$$= 2^{-i-1}(b-a)(y_1-y_2)\left(2^{-j-1}(b-a)\frac{\partial f}{\partial y}(x_1,y^*) - 2^{-j-1}(b-a)\frac{\partial f}{\partial y}(x_2,y^*) \right).$$

Now, we use the mean value theorem of integral calculus

$$a+k\frac{(b-a)}{2^j}\le x_1\le a+\left(k+\frac{1}{2}\right)\frac{(b-a)}{2^j},\quad a+\left(k+\frac{1}{2}\right)\frac{(b-a)}{2^j}\le x_2\le a+(k+1)\frac{(b-a)}{2^j}$$

$$c_{ij}\le 2^{-i-j-2}(b-a)^2(y_1-y_2)(x_1-x_2)\frac{\partial^2 f(x^*,y^*)}{\partial x\partial y}.$$

But for $x_1\le x^*\le x_2$, $(y_1-y_2)\le(b-a)$, and $(x_1-x_2)\le(b-a)$, we obtain

$$c_{ij}\le\frac{(b-a)^4}{2^{i+j+2}}M,\quad \text{if}\left|\frac{\partial^2 f(x^*,y^*)}{\partial x\partial y}\right|\le M.$$

Therefore, $c_{ij}'^2=c_{ij}^2\dfrac{(b-a)^2}{2^{i+j}}\le\dfrac{(b-a)^{10}}{2^{3i+3j+4}}M^2$

$$\sum_{n=k}^{\infty}\sum_{m=l}^{\infty}c_{nm}'^2\le\sum_{n=2^{\alpha+1}}^{\infty}\sum_{n=2^{\beta+1}}^{\infty}\frac{(b-a)^{10}}{2^{3i+3j+4}}M^2,\quad \alpha,\beta=0,1,2,\ldots$$

$$\le(b-a)^{10}M^2\sum_{n=2^{\alpha+1}}^{\infty}\sum_{i=\beta+1}^{\infty}\sum_{m=2^i}^{2^{i+1}-1}2^{-3i-3j-4}$$

$$\le(b-a)^{10}M^2\sum_{n=2^{\alpha+1}}^{\infty}2^{-3j-4}\sum_{i=\beta+1}^{\infty}\left(2^{i+1}-1-2^i+1\right)2^{-3i}$$

$$\le(b-a)^{10}M^2\sum_{n=2^{\alpha+1}}^{\infty}2^{-3j-4}\sum_{i=\beta+1}^{\infty}2^{-2i}$$

$$\le(b-a)^{10}M^2\sum_{n=2^{\alpha+1}}^{\infty}2^{-3j-4}2^{-2(\beta+1)}\frac{1}{\left(1-\frac{1}{2^?}\right)}$$

$$\le\frac{4(b-a)^{10}}{3l^2}M^2 2^{-4}\sum_{j=\alpha+1}^{\infty}\sum_{n=2^j}^{2^{j+1}-1}2^{-3j}$$

$$\le\frac{4(b-a)^{10}}{3l^2}M^2 2^{-4}\sum_{j=\alpha+1}^{\infty}2^{-2j}$$

$$\le\frac{4(b-a)^{10}}{3l^2}M^2 2^{-4}\left(\frac{4}{3}\right)2^{-2(\alpha+1)}$$

$$\le\left(\frac{16}{9}\right)\frac{(b-a)^{10}}{l^2k^2}M^2 2^{-4}$$

$$\le\left(\frac{16}{144}\right)\frac{(b-a)^{10}}{l^2k^2}M^2.65.$$

Next

$$\sum_{n=k}^{\infty}\sum_{m=0}^{l-1} c_{nm}'^2 \le \sum_{n=k}^{\infty}\sum_{m=0}^{l-1} \frac{(b-a)^{10} M^2}{2^{3i+3j+4}} \le \sum_{n=2^{\alpha+1}}^{\infty} \frac{(b-a)^{10} M^2}{2^{3j+4}} \sum_{i=0}^{\beta}\sum_{m=2^i-1}^{2^{i+1}-1} 2^{-3i}$$

$$\le \sum_{n=2^{\alpha+1}}^{\infty} \frac{(b-a)^{10} M^2}{2^{3j+4}} \sum_{i=0}^{\beta}\left(2^{-2i}+2^{-3i}\right)$$

$$\le \left(\frac{52}{21}\right) 2^{-4}(b-a)^{10} M^2 \sum_{j=\alpha+1}^{\infty}\sum_{n=2^j}^{2^{j+1}-1} 2^{-3j}$$

$$\le \left(\frac{52}{336}\right)(b-a)^{10} M^2 \sum_{j=\alpha+1}^{\infty} 2^{-2j}$$

$$\le \left(\frac{52}{336}\right)(b-a)^{10} M^2 \left(\frac{2^{-2(\alpha+1)}}{\left(1-\frac{1}{2^2}\right)}\right)$$

$$= \frac{52(b-a)^{10} M^2}{252k^2}.$$

Similarly, we have $\sum_{n=0}^{k-1}\sum_{m=l}^{\infty} c_{nm}'^2 \le \frac{52(b-a)^{10} M^2}{252l^2}$.

Then

$$\sum_{n=k}^{\infty}\sum_{m=l}^{\infty} c_{nm}'^2 + \sum_{n=k}^{\infty}\sum_{m=0}^{l-1} c_{nm}'^2 + \sum_{n=0}^{k-1}\sum_{m=l}^{\infty} c_{nm}'^2 \le \left(\frac{16}{144}\right)\frac{(b-a)^{10}}{l^2k^2} M^2 + \frac{52(b-a)^{10} M^2}{252k^2} + \frac{52(b-a)^{10} M^2}{252l^2}.$$

Hence, we obtain $\left\| f(x,y) - f_{kl}(x,y)\right\| \le \frac{(b-a)^{10} M^2}{3}\left(\frac{1}{3l^2k^2}+\frac{13}{21k^2}+\frac{13}{21l^2}\right)$.

As $l \to \infty$ and $k \to \infty$, we can get $\|f(x,y) - f_{kl}(x,y)\| \to 0$.

4.11 Conclusion

The numerical solutions of fractional order Burgers–Fisher equation, generalized Fisher's type equation, and the time- and space-fractional FPEs have been analyzed in this chapter by utilizing the two-dimensional Haar wavelet method. The obtained results are then compared with OHAM solutions, exact solutions, and results available in the literature. The Haar wavelet technique provides quite satisfactory results for the fractional order Burgers–Fisher (4.27) and generalized Fisher equations (4.52). The main advantage of this Haar wavelet method is that they transfer the whole scheme into a system of algebraic equations for which the computation is easy and simple. OHAM allows fine tuning of the convergence region and rate of convergence by suitably identifying convergence control parameters $C_1, C_2, C_3, \ldots$. It has been observed that for the Burgers–Fisher equation, OHAM provides more accurate results than the Haar wavelet method as presented in the tables. But in the case of the generalized Fisher's equation, both methods are competitive. The results obtained by OHAM are slightly more accurate than the results obtained by the Haar wavelet method.

The solutions of time- and space-fractional FPEs have been compared with exact solutions as well as results obtained by ADM, VIM, and OTM that are available in open literature. These results have been cited in the tables in order to justify the accuracy and efficiency of the proposed scheme based on the two-dimensional Haar wavelet method. It can be noticed that the Haar wavelet technique provides quite satisfactory results in comparison to results obtained by ADM, VIM, and OTM [103–105] for the fractional order FPEs as demonstrated in the tables.

5

Application of Legendre Wavelet Methods for the Numerical Solution of Fractional Differential Equations

5.1 Introduction

In contemporary years, fractional calculus has become the focus of curiosity for many researchers in exclusive disciplines of applied science and engineering because of the fact that a realistic modeling of a physical phenomenon can be efficiently executed by way of utilizing fractional calculus. For the intent of this chapter, Caputo's definition of fractional derivative will be used. The advantage of Caputo's approach is that the initial conditions for fractional differential equations with Caputo's derivatives tackle the normal kind as for integer-order differential equations.

The investigation of traveling wave solutions for nonlinear fractional order partial differential equations (PDEs) plays an important role in the study of nonlinear physical phenomena. It is important to find new solutions since either new exact solutions or numerical approximate solutions may provide more information for understanding physical phenomena. In this chapter, the numerical solutions of fractional order PDEs comprising the Caputo fractional derivative are discussed. Fractional differential equations such as the KdV–Burger–Kuramoto (KBK) equation, seventh-order KdV (sKdV) equation, and Kaup–Kupershmidt (KK) equation have been solved using two-dimensional Legendre wavelet and Legendre multiwavelet methods.

The main focus of the present chapter is the application of the two-dimensional Legendre wavelet technique for solving nonlinear fractional differential equations like the time-fractional KBK equation and time-fractional sKdV equation in order to demonstrate the efficiency and accuracy of the proposed wavelet method. Also, the time-fractional KK equation has been solved by using the two-dimensional Legendre multiwavelet method. The obtained numerical approximate results of the proposed Legendre wavelet methods are then compared with the exact solutions and those available in literature.

5.2 Outline of the Present Study

Consider a nonlinear time-fractional parabolic PDE of the form

$$\frac{\partial^\alpha u}{\partial t^\alpha} = -u\frac{\partial}{\partial x}\left(\frac{2u}{x} - x\right) + u\frac{\partial^2 u}{\partial x^2}, \quad t > 0,\ x > 0 \tag{5.1}$$

with Dirichlet boundary conditions

$$u(0,t)=0 \quad \text{and} \quad u(1,t)=E_\alpha\left(t^\alpha\right),$$

where

α is the parameter describing the order of the fractional time derivative

E_α is the Mittag-Leffler function

Fractional diffusion equations like Equation 5.1 arise in continuous-time random walks, modeling of anomalous diffusive and subdiffusive systems, unification of diffusion and wave propagation phenomena, etc. The motivation of the present work is the application of the two-dimensional Legendre wavelet method for solving the problem of fractional PDEs with Dirichlet boundary conditions. To exhibit the effectiveness, the obtained numerical approximate results of the two-dimensional Legendre wavelet method are compared with that of the Haar wavelet method as well as with the exact solution derived by using the homotopy perturbation method (HPM).

Next, we consider the following time-fractional KBK equation [109]:

$$\frac{\partial^\alpha u}{\partial t^\alpha}+u\frac{\partial u}{\partial x}-\lambda_1\frac{\partial^2 u}{\partial x^2}+\lambda_2\frac{\partial^3 u}{\partial x^3}+\lambda_3\frac{\partial^4 u}{\partial x^4}=f(x,t), \quad t>0,\ x>0 \tag{5.2}$$

where

α is the order of the fractional time derivative

$\lambda_1,\lambda_2,\lambda_3\geq 0$ are parameters characterizing instability, dispersion, and dissipation, respectively [110]

The classical KBK equation is an important mathematical model arising in many different physical contexts to describe some physical processes in the motion of turbulence and other unstable process systems. It can also be used to describe long waves on a viscous fluid flowing down along an inclined plane [111], unstable drift waves in plasma [112], and a turbulent cascade model in a barotropic atmosphere [113].

There are a lot of studies available for the classical KBK equation, and some profound results have been established. But according to the best possible information of the authors, the detailed study of the nonlinear fractional order KBK equation is only beginning. Very few mathematical methods such as the homotopy analysis method (HAM) [114], He's variational iteration method, and Adomian's decomposition method [115] are available in open literature for the numerical solution of fractional KBK equation.

The KdV type of equations, which were first derived by Korteweg and de Vries in 1895 [172] and are used to describe weakly nonlinear shallow water waves, have emerged as an important class of nonlinear evolution equation and are often used in practical applications. In reality, the next state of a physical phenomenon might depend not only on its current state but also on its historical states (non-local property), which can be successfully modeled by using the theory of derivatives and integrals of fractional order. The sKdV equation was first introduced by Pomeau et al. [116] in order to discuss the structural stability of the KdV equation under singular perturbation.

The time-fractional generalized sKdV equation is given by [117]

$$\frac{\partial^\alpha u}{\partial t^\alpha}+\frac{\partial}{\partial x}\left(g(u)\right)+\frac{\partial^3 u}{\partial x^3}-\frac{\partial^5 u}{\partial x^5}+\sigma\frac{\partial^7 u}{\partial x^7}=f(x,t), \quad t>0,\ x>0 \tag{5.3}$$

where

σ is a constant

$\alpha(0<\alpha\leq 1)$ is the parameter describing the order of the fractional time derivative

$f(x,t)$ is the forcing term

The sKdV equation arises in fluid flow through porous media, fluid dynamics, plasma physics, optical fibers, elasticity, economics, optimization, hydrodynamics, hydro-magnetic stability, structural dynamics, medical imaging, and pure and applied sciences [118–120].

Next, we consider fractional order KK equation. Here, we compare two different methods, one numerical technique, namely, the Legendre multiwavelet method, and the other an analytical technique, namely, the optimal homotopy asymptotic method (OHAM), for solving a fractional order KK equation. Two-dimensional Legendre multiwavelet expansion together with operational matrices of fractional integration and derivative of wavelet functions is used to compute the numerical solution of a nonlinear time-fractional KK equation. The approximate solutions of a time-fractional KK equation thus obtained by the Legendre multiwavelet method are compared with the exact solutions as well as with OHAM.

Consider the following time-fractional KK equation [121]:

$$\frac{\partial^\alpha u}{\partial t^\alpha}+45u^2\frac{\partial u}{\partial x}-15p\frac{\partial u}{\partial x}\frac{\partial^2 u}{\partial x^2}-15u\frac{\partial^3 u}{\partial x^3}+\frac{\partial^5 u}{\partial x^5}=0,\quad t>0,\ x>0. \tag{5.4}$$

Equation 5.4 is a variation of the following KK equation [122–125]:

$$u_t+45u^2u_x-15pu_xu_{xx}-15uu_{xxx}+u_{xxxxx}=0$$

Here, $0<\alpha\leq 1$ is the parameter describing the order of the fractional time derivative. The classical KK equation is an important dispersive equation proposed first by Kaup in 1980 [122] and developed by Kupershmidt in 1994 [123]. This equation arises in the study of capillary gravity waves. The classical KK equation is known to be integrable [124] for $p=5/2$ and to have bilinear representations [125], but the explicit form of its N-soliton solution is apparently not known. A great deal of research work has been invested in recent years for the study of classical KK equations. Various methods have been developed independently by which soliton and solitary wave solutions may be obtained for nonlinear evolution equations. Our aim in the present work is to implement two-dimensional Legendre multiwavelet and OHAM in order to exhibit the capabilities of these methods in handling nonlinear equations like the fractional order KK equation.

5.3 Solution of a Time-Fractional Parabolic Partial Differential Equation

5.3.1 Application of HPM to Find the Exact Solution of Fractional Order Parabolic PDE

Consider a nonlinear time-fractional PDE of the form

$$\frac{\partial^\alpha u}{\partial t^\alpha}=-u\frac{\partial}{\partial x}\left(\frac{2u}{x}-x\right)+u\frac{\partial^2 u}{\partial x^2},\quad t>0,\ x>0.$$

This implies

$$\frac{\partial^\alpha u}{\partial t^\alpha} = \frac{2u^2}{x^2} - \frac{2u}{x}\frac{\partial u}{\partial x} + u + u\frac{\partial^2 u}{\partial x^2}, \quad t > 0,\ x > 0, \tag{5.5}$$

where $0 < \alpha \leq 1$, subject to the initial condition

$$u(x,0) = x^2. \tag{5.6}$$

To solve Equation 5.5 by HPM [25,26], we construct the following homotopy:

$$(1-p)\frac{\partial^\alpha u}{\partial t^\alpha} + p\left(\frac{\partial^\alpha u}{\partial t^\alpha} - \frac{2u^2}{x^2} + \frac{2u}{x}\frac{\partial u}{\partial x} - u - u\frac{\partial^2 u}{\partial x^2}\right) = 0. \tag{5.7}$$

This implies

$$\frac{\partial^\alpha u}{\partial t^\alpha} = p\left(\frac{2u^2}{x^2} - \frac{2u}{x}\frac{\partial u}{\partial x} + u + u\frac{\partial^2 u}{\partial x^2}\right). \tag{5.8}$$

By substituting $u(x,t) = \sum_{n=0}^{\infty} p^n u_n(x,t)$ in Equation 5.8, we get

$$\frac{\partial^\alpha}{\partial t^\alpha}\left(\sum_{n=0}^{\infty} p^n u_n(x,t)\right) = p\left[\frac{2}{x^2}\left(\sum_{n=0}^{\infty} p^n u_n(x,t)\right)^2 - \frac{2}{x}\left(\sum_{n=0}^{\infty} p^n u_n(x,t)\right)\frac{\partial}{\partial x}\left(\sum_{n=0}^{\infty} p^n u_n(x,t)\right)\right.$$
$$\left. + \sum_{n=0}^{\infty} p^n u_n(x,t) + \left(\sum_{n=0}^{\infty} p^n u_n(x,t)\right)\frac{\partial^2}{\partial x^2}\left(\sum_{n=0}^{\infty} p^n u_n(x,t)\right)\right]. \tag{5.9}$$

Collecting the coefficients of different powers of p for Equation 5.9, we have the following equations:

$$\text{Coefficients of } p^0: \frac{\partial^\alpha u_0}{\partial t^\alpha} = 0 \tag{5.10}$$

$$\text{Coefficients of } p^1: \frac{\partial^\alpha u_1}{\partial t^\alpha} = \frac{2}{x^2}u_0^2 - \frac{2}{x}u_0\frac{\partial u_0}{\partial x} + u_0 + u_0\frac{\partial^2 u_0}{\partial x^2} \tag{5.11}$$

$$\text{Coefficients of } p^2: \frac{\partial^\alpha u_2}{\partial t^\alpha} = \frac{2}{x^2}(2u_0u_1) - \frac{2}{x}\left(u_0\frac{\partial u_1}{\partial x} + u_1\frac{\partial u_0}{\partial x}\right) + u_1 + \left(\frac{\partial^2 u_0}{\partial x^2}u_1 + u_0\frac{\partial^2 u_1}{\partial x^2}\right) \tag{5.12}$$

and so on.

By putting $u(x,0) = u_0$ in Equations 5.10 through 5.12 and solving them, we obtain

$$u_1(x,t) = \frac{x^2 t^\alpha}{\Gamma(1+\alpha)}, \tag{5.13}$$

$$u_2(x,t) = \frac{x^2 t^{2\alpha}}{\Gamma(1+2\alpha)}, \tag{5.14}$$

and so on.

Finally, the approximate solution for Equation 5.5 is given by

$$u = u_0(x,t) + u_1(x,t) + u_2(x,t) + \cdots. \tag{5.15}$$

Thus, we have

$$\begin{aligned} u(x,t) &= x^2 + \frac{x^2 t^{\alpha}}{\Gamma(1+\alpha)} + \frac{x^2 t^{2\alpha}}{\Gamma(1+2\alpha)} + \cdots \\ &= x^2 E_\alpha(t^\alpha), \end{aligned} \tag{5.16}$$

where E_α is the Mittag-Leffler function.

Taking $\alpha = 1$ in Equation 5.16, we reproduce the solution of the problem as follows:

$$u(x,t) = x^2\left[1 + \frac{t}{\Gamma(2)} + \frac{t^2}{\Gamma(3)} + \frac{t^3}{\Gamma(4)} + \cdots\right].$$

The solution is equivalent to the exact solution in a closed form:

$$u(x,t) = x^2 e^t.$$

It is clear that a closed form of solution is obtainable by adding more terms to the homotopy perturbation series.

5.3.2 Application of a Two-Dimensional Haar Wavelet for the Numerical Solution of a Fractional PDE

To solve the nonlinear fractional PDE considered in Equations 5.5 and 5.6, let us divide both the space and time intervals [0, 1] into m equal subintervals, each of width $\Delta = 1/m$.

The Haar wavelet solution of $u(x,t)$ is sought by assuming that $\partial^2 u(x,t)/\partial x^2$ can be expanded in terms of Haar wavelets as

$$\frac{\partial^2 u(x,t)}{\partial x^2} = \sum_{i=1}^{m}\sum_{j=1}^{m} c_{ij} h_i(x) h_j(t). \tag{5.17}$$

Integrating Equation 5.17 with respect to x from 0 to x, we get

$$\frac{\partial u(x,t)}{\partial x} - p(t) = \sum_{i=1}^{m}\sum_{j=1}^{m} c_{ij} Q h_i(x) h_j(t). \tag{5.18}$$

Again, integrating Equation 5.18 with respect to x from 0 to x, we get

$$u(x,t)=\sum_{i=1}^{m}\sum_{j=1}^{m}c_{ij}Q^2h_i(x)h_j(t)+q(t)+xp(t). \tag{5.19}$$

Putting $x=0$ in Equation 5.19, we get

$$q(t)=u(0,t)=0. \tag{5.20}$$

Putting $x=1$ in Equation 5.19, we get

$$\begin{aligned}p(t)&=u(1,t)-u(0,t)-\sum_{i=1}^{m}\sum_{j=1}^{m}c_{ij}\left[Q^2h_i(x)\right]_{x=1}h_j(t)\\&=u(1,t)-\sum_{i=1}^{m}\sum_{j=1}^{m}c_{ij}\left[Q^2h_i(x)\right]_{x=1}h_j(t).\end{aligned} \tag{5.21}$$

Putting Equation 5.21 in Equations 5.18 and 5.19, we have

$$\frac{\partial u(x,t)}{\partial x}=\sum_{i=1}^{m}\sum_{j=1}^{m}c_{ij}Qh_i(x)h_j(t)+u(1,t)-\sum_{i=1}^{m}\sum_{j=1}^{m}c_{ij}\left[Q^2h_i(x)\right]_{x=1}h_j(t); \tag{5.22}$$

$$u(x,t)=\sum_{i=1}^{m}\sum_{j=1}^{m}c_{ij}Q^2h_i(x)h_j(t)+x\left[u(1,t)-\sum_{i=1}^{m}\sum_{j=1}^{m}c_{ij}\left[Q^2h_i(x)\right]_{x=1}h_j(t)\right]. \tag{5.23}$$

The nonlinear term presented in Equation 5.5 can be approximated using the Haar wavelet function as

$$u\frac{\partial^2 u}{\partial x^2}-u\frac{\partial}{\partial x}\left(\frac{2u}{x}\right)=\sum_{i=1}^{m}\sum_{j=1}^{m}d_{ij}h_i(x)h_j(t),$$

which yields

$$u\frac{\partial^2 u}{\partial x^2}+\frac{2u}{x}\left(\frac{u}{x}-\frac{\partial u}{\partial x}\right)=\sum_{i=1}^{m}\sum_{j=1}^{m}d_{ij}h_i(x)h_j(t). \tag{5.24}$$

Therefore, substituting Equations 5.17, 5.22, and 5.23 in Equation 5.24, we have

$$\left(\sum_{i=1}^{m}\sum_{j=1}^{m}c_{ij}Q^2h_i(x)h_j(t)+x\left[u(1,t)-\sum_{i=1}^{m}\sum_{j=1}^{m}c_{ij}\left[Q^2h_i(x)\right]_{x=1}h_j(t)\right]\right)\left(\sum_{i=1}^{m}\sum_{j=1}^{m}c_{ij}h_i(x)h_j(t)\right)$$
$$+\frac{2}{x}\left(\sum_{i=1}^{m}\sum_{j=1}^{m}c_{ij}Q^2h_i(x)h_j(t)+x\left[u(1,t)-\sum_{i=1}^{m}\sum_{j=1}^{m}c_{ij}\left[Q^2h_i(x)\right]_{x=1}h_j(t)\right]\right)$$
$$\times\left[\frac{1}{x}\left(\sum_{i=1}^{m}\sum_{j=1}^{m}c_{ij}Q^2h_i(x)h_j(t)+x\left[u(1,t)-\sum_{i=1}^{m}\sum_{j=1}^{m}c_{ij}\left[Q^2h_i(x)\right]_{x=1}h_j(t)\right]\right)\right.$$
$$\left.-\left(\sum_{i=1}^{m}\sum_{j=1}^{m}c_{ij}Qh_i(x)h_j(t)+u(1,t)-\sum_{i=1}^{m}\sum_{j=1}^{m}c_{ij}\left[Q^2h_i(x)\right]_{x=1}h_j(t)\right)\right]=\sum_{i=1}^{m}\sum_{j=1}^{m}d_{ij}h_i(x)h_j(t). \quad (5.25)$$

Substituting Equation 5.24 in Equation 5.5, we will have

$$\frac{\partial^\alpha u}{\partial t^\alpha}=\sum_{i=1}^{m}\sum_{j=1}^{m}d_{ij}h_i(x)h_j(t)+u. \quad (5.26)$$

Now, applying J^α to both sides of Equation 5.26 yields

$$u(x,t)-u(x,0)=J^\alpha\left(\sum_{i=1}^{m}\sum_{j=1}^{m}d_{ij}h_i(x)h_j(t)+u\right). \quad (5.27)$$

Substituting Equations 5.6 and 5.23 in Equation 5.27, we get

$$\sum_{i=1}^{m}\sum_{j=1}^{m}c_{ij}Q^2h_i(x)h_j(t)+x\left[u(1,t)-\sum_{i=1}^{m}\sum_{j=1}^{m}c_{ij}\left[Q^2h_i(x)\right]_{x=1}h_j(t)\right]-x^2=\sum_{i=1}^{m}\sum_{j=1}^{m}d_{ij}h_i(x)Q_t^\alpha h_j(t)$$
$$+\left(\sum_{i=1}^{m}\sum_{j=1}^{m}c_{ij}Q^2h_i(x)Q_t^\alpha h_j(t)+x\left[J^\alpha u(1,t)-\sum_{i=1}^{m}\sum_{j=1}^{m}c_{ij}\left[Q^2h_i(x)\right]_{x=1}Q_t^\alpha h_j(t)\right]\right). \quad (5.28)$$

Now, substituting the collocation points $x_l=(l-0.5)/m$ and $t_k=(k-0.5)/m$ for $l,k=1,2,\ldots,m$ in Equations 5.25 and 5.28, we have $2m^2$ equations in $2m^2$ unknowns in c_{ij} and d_{ij}. By solving these system of equations using mathematical software, the Haar wavelet coefficients c_{ij} and d_{ij} can be obtained.

5.3.3 Application of Two-Dimensional Legendre Wavelet for Solving Fractional PDE

The Legendre wavelet solution of $u(x,t)$ for the nonlinear fractional PDE considered in Equations 5.5 and 5.6 is sought by assuming that $u(x,t)$ can be expanded in terms of Legendre wavelets as

$$u(x,t)=\sum_{n=1}^{2^{k_1-1}}\sum_{i=0}^{M_1-1}\sum_{l=1}^{2^{k_2-1}}\sum_{j=0}^{M_2-1}c_{n,i,l,j}\psi_{n,i,l,j}(x,t), \tag{5.29}$$

where $n=1,\ldots,2^{k_1-1}$, $i=0,\ldots,M_1-1$, $l=1,\ldots,2^{k_2-1}$, $j=0,\ldots,M_2-1$.

Applying J_t^α on both sides of Equation 5.5, we have

$$u(x,t)-u(x,0)=J_t^\alpha\left[-u\frac{\partial}{\partial x}\left(\frac{2u}{x}-x\right)+u\frac{\partial^2 u}{\partial x^2}\right].$$

This implies

$$\begin{aligned}u(x,t)-u(x,0)=J_t^\alpha&\left[-\left(\sum_{n=1}^{2^{k_1-1}}\sum_{i=0}^{M_1-1}\sum_{l=1}^{2^{k_2-1}}\sum_{j=0}^{M_2-1}c_{n,i,l,j}\psi_{n,i,l,j}(x,t)\right)\right.\\&\times\frac{\partial}{\partial x}\left(\frac{2}{x}\left(\sum_{n=1}^{2^{k_1-1}}\sum_{i=0}^{M_1-1}\sum_{l=1}^{2^{k_2-1}}\sum_{j=0}^{M_2-1}c_{n,i,l,j}\psi_{n,i,l,j}(x,t)\right)-x\right)\\&+\left(\sum_{n=1}^{2^{k_1-1}}\sum_{i=0}^{M_1-1}\sum_{l=1}^{2^{k_2-1}}\sum_{j=0}^{M_2-1}c_{n,i,l,j}\psi_{n,i,l,j}(x,t)\right)\\&\left.\times\frac{\partial^2}{\partial x^2}\left(\sum_{n=1}^{2^{k_1-1}}\sum_{i=0}^{M_1-1}\sum_{l=1}^{2^{k_2-1}}\sum_{j=0}^{M_2-1}c_{n,i,l,j}\psi_{n,i,l,j}(x,t)\right)\right].\end{aligned} \tag{5.30}$$

Now, substituting the collocation points $x_l=(l-0.5)/(2^{k_1-1}M_1)$ and $t_r=(r-0.5)/(2^{k_2-1}M_2)$ for $l=1,2,\ldots,2^{k_1-1}M_1$ and $r=1,2,\ldots,2^{k_2-1}M_2$ in Equation 5.30, we have $(2^{k_1-1}M_1)(2^{k_2-1}M_2)$ equations in $(2^{k_1-1}M_1)(2^{k_2-1}M_2)$ unknowns involving $c_{n,i,l,j}$. By solving this system of equations using mathematical software, the Legendre wavelet coefficients $c_{n,i,l,j}$ can be obtained.

5.4 Numerical Results of Fractional Order PDE

Tables 5.1 through 5.3 show the comparison of the absolute errors for fractional order PDEs obtained by using the Haar wavelet method and Legendre wavelet method at different values of x, t, and α. Tables 5.4 and 5.5, respectively, show the approximate solutions of fractional order PDEs (5.5) obtained by using the Haar wavelet method, Legendre wavelet method, and HPM at various points of x and t, taking $\alpha=0.5$ and 0.75. The results thus obtained have been cited in the tables in order to justify the accuracy and efficiency of the proposed schemes.

TABLE 5.1

Comparison of Absolute Errors Obtained by the Haar Wavelet Method and Legendre Wavelet Method for a Classical Order PDE (5.5) at Various Points of x and t for $\alpha=1$

	$t=0.2$		$t=0.4$		$t=0.6$		$t=0.8$	
x	$\|u_{Exact}-u_{Haar}\|$	$\|u_{Exact}-u_{LW}\|$	$\|u_{Exact}-u_{Haar}\|$	$\|u_{Exact}-u_{LW}\|$	$\|u_{Exact}-u_{Haar}\|$	$\|u_{Exact}-u_{LW}\|$	$\|u_{Exact}-u_{Haar}\|$	$\|u_{Exact}-u_{LW}\|$
0	0	0	0	0	0	0	0	0
0.1	2.14328E−3	3.92494E−8	9.25684E−4	1.09313E−7	9.10577E−4	1.33628E−7	3.57450E−3	1.27536E−7
0.2	3.81027E−3	1.56998E−7	1.64566E−3	4.37251E−7	1.61880E−3	5.34510E−7	6.35467E−3	5.10146E−7
0.3	5.00098E−3	3.53244E−7	2.15993E−3	9.83815E−7	2.12468E−3	1.20265E−6	8.34051E−3	1.14783E−6
0.4	5.71541E−3	6.27990E−7	2.46849E−3	1.74900E−6	2.42821E−3	2.13804E−6	9.53201E−3	2.04058E−6
0.5	5.95355E−3	9.81235E−7	2.57135E−3	2.73282E−6	2.52938E−3	3.34069E−6	9.92917E−3	3.18841E−6
0.6	5.71541E−3	1.41298E−6	2.46849E−3	3.93526E−6	2.42821E−3	4.81059E−6	9.53201E−3	4.59131E−6
0.7	5.00098E−3	1.92322E−6	2.15993E−3	5.35632E−6	2.12468E−3	6.54775E−6	8.34051E−3	6.24929E−6
0.8	3.81027E−3	2.51196E−6	1.64566E−3	6.99601E−6	1.61880E−3	8.55216E−6	6.35467E−3	8.16234E−6
0.9	2.14328E−3	3.17920E−6	9.25684E−4	8.85433E−6	9.10577E−4	1.08238E−5	3.57450E−3	1.03305E−5
1.0	0	6.16414E−6	0	2.46565E−5	0	5.54772E−5	0	9.86262E−5

TABLE 5.2

Comparison of Absolute Errors Obtained by the Haar Wavelet Method and Legendre Wavelet Method for a Fractional Order PDE (5.5) at Various Points of x and t for $\alpha = 0.75$

	$t = 0.2$		$t = 0.4$		$t = 0.6$		$t = 0.8$	
x	$\lvert u_{Exact} - u_{Haar} \rvert$	$\lvert u_{Exact} - u_{LW} \rvert$	$\lvert u_{Exact} - u_{Haar} \rvert$	$\lvert u_{Exact} - u_{LW} \rvert$	$\lvert u_{Exact} - u_{Haar} \rvert$	$\lvert u_{Exact} - u_{LW} \rvert$	$\lvert u_{Exact} - u_{Haar} \rvert$	$\lvert u_{Exact} - u_{LW} \rvert$
0	0	0	0	0	0	0	0	0
0.1	3.24200E−4	8.05544E−6	1.64793E−3	4.26508E−6	3.90569E−3	1.02880E−5	6.80080E−3	1.20536E−5
0.2	2.90100E−3	3.22218E−5	6.85948E−3	1.70603E−5	1.06992E−2	4.11520E−5	1.53909E−2	4.82144E−5
0.3	8.26860E−3	7.24990E−5	1.34063E−2	3.83857E−5	1.79335E−2	9.25921E−5	2.34600E−2	1.08482E−4
0.4	1.43532E−2	1.28887E−4	1.97217E−2	6.82413E−5	2.41796E−2	1.64608E−4	2.97960E−2	1.92858E−4
0.5	1.97971E−2	2.01386E−4	2.45854E−2	1.06627E−4	2.84360E−2	2.57200E−4	3.36037E−2	3.01340E−4
0.6	2.34026E−2	2.89996E−4	2.70595E−2	1.53543E−4	2.99891E−2	3.70368E−4	3.43426E−2	4.33929E−4
0.7	2.41057E−2	3.94717E−4	2.63927E−2	2.08989E−4	2.82977E−2	5.04113E−4	3.16167E−2	5.90626E−4
0.8	2.09565E−2	5.15548E−4	2.19721E−2	2.72965E−4	2.29403E−2	6.58433E−4	2.51261E−2	7.71430E−4
0.9	1.31498E−2	6.52491E−4	1.33111E−2	3.45472E−4	1.35947E−2	8.33329E−4	1.46467E−2	9.76341E−4
1.0	0	8.05544E−4	0	4.26508E−4	0	0.00102880	0	0.00120536

TABLE 5.3

Comparison of Absolute Errors Obtained by the Haar Wavelet Method and Legendre Wavelet Method for a Fractional Order PDE (5.5) at Various Points of x and t for $\alpha=0.5$

	$t=0.2$		$t=0.4$		$t=0.6$		$t=0.8$	
x	$\lvert u_{Exact}-u_{Haar}\rvert$	$\lvert u_{Exact}-u_{LW}\rvert$	$\lvert u_{Exact}-u_{Haar}\rvert$	$\lvert u_{Exact}-u_{LW}\rvert$	$\lvert u_{Exact}-u_{Haar}\rvert$	$\lvert u_{Exact}-u_{LW}\rvert$	$\lvert u_{Exact}-u_{Haar}\rvert$	$\lvert u_{Exact}-u_{LW}\rvert$
0	0	0	0	0	0	0	0	0
0.1	2.50310E−3	4.50884E−5	4.78710E−3	3.56336E−5	7.30539E−3	5.72330E−5	1.05971E−2	6.74305E−5
0.2	1.15518E−2	1.80354E−4	1.55024E−2	1.42534E−4	1.92456E−2	2.28932E−4	2.41756E−2	2.69722E−4
0.3	2.34904E−2	4.05796E−4	2.80273E−2	3.20702E−4	3.18941E−2	5.15097E−4	3.72663E−2	6.06875E−4
0.4	3.55962E−2	7.21415E−4	3.97116E−2	5.70137E−4	4.29010E−2	9.15729E−4	4.78790E−2	1.07889E−3
0.5	4.55434E−2	1.12721E−3	4.85005E−2	8.90839E−4	5.05414E−2	1.43083E−3	5.46027E−2	1.68576E−3
0.6	5.13466E−2	1.62318E−3	5.27682E−2	1.28281E−3	5.35106E−2	2.06039E−3	5.64007E−2	2.42750E−3
0.7	5.12355E−2	2.20933E−3	5.11514E−2	1.74604E−3	5.07538E−2	2.80442E−3	5.24535E−2	3.30410E−3
0.8	4.35902E−2	2.88566E−3	4.24697E−2	2.28055E−3	4.13874E−2	3.66291E−3	4.20889E−2	4.31555E−3
0.9	2.69659E−2	3.65216E−3	2.57206E−2	2.88632E−3	2.46829E−2	4.63588E−3	2.47617E−2	5.46187E−3
1.0	0	4.50884E−3	0	3.56336E−3	0	5.72330E−3	0	6.74305E−3

TABLE 5.4

Approximate Solutions of a Fractional Order PDE (5.5) Using the Haar Wavelet Method, HPM, and Legendre Wavelet Method at Various Points of x and t for $\alpha = 0.5$

x	$t = 0.2$			$t = 0.4$			$t = 0.6$			$t = 0.8$		
	u_{Haar}	$u_{Legendre}$	u_{HPM}	u_{Haar}	$u_{Legendre}$	u_{HPM}	u_{Haar}	$u_{Legendre}$	u_{HPM}	u_{Haar}	$u_{Legendre}$	u_{HPM}
0	0	0	0	0	0	0	0	0	0	0	0	0
0.1	0.014717	0.018035	0.017990	0.019705	0.024336	0.024300	0.025527	0.031519	0.03146	0.032853	0.039996	0.03992
0.2	0.060408	0.072141	0.071961	0.075175	0.097344	0.097202	0.092130	0.126077	0.12585	0.113197	0.159983	0.15971
0.3	0.133417	0.162317	0.16191	0.162291	0.219025	0.21870	0.195885	0.283674	0.28316	0.237565	0.359962	0.35935
0.4	0.231021	0.288564	0.28784	0.278404	0.389377	0.38881	0.334440	0.504310	0.50339	0.403966	0.639933	0.63885
0.5	0.350894	0.450882	0.44975	0.421457	0.608402	0.60751	0.506071	0.787984	0.78655	0.610988	0.999895	0.99821
0.6	0.491052	0.649269	0.64765	0.589825	0.876098	0.87482	0.709473	1.134700	1.13264	0.857595	1.439850	1.43742
0.7	0.649723	0.883728	0.88152	0.782146	1.192470	1.19072	0.943592	1.544450	1.54164	1.142970	1.959790	1.95649
0.8	0.825288	1.154260	1.15137	0.997238	1.557510	1.55523	1.207540	2.017240	2.01358	1.466440	2.559730	2.55541
0.9	1.016300	1.460860	1.45720	1.234100	1.971220	1.96833	1.500600	2.553070	2.54843	1.827450	3.239660	3.23420
1.0	1.22140	1.803530	1.79902	1.49182	2.433610	2.43004	1.82212	3.151940	3.14621	2.22554	3.999580	3.99284

TABLE 5.5

Approximate Solutions of a Fractional Order PDE (5.5) Using the Haar Wavelet Method, HPM, and Legendre Wavelet Method at Various Points of x and t for $\alpha = 0.75$

	$t=0.2$			$t=0.4$			$t=0.6$			$t=0.8$		
x	u_{Haar}	$u_{Legendre}$	u_{HPM}	u_{Haar}	$u_{Legendre}$	u_{HPM}	u_{Haar}	$u_{Legendre}$	u_{HPM}	u_{Haar}	$u_{Legendre}$	u_{HPM}
0	0	0	0	0	0	0	0	0	0	0	0	0
0.1	0.01189	0.01406	0.01405	0.016566	0.018009	0.018004	0.022127	0.022639	0.022629	0.02906	0.028179	0.028166
0.2	0.05176	0.05622	0.05619	0.066533	0.072035	0.072018	0.083584	0.090558	0.090516	0.10441	0.112715	0.112667
0.3	0.11819	0.12649	0.12642	0.147670	0.162079	0.162041	0.181924	0.203755	0.203662	0.22376	0.253608	0.253500
0.4	0.20978	0.22488	0.22475	0.258414	0.288141	0.288073	0.315719	0.362231	0.362066	0.38588	0.450859	0.450666
0.5	0.32515	0.35137	0.35117	0.397542	0.450221	0.450114	0.483966	0.565986	0.565729	0.58999	0.704468	0.704166
0.6	0.46311	0.50597	0.50568	0.564116	0.648318	0.648164	0.685952	0.815020	0.814649	0.83554	1.014430	1.014000
0.7	0.62259	0.68869	0.68829	0.757387	0.882432	0.882223	0.921136	1.109330	1.108830	1.12213	1.380760	1.380170
0.8	0.80265	0.89951	0.89899	0.976740	1.152560	1.152290	1.189100	1.448920	1.448270	1.44947	1.803440	1.802670
0.9	1.00249	1.13844	1.13779	1.221690	1.458710	1.458370	1.489510	1.833790	1.832960	1.81733	2.282470	2.281500
1.0	1.22140	1.40548	1.40468	1.491820	1.800880	1.800460	1.822120	2.263940	2.262910	2.22554	2.817870	2.816660

5.5 Implementation of Legendre Wavelets for Solving a Fractional KBK Equation

Consider the nonlinear time-fractional KBK equation [107]

$$\frac{\partial^\alpha u}{\partial t^\alpha}+u\frac{\partial u}{\partial x}-\lambda_1\frac{\partial^2 u}{\partial x^2}+\lambda_2\frac{\partial^3 u}{\partial x^3}+\lambda_3\frac{\partial^4 u}{\partial x^4}=f(x,t), \quad t>0,\ x>0 \tag{5.31}$$

where

$0<\alpha\leq 1$, subject to the initial condition $u(x,0)=0$

$f(x,t)$ is the forcing term

To show the effectiveness and accuracy of the proposed scheme, two test examples have been considered. The numerical solutions thus obtained are compared with the exact solutions.

The Legendre wavelet solution of $u(x,t)$ is sought by assuming that $u(x,t)$ can be expanded in terms of Legendre wavelets as

$$u(x,t)=\sum_{n=1}^{2^{k_1-1}}\sum_{i=0}^{M_1-1}\sum_{l=1}^{2^{k_2-1}}\sum_{j=0}^{M_2-1}c_{n,i,l,j}\psi_{n,i,l,j}(x,t), \tag{5.32}$$

where $n=1,\ldots,2^{k_1-1}$, $i=0,\ldots,M_1-1$, $l=1,\ldots,2^{k_2-1}$, $j=0,\ldots,M_2-1$.

The nonlinear term presented in Equation 5.31 can be approximated using the Legendre wavelet function as

$$u\frac{\partial u}{\partial x}=\sum_{n=1}^{2^{k_1-1}}\sum_{i=0}^{M_1-1}\sum_{l=1}^{2^{k_2-1}}\sum_{j=0}^{M_2-1}a_{n,i,l,j}\psi_{n,i,l,j}(x,t). \tag{5.33}$$

This implies

$$\left(\sum_{n=1}^{2^{k_1-1}}\sum_{i=0}^{M_1-1}\sum_{l=1}^{2^{k_2-1}}\sum_{j=0}^{M_2-1}c_{n,i,l,j}\psi_{n,i,l,j}(x,t)\right)\left[\sum_{n=1}^{2^{k_1-1}}\sum_{i=0}^{M_1-1}\sum_{l=1}^{2^{k_2-1}}\sum_{j=0}^{M_2-1}c_{n,i,l,j}\frac{\partial\psi_{n,i,l,j}(x,t)}{\partial x}\right]$$
$$=\sum_{n=1}^{2^{k_1-1}}\sum_{i=0}^{M_1-1}\sum_{l=1}^{2^{k_2-1}}\sum_{j=0}^{M_2-1}a_{n,i,l,j}\psi_{n,i,l,j}(x,t). \tag{5.34}$$

Again, applying J_t^α on both sides of Equation 5.31, we have

$$u(x,t)-u(x,0)=J_t^\alpha\left[f(x,t)-u\frac{\partial u}{\partial x}+\lambda_1\frac{\partial^2 u}{\partial x^2}-\lambda_2\frac{\partial^3 u}{\partial x^3}-\lambda_3\frac{\partial^4 u}{\partial x^4}\right]. \tag{5.35}$$

Putting Equations 5.32 and 5.33 in Equation 5.35, we have

$$u(x,t)-u(x,0)=J_t^{\alpha}\left[f(x,t)-\sum_{n=1}^{2^{k_1-1}}\sum_{i=0}^{M_1-1}\sum_{l=1}^{2^{k_2-1}}\sum_{j=0}^{M_2-1}a_{n,i,l,j}\psi_{n,i,l,j}(x,t)\right.$$
$$+\lambda_1\left(\sum_{n=1}^{2^{k_1-1}}\sum_{i=0}^{M_1-1}\sum_{l=1}^{2^{k_2-1}}\sum_{j=0}^{M_2-1}c_{n,i,l,j}\frac{\partial^2\psi_{n,i,l,j}(x,t)}{\partial x^2}\right)$$
$$-\lambda_2\left(\sum_{n=1}^{2^{k_1-1}}\sum_{i=0}^{M_1-1}\sum_{l=1}^{2^{k_2-1}}\sum_{j=0}^{M_2-1}c_{n,i,l,j}\frac{\partial^3\psi_{n,i,l,j}(x,t)}{\partial x^3}\right)$$
$$\left.-\lambda_3\left(\sum_{n=1}^{2^{k_1-1}}\sum_{i=0}^{M_1-1}\sum_{l=1}^{2^{k_2-1}}\sum_{j=0}^{M_2-1}c_{n,i,l,j}\frac{\partial^4\psi_{n,i,l,j}(x,t)}{\partial x^4}\right)\right]. \tag{5.36}$$

Now, substituting the collocation points $x_l=(l-0.5)/(2^{k_1-1}M_1)$ and $t_r=(r-0.5)/(2^{k_2-1}M_2)$ for $l=1,2,\ldots,2^{k_1-1}M_1$ and $r=1,2,\ldots,2^{k_2-1}M_2$ in Equations 5.34 and 5.36, we have $2(2^{k_1-1}M_1)(2^{k_2-1}M_2)$ equations in $2(2^{k_1-1}M_1)(2^{k_2-1}M_2)$ unknowns in $a_{n,i,l,j}$ and $c_{n,i,l,j}$. By solving this system of equations using mathematical software, the Legendre wavelet coefficients $a_{n,i,l,j}$ and $c_{n,i,l,j}$ can be obtained.

Example 5.1

Consider the time-fractional KBK equation (5.31) with the following forcing term:

$$f(x,t)=\frac{t^{\alpha}\cos x}{\Gamma(1+\alpha)}-\frac{t^{4\alpha}\cos(x)\sin(x)}{\left(\Gamma(1+2\alpha)\right)^2}+\lambda_1\frac{t^{2\alpha}\cos(x)}{\Gamma(1+2\alpha)}+\lambda_2\frac{t^{2\alpha}\sin(x)}{\Gamma(1+2\alpha)}+\lambda_3\frac{t^{2\alpha}\cos(x)}{\Gamma(1+2\alpha)}. \tag{5.37}$$

The exact solution of Equation 5.31 is given by

$$u(x,t)=\frac{t^{2\alpha}\cos(x)}{\Gamma(1+2\alpha)}. \tag{5.38}$$

In the case of Example 5.1, Table 5.6 shows the comparison of absolute errors obtained by the Legendre wavelet method for Equation 5.31, taking $\lambda_1=\lambda_2=\lambda_3=1$ and $\alpha=1$. Similarly, Table 5.7 shows the comparison of absolute errors for Equation 5.31, taking $\alpha=0.75$. Table 5.8 shows the L_2 and L_∞ errors at different values of α and x. Consequently, it is observed from Tables 5.6 through 5.8 that the solutions obtained by the present method are in good agreement with the exact results.

Example 5.2

Consider the time-fractional KBK equation (5.31) with the following forcing term:

$$f(x,t)=\frac{t^{\alpha}\sin(x)}{\Gamma(1+\alpha)}+\frac{t^{4\alpha}\cos(x)\sin(x)}{\left(\Gamma(1+2\alpha)\right)^2}+\lambda_1\frac{t^{2\alpha}\sin(x)}{\Gamma(1+2\alpha)}-\lambda_2\frac{t^{2\alpha}\cos(x)}{\Gamma(1+2\alpha)}+\lambda_3\frac{t^{2\alpha}\sin(x)}{\Gamma(1+2\alpha)}. \tag{5.39}$$

TABLE 5.6

Comparison of Absolute Errors Obtained by the Legendre Wavelet Method for the KBK Equation (5.31) Given in Example 5.1 at Various Points of x and t, Taking $\alpha = 1$

	$\vert u_{Exact} - u_{Legendre} \vert$					
x	$t=0$	$t=0.1$	$t=0.2$	$t=0.3$	$t=0.4$	$t=0.5$
0.1	6.95349E−6	1.84959E−4	2.90830E−4	9.67277E−5	7.19605E−4	2.50855E−3
0.2	3.73063E−4	2.97674E−4	4.23930E−4	2.76159E−4	5.99905E−4	2.64995E−3
0.3	8.23433E−4	4.25867E−4	5.67587E−4	4.61419E−4	4.94075E−4	2.83828E−3
0.4	1.34267E−3	5.68389E−4	7.19135E−4	6.46241E−4	4.14924E−4	3.09695E−3
0.5	1.92911E−3	7.24113E−4	8.76214E−4	8.25379E−4	3.72850E−4	3.44474E−3
0.6	2.58055E−3	8.91885E−4	1.03672E−3	9.94495E−4	3.75989E−4	3.89591E−3
0.7	3.29387E−3	1.07048E−3	1.19879E−3	1.15008E−3	4.30298E−4	4.46022E−3
0.8	4.06469E−3	1.25857E−3	1.36074E−3	1.28944E−3	5.39558E−4	5.14289E−3
0.9	4.88696E−3	1.45468E−3	1.52113E−3	1.41067E−3	7.05318E−4	5.94439E−3
1.0	5.75262E−3	1.65715E−3	1.67872E−3	1.5127E−3	9.26776E−4	6.86019E−3

TABLE 5.7

Comparison of Absolute Errors Obtained by the Legendre Wavelet Method for the Fractional Order KBK Equation (5.31) Given in Example 5.1 at Various Points of x and t, Taking $\alpha = 0.75$

	$\vert u_{Exact} - u_{Legendre} \vert$							
x	$t=0$	$t=0.1$	$t=0.2$	$t=0.3$	$t=0.4$	$t=0.5$	$t=0.6$	$t=0.7$
0.1	9.99667E−4	2.12737E−4	2.28422E−4	7.56638E−3	1.76708E−2	9.57606E−2	2.20683E−2	1.05767E−2
0.2	9.08162E−4	2.41145E−4	2.65580E−3	9.06168E−3	2.18543E−2	5.85324E−2	2.74249E−2	1.98501E−2
0.3	7.96724E−4	2.86820E−4	3.10412E−3	1.07513E−2	2.64650E−2	1.30486E−2	3.19023E−2	2.75266E−2
0.4	6.65884E−4	3.51375E−4	3.63645E−3	1.26528E−2	3.15340E−2	4.06402E−2	3.56728E−2	3.38164E−2
0.5	8.39996E−4	3.19966E−4	2.78217E−3	8.89023E−3	2.04192E−2	3.80953E−2	7.45291E−3	8.48950E−3
0.6	7.41453E−4	3.22493E−4	2.93705E−3	9.67846E−3	2.29045E−2	6.76929E−2	8.91490E−3	4.26215E−3
0.7	6.31244E−4	3.35989E−4	3.12840E−3	1.05325E−2	2.54926E−2	1.00724E−1	9.53959E−3	1.40098E−3
0.8	5.10374E−4	3.61167E−4	3.36005E−3	1.14623E−2	2.82012E−2	1.36832E−1	9.45691E−3	2.80664E−4
0.9	3.80015E−4	3.98257E−4	3.63348E−3	1.24718E−2	3.10362E−2	1.75597E−1	8.77358E−3	9.43238E−4
1.0	2.41511E−4	4.47018E−4	3.94789E−3	1.35589E−2	3.39907E−2	2.16518E−1	7.57357E−3	7.23009E−4

The exact solution of Equation 5.31 is given by

$$u(x,t) = \frac{t^{2\alpha}\sin(x)}{\Gamma(1+2\alpha)}. \tag{5.40}$$

In the case of Example 5.2, Table 5.9 shows the comparison of absolute errors obtained by the Legendre wavelet method for Equation 5.31, taking $\lambda_1 = \lambda_2 = \lambda_3 = 1$ and $\alpha = 1$. Similarly, Table 5.10 shows the comparison of absolute errors for Equation 5.31, taking $\alpha = 0.75$. Consequently, it is observed from Tables 5.9 and 5.10 that the solutions obtained by the present method are in good agreement with the exact results.

TABLE 5.8

L_2 and L_∞ Error Norms for the Nonlinear Time-Fractional KBK Equation (5.31) Given in Example 5.1 Using the Two-Dimensional Legendre Wavelet Method at Various Points of x

	$\alpha=1$		$\alpha=0.75$	
x	L_2	L_∞	L_2	L_∞
0.1	1.07540E−3	2.50855E−3	3.56012E−2	9.57606E−2
0.2	1.14498E−3	2.64995E−3	2.53472E−2	5.85324E−2
0.3	1.27112E−3	2.83828E−3	1.86148E−2	3.19023E−2
0.4	1.46195E−3	3.09695E−3	2.55832E−2	4.06402E−2
0.5	1.71756E−3	3.44474E−3	1.61377E−2	3.80953E−2
0.6	2.03462E−3	3.89591E−3	2.75358E−2	6.76929E−2
0.7	2.40950E−3	4.46022E−3	3.70968E−2	1.00724E−1
0.8	2.83903E−3	5.14289E−3	4.96876E−2	1.36832E−1
0.9	3.32003E−3	5.94439E−3	6.32895E−2	1.75597E−1
1.0	3.84851E−3	6.86019E−3	7.76957E−2	2.16518E−1

TABLE 5.9

Comparison of Absolute Errors Obtained by the Legendre Wavelet Method for the Fractional Order KBK Equation (5.31) Given in Example 5.2 at Various Points of x and t, Taking $\alpha=1$

				$\lvert u_{Exact}-u_{Legendre}\rvert$			
x	$t=0$	$t=0.1$	$t=0.2$	$t=0.3$	$t=0.4$	$t=0.5$	$t=0.6$
0.1	4.98657E−5	5.68684E−5	1.28221E−4	4.93990E−4	1.48989E−3	3.49306E−3	6.95761E−3
0.2	1.04370E−4	5.39855E−5	1.19191E−4	4.39997E−4	1.31699E−3	3.07391E−3	6.12894E−3
0.3	1.69932E−4	4.85989E−5	1.04660E−4	3.69101E−4	1.10257E−3	2.56670E−3	5.13534E−3
0.4	2.46500E−4	4.05274E−5	8.31235E−5	2.76311E−4	8.35246E−4	1.95025E−3	3.94182E−3
0.5	3.34173E−4	2.95642E−5	5.30284E−5	1.56464E−4	5.03251E−4	1.20250E−3	2.51167E−3
0.6	4.33146E−4	1.54809E−5	1.27887E−5	4.30468E−6	9.45733E−5	3.00882E−4	8.06986E−4
0.7	5.43658E−4	1.96621E−6	3.91878E−5	1.85428E−4	4.02820E−4	7.77358E−4	1.21073E−3
0.8	6.65935E−4	2.30271E−5	1.04464E−4	4.17885E−4	1.00074E−3	2.05467E−3	3.57988E−3
0.9	8.00141E−4	4.79469E−5	1.84526E−4	6.97968E−4	1.71050E−3	3.55273E−3	6.33795E−3
1.0	9.46323E−4	7.69510E−5	2.80724E−4	1.03017E−3	2.54254E−3	5.29182E−3	9.52067E−3

5.6 Numerical Results and Discussion of a Time-Fractional KBK Equation

In order to measure the accuracy of the numerical scheme, L_2 and L_∞ error norms are calculated using the following formulae:

$$L_2 = R.M.S.\ error = \frac{1}{\sqrt{N}}\sqrt{\sum_{k=1}^{N}\left(\left|u_{approx}(x,t_k)-u_{exact}(x,t_k)\right|\right)^2}, \tag{5.41}$$

$$L_\infty = \max\left|u_{approx}(x,t_k)-u_{exact}(x,t_k)\right|. \tag{5.42}$$

TABLE 5.10

Comparison of Absolute Errors Obtained by the Legendre Wavelet Method for the Fractional Order KBK Equation (5.31) Given in Example 5.2 at Various Points of x and t, Taking $\alpha=0.75$

	$\|u_{Exact}-u_{Legendre}\|$							
x	$t=0.05$	$t=0.1$	$t=0.15$	$t=0.2$	$t=0.25$	$t=0.3$	$t=0.35$	$t=0.4$
0.1	2.27075E−5	2.46815E−4	7.82738E−4	1.72484E−3	3.10534E−3	4.90591E−3	7.05637E−3	9.42941E−3
0.2	2.61128E−5	2.14264E−4	6.52493E−4	1.40683E−3	2.46587E−3	3.76612E−3	5.19167E−3	6.56591E−3
0.3	2.80289E−5	1.72666E−4	4.94899E−4	1.02750E−3	1.71007E−3	2.42812E−3	3.01363E−3	3.23418E−3
0.4	2.78829E−5	1.19612E−4	3.04250E−4	5.76276E−4	8.21037E−4	8.67143E−4	4.88257E−4	6.10355E−4
0.5	3.15224E−5	1.14364E−4	3.08178E−4	6.41627E−4	1.04190E−3	1.40877E−3	1.62634E−3	1.55604E−3
0.6	3.54238E−5	8.08914E−5	1.61045E−4	2.66399E−4	2.74380E−4	3.61691E−4	6.09843E−4	1.84669E−3
0.7	3.64308E−5	3.35835E−5	2.33063E−5	1.85579E−4	6.28651E−4	1.55294E−3	3.16851E−3	5.70490E−3
0.8	3.38764E−5	3.04752E−5	2.51819E−4	7.27157E−4	1.68784E−3	3.38895E−3	6.09183E−3	1.00748E−2
0.9	2.71227E−5	1.14065E−4	5.31069E−4	1.37042E−3	2.92247E−3	5.49997E−3	9.41849E−3	1.50075E−2
1.0	1.55712E−5	2.19782E−4	8.67153E−4	2.12648E−3	4.35005E−3	7.91130E−3	1.31829E−2	2.05479E−2

The comparison of absolute errors for the time-fractional KBK equation (5.31) given in Example 5.1 has been exhibited in Tables 5.6 and 5.7, which are constructed using the results obtained by the Legendre wavelet method at different values of x and t, taking $\alpha=1$ and 0.75, respectively. Similarly, Tables 5.9 and 5.10 show absolute errors of the fractional order KBK equation (5.31) in Example 5.2 at various points of x and t, taking $\alpha=1$ and 0.75, respectively. To show the accuracy of the proposed method, L_2 and L_∞ error norms for the fractional order nonlinear KBK equation given in Example 5.1 have been presented in Table 5.8. From Tables 5.6 through 5.10, one can see a pretty good agreement between the exact solutions and the solutions acquired by the two-dimensional Legendre wavelet method.

5.7 Application of Analytical and Numerical Methods for Solving the Time-Fractional sKdV Equation

5.7.1 Implementation of the Legendre Wavelet Method for a Numerical Solution of the Fractional sKdV Equation

To show the effectiveness and accuracy of the proposed scheme, we consider two test examples. The numerical solutions thus obtained are compared with the exact solutions and the solutions obtained by HAM.

Consider the nonlinear time-fractional generalized sKdV equation [117]

$$\frac{\partial^\alpha u}{\partial t^\alpha}+u\frac{\partial u}{\partial x}+\frac{\partial^3 u}{\partial x^3}-\frac{\partial^5 u}{\partial x^5}+\sigma\frac{\partial^7 u}{\partial x^7}=f(x,t),\quad 0<\alpha\le 1 \tag{5.43}$$

with initial condition $u(x,t_0)=g(x)$.

The Legendre wavelet solution of $u(x,t)$ is sought by assuming that $u(x,t)$ can be expanded in terms of a Legendre wavelet as

$$u(x,t)=\sum_{n=1}^{2^{k_1-1}}\sum_{i=0}^{M_1-1}\sum_{l=1}^{2^{k_2-1}}\sum_{j=0}^{M_2-1}c_{n,i,l,j}\psi_{n,i,l,j}(x,t), \tag{5.44}$$

where $n=1,\ldots,2^{k_1-1}$, $i=0,\ldots,M_1-1$, $l=1,\ldots,2^{k_2-1}$, $j=0,\ldots,M_2-1$.

The nonlinear term presented in Equation 5.43 can be approximated using the Legendre wavelet function as

$$u\frac{\partial u}{\partial x}=\sum_{n=1}^{2^{k_1-1}}\sum_{i=0}^{M_1-1}\sum_{l=1}^{2^{k_2-1}}\sum_{j=0}^{M_2-1}a_{n,i,l,j}\psi_{n,i,l,j}(x,t). \tag{5.45}$$

This implies

$$\left(\sum_{n=1}^{2^{k_1-1}}\sum_{i=0}^{M_1-1}\sum_{l=1}^{2^{k_2-1}}\sum_{j=0}^{M_2-1}c_{n,i,l,j}\psi_{n,i,l,j}(x,t)\right)\left[\sum_{n=1}^{2^{k_1-1}}\sum_{i=0}^{M_1-1}\sum_{l=1}^{2^{k_2-1}}\sum_{j=0}^{M_2-1}c_{n,i,l,j}\frac{\partial\psi_{n,i,l,j}(x,t)}{\partial x}\right]$$
$$=\sum_{n=1}^{2^{k_1-1}}\sum_{i=0}^{M_1-1}\sum_{l=1}^{2^{k_2-1}}\sum_{j=0}^{M_2-1}a_{n,i,l,j}\psi_{n,i,l,j}(x,t). \tag{5.46}$$

Again, applying J_t^α on both sides of Equation 5.43, we have

$$u(x,t)-u(x,0)=J_t^\alpha\left[f(x,t)-u\frac{\partial u}{\partial x}-\frac{\partial^3 u}{\partial x^3}+\frac{\partial^5 u}{\partial x^5}-\sigma\frac{\partial^7 u}{\partial x^7}\right]. \tag{5.47}$$

Putting Equations 5.44 and 5.45 in Equation 5.47, we have

$$\begin{aligned}\sum_{n=1}^{2^{k_1-1}}\sum_{i=0}^{M_1-1}\sum_{l=1}^{2^{k_2-1}}\sum_{j=0}^{M_2-1}c_{n,i,l,j}\psi_{n,i,l,j}(x,t)-u(x,0)=J_t^\alpha\Bigg[&f(x,t)-\sum_{n=1}^{2^{k_1-1}}\sum_{i=0}^{M_1-1}\sum_{l=1}^{2^{k_2-1}}\sum_{j=0}^{M_2-1}a_{n,i,l,j}\psi_{n,i,l,j}(x,t)\\&-\sum_{n=1}^{2^{k_1-1}}\sum_{i=0}^{M_1-1}\sum_{l=1}^{2^{k_2-1}}\sum_{j=0}^{M_2-1}c_{n,i,l,j}\frac{\partial^3}{\partial x^3}\psi_{n,i,l,j}(x,t)\\&+\sum_{n=1}^{2^{k_1-1}}\sum_{i=0}^{M_1-1}\sum_{l=1}^{2^{k_2-1}}\sum_{j=0}^{M_2-1}c_{n,i,l,j}\frac{\partial^5}{\partial x^5}\psi_{n,i,l,j}(x,t)\\&-\sigma\sum_{n=1}^{2^{k_1-1}}\sum_{i=0}^{M_1-1}\sum_{l=1}^{2^{k_2-1}}\sum_{j=0}^{M_2-1}c_{n,i,l,j}\frac{\partial^7}{\partial x^7}\psi_{n,i,l,j}(x,t)\Bigg].\end{aligned} \tag{5.48}$$

Now, substituting the collocation points $x_l=(l-0.5)/(2^{k_1-1}M_1)$ and $t_r=(r-0.5)/(2^{k_1-1}M_2)$ for $l=1,2,\ldots,2^{k_1-1}M_1$ and $r=1,2,\ldots,2^{k_2-1}M_2$ in Equations 5.46 and 5.48, we have $2(2^{k_1-1}M_1)(2^{k_2-1}M_2)$ equations in $2(2^{k_1-1}M_1)(2^{k_2-1}M_2)$ unknowns in $a_{n,i,l,j}$ and $c_{n,i,l,j}$. By solving this system of equations using mathematical software, the Legendre wavelet coefficients $a_{n,i,l,j}$ and $c_{n,i,l,j}$ can be obtained.

5.7.2 Comparison with HAM for a Solution of the Time-Fractional sKdV Equation

Consider the nonlinear time-fractional generalized sKdV equation [117]

$$\frac{\partial^\alpha u}{\partial t^\alpha}+u\frac{\partial u}{\partial x}+\frac{\partial^3 u}{\partial x^3}-\frac{\partial^5 u}{\partial x^5}+\sigma\frac{\partial^7 u}{\partial x^7}=0, \quad 0<\alpha\le 1, \tag{5.49}$$

subject to the initial condition [120]

$$u(x,0)=a_0+a_6\text{sech}^6(kx),$$

where $k=\frac{5}{\sqrt{1538}}$, $a_0=c-\frac{18{,}000}{769^2}$, $a_6=\frac{519{,}750}{769^2}$ and c is an arbitrary parameter.

To obtain the approximate solution of the time-fractional sKdV equation (5.49), we choose the linear operator

$$\mathcal{L}\left[\phi(x,t;p)\right]=D_t^\alpha\phi(x,t;p). \tag{5.50}$$

Let us construct the mth-order deformation equation for Equation 5.49 as follows:

$$\mathcal{L}\left[u_m(x,t)-\chi_m u_{m-1}(x,t)\right]=\hbar\mathfrak{R}_m(u_0,u_1,\ldots,u_{m-1}), \tag{5.51}$$

where

$$\mathfrak{R}_m(u_0,u_1,\ldots,u_{m-1})=\frac{1}{(m-1)!}\left.\frac{\partial^{m-1}\mathcal{N}\left[\phi(x,t;p)\right]}{\partial p^{m-1}}\right|_{p=0}, \tag{5.52}$$

$$\mathcal{N}\left[\phi(x,t;p)\right]=D_t^\alpha\phi(x,t;p)+\phi(x,t;p)\frac{\partial\phi(x,t;p)}{\partial x}+\frac{\partial^3\phi(x,t;p)}{\partial x^3}-\frac{\partial^5\phi(x,t;p)}{\partial x^5}+\sigma\frac{\partial^7\phi(x,t;p)}{\partial x^7},$$

auxiliary function $H(x,t)=1$, and

$$\phi(x,t;p)=u_0(x,t)+\sum_{m=1}^{\infty}p^m u_m(x,t),$$

$$u_m(x,t)=\frac{1}{m!}\left.\frac{\partial^m\phi(x,t;p)}{\partial p^m}\right|_{p=0}.$$

Now, the solution of the first deformation equation of Equation 5.51 is given by

$$u_1(x,t)=\hbar J_t^\alpha\left[\frac{\partial^\alpha u_0}{\partial t^\alpha}+u_0\frac{\partial u_0}{\partial x}+\frac{\partial^3 u_0}{\partial x^3}-\frac{\partial^5 u_0}{\partial x^5}+\sigma\frac{\partial^7 u_0}{\partial x^7}\right]. \tag{5.53}$$

Similarly, the solutions of second- and third-order deformation equations are

$$u_2(x,t) = u_1(x,t) + \hbar J_t^{\alpha}\left[\frac{\partial^{\alpha} u_1}{\partial t^{\alpha}} + u_0 \frac{\partial u_1}{\partial x} + u_1 \frac{\partial u_0}{\partial x} + \frac{\partial^3 u_1}{\partial x^3} - \frac{\partial^5 u_1}{\partial x^5} + \sigma \frac{\partial^7 u_1}{\partial x^7}\right], \tag{5.54}$$

$$u_3(x,t) = u_2(x,t) + \hbar J_t^{\alpha}\left[\frac{\partial^{\alpha} u_2}{\partial t^{\alpha}} + u_0 \frac{\partial u_2}{\partial x} + u_1 \frac{\partial u_1}{\partial x} + u_2 \frac{\partial u_0}{\partial x} + \frac{\partial^3 u_2}{\partial x^3} - \frac{\partial^5 u_2}{\partial x^5} + \sigma \frac{\partial^7 u_2}{\partial x^7}\right]. \tag{5.55}$$

By putting the initial condition $u_0 = u(x,0)$ in Equations 5.53 through 5.55 and solving them, we obtain the expressions for u_1, u_2, u_3.

Finally, the approximate solution for the time-fractional sKdV equation is given by

$$u = u_1(x,t) + u_2(x,t) + u_3(x,t) + \cdots. \tag{5.56}$$

Example 5.3

Consider the nonlinear time-fractional generalized sKdV equation (5.49) subject to the following initial condition $u(x,0) = 0$ and the forcing term

$$f(x,t) = \frac{t^{\alpha}\cos(x)}{\Gamma(1+\alpha)} - \frac{t^{4\alpha}\sin(x)\cos(x)}{\left(\Gamma(1+2\alpha)\right)^2} + \frac{3t^{2\alpha}\sin(x)}{\Gamma(1+2\alpha)}. \tag{5.57}$$

The exact solution of Equation 5.49 is given by

$$u(x,t) = \frac{t^{2\alpha}\cos(x)}{\Gamma(1+2\alpha)}.$$

The numerical solutions of Example 5.3 are presented for $\sigma = 1$ in Tables 5.11 and 5.12. The results are compared with the exact solutions. It has been observed from Tables 5.11 and 5.12 that the solutions obtained by the present method are in good agreement with that of the exact solutions.

Example 5.4

Consider the nonlinear homogeneous time-fractional sKdV equation (5.49) subject to the following initial condition [120]:

$$u(x,0) = a_0 + a_6 \operatorname{sech}^6(kx), \tag{5.58}$$

where $k = \frac{5}{\sqrt{1538}}$, $a_0 = c - \frac{18{,}000}{769^2}$, $a_6 = \frac{519{,}750}{769^2}$, and c is an arbitrary parameter.

The numerical solutions of Example 5.4 are presented for $\sigma = 0.01$ in Tables 5.13 and 5.14. The results are compared with the exact solutions as well as solutions obtained by HAM. It has been observed from Tables 5.13 and 5.14 that the solutions obtained by the present method are in good agreement with the exact solutions and the solutions obtained by HAM.

TABLE 5.11

Comparison of L_2 and L_∞ Error Norms Obtained by the Two-Dimensional Legendre Wavelet Method for the Nonlinear sKdV Equation (5.49) Given in Example 5.3 at Various Points of t Taking M = 6 and 8 when $\alpha = 1$

	$M=8, k=2$		$M=6, k=2$	
t	L_2	L_∞	L_2	L_∞
0.05	9.16437E−6	5.3919E−6	4.18475E−5	2.13437E−5
0.10	1.01505E−4	5.5779E−5	3.62294E−4	1.82466E−4
0.15	4.68588E−4	2.3893E−4	1.31216E−3	6.41485E−4
0.20	1.43992E−3	6.9125E−4	3.34637E−3	1.58899E−3
0.25	3.50308E−3	1.6015E−3	7.00733E−3	3.23727E−3
0.30	7.34409E−3	3.2189E−3	1.29496E−2	5.83178E−3
0.35	1.39175E−2	5.8669E−3	2.19626E−2	9.66262E−3
0.40	2.45602E−2	9.9599E−3	3.49928E−2	1.50760E−2
0.45	4.11652E−2	1.6021E−2	5.31644E−2	2.24858E−2
0.50	4.86290E−2	2.6175E−2	8.00433E−2	3.34444E−2

TABLE 5.12

Comparison of Approximate Solutions Obtained by the Two-Dimensional Legendre Wavelet Method for the Fractional Order Nonlinear sKdV Equation (5.49) Given in Example 5.3 at Various Points of x and t, Taking $\alpha = 0.75$

	$\lvert u_{Exact} - u_{LW} \rvert$					
x	$t=0.05$	$t=0.1$	$t=0.15$	$t=0.2$	$t=0.25$	$t=0.3$
0.1	1.48844E−3	1.02439E−3	1.34522E−3	6.84404E−3	1.65336E−2	3.16301E−2
0.2	1.44428E−3	8.04932E−4	1.87684E−3	7.87671E−3	1.83517E−2	3.47030E−2
0.3	1.38657E−3	5.78750E−4	2.38982E−3	8.83756E−3	2.00105E−2	0.37488E−2
0.4	1.31595E−3	3.49808E−4	2.87396E−3	9.70634E−3	2.14745E−2	3.99294E−2
0.5	3.14870E−3	5.32314E−3	8.43096E−3	1.46331E−2	2.77386E−2	5.23307E−2
0.6	3.56192E−3	5.91560E−3	9.27444E−3	1.57778E−2	2.93468E−2	5.46608E−2
0.7	3.94745E−3	6.45309E−3	1.00217E−2	1.67499E−2	3.06354E−2	5.64109E−2
0.8	4.30041E−3	6.92704E−3	1.06584E−2	1.75272E−2	3.15695E−2	5.75265E−2
0.9	4.61606E−3	7.32937E−3	1.11714E−2	1.80894E−2	3.21175E−2	5.79590E−2
1	4.88988E−3	7.65262E−3	1.15487E−2	1.84184E−2	3.22520E−2	5.76668E−2

5.8 Numerical Results and Discussion of the Time-Fractional sKdV Equation

The comparison of the L_2 and L_∞ error norms for the time-fractional sKdV equation (5.49) given in Examples 5.3 and 5.4 has been exhibited in Tables 5.11 and 5.14, which are constructed using the results obtained by Legendre wavelet method at different values of t, taking M = 6, 8, and 4 when $\alpha = 1$. Tables 5.12 and 5.13 show the absolute errors of the time-fractional sKdV equation (5.49) for Example 5.3 at various points of x and t, taking $\alpha = 0.75$, 0.9, and 0.85, respectively. Similarly, Table 5.15 illustrates the absolute errors of the fractional order sKdV equation (5.49) for Example 5.4 at various points of x and t, taking $\alpha = 0.75$. Agreement between present numerical results and exact solutions appears very satisfactory through illustration in Tables 5.11 through 5.15. To show the accuracy of the

TABLE 5.13

Absolute Errors Obtained by the Two-Dimensional Legendre Wavelet Method for the Fractional Order Nonlinear sKdV Equation (5.49) Given in Example 5.3 at Various Points of x and t, Taking $\alpha = 0.9$ and 0.85

	$\lvert u_{Exact} - u_{LW} \rvert$									
	$\alpha = 0.9$					$\alpha = 0.85$				
x	$t = 0.05$	$t = 0.1$	$t = 0.15$	$t = 0.2$	$t = 0.25$	$t = 0.05$	$t = 0.1$	$t = 0.15$	$t = 0.2$	$t = 0.25$
0.1	1.9246E−2	1.3604E−2	4.4142E−3	3.8954E−2	7.4497E−2	6.9542E−4	7.6950E−3	1.5974E−2	2.5318E−2	3.8521E−2
0.2	1.7746E−2	1.2199E−2	5.2290E−3	3.9733E−2	7.7033E−2	7.1046E−5	1.0510E−2	1.8741E−2	2.8015E−2	4.4560E−2
0.3	1.6012E−2	1.0617E−2	6.0247E−3	4.0160E−2	7.8958E−2	5.7288E−4	1.3190E−2	2.1255E−2	3.0330E−2	5.0048E−2
0.4	1.4047E−2	8.8587E−3	6.7945E−3	4.0212E−2	8.0217E−2	1.2304E−3	1.5746E−2	2.3537E−2	3.2285E−2	5.5001E−2
0.5	1.1856E−2	6.9277E−3	7.5331E−3	3.9867E−2	8.0759E−2	2.1161E−3	6.6038E−3	2.9885E−3	2.1245E−2	3.8269E−2
0.6	9.4467E−3	4.8277E−3	8.2358E−3	3.9110E−2	8.0539E−2	3.1420E−4	6.6306E−5	1.1018E−2	2.9948E−2	5.0558E−2
0.7	6.8278E−3	2.5638E−3	8.8993E−3	3.7927E−2	7.9514E−2	1.4167E−3	6.3441E−3	1.8809E−2	3.8240E−2	6.2237E−2
0.8	4.0113E−3	1.4338E−4	9.5203E−3	3.6306E−2	7.7648E−2	3.1190E−3	1.2661E−2	2.6382E−2	4.6132E−2	7.3303E−2
0.9	1.0121E−3	2.4241E−3	1.0096E−2	3.4243E−2	7.4911E−2	4.7975E−3	1.8884E−2	3.3728E−2	5.3601E−2	8.3721E−2
1	2.1510E−3	5.1261E−3	1.0624E−2	3.1738E−2	7.1281E−2	6.4193E−3	2.4976E−2	4.0805E−2	6.0597E−2	9.3426E−2

TABLE 5.14

Comparison of L_2 and L_∞ Error Norms Obtained by the Two-Dimensional Legendre Wavelet Method for the Nonlinear sKdV Equation (5.49) Given in Example 5.4 at Various Points t, Taking $M=4$ and 8 when $\alpha=1$

	$M=8, k=1$		$M=4, k=1$	
t	L_2	L_∞	L_2	L_∞
0	1.13870E−5	4.68205E−6	6.44281E−5	3.33213E−5
0.1	5.72636E−3	2.66655E−3	9.28479E−3	5.07039E−3
0.2	1.27713E−2	5.63322E−3	1.78148E−2	9.90263E−3
0.3	2.15239E−2	9.08072E−3	2.55958E−2	1.44389E−2
0.4	3.23356E−2	1.32161E−2	3.26941E−2	1.86549E−2
0.5	4.55238E−2	1.82721E−2	3.91898E−2	2.25262E−2
0.6	6.13666E−2	2.45038E−2	4.51782E−2	2.60282E−2
0.7	8.35997E−2	3.85932E−2	5.07729E−2	2.91367E−2

TABLE 5.15

Comparison of Approximate Solutions Obtained by the Two-Dimensional Legendre Wavelet Method and HAM for the Fractional Order Nonlinear sKdV Equation (5.49) Given in Example 5.4 at Various Points of x and t, Taking $\hbar=-1.45$ and $\alpha=0.75$

	$t=0.1$		$t=0.2$		$t=0.3$		$t=0.4$		$t=0.5$	
x	$u_{Legendre}$	u_{HAM}	$u_{Legendre}$	u_{HAM}	$u_{Legendre}$	u_{HAM}	$u_{Legendre}$	u_{HAM}	$u_{Legendre}$	u_{HAM}
0.1	0.875221	0.878962	0.877833	0.878771	0.880844	0.877951	0.884388	0.876460	0.888389	0.874258
0.2	0.872626	0.878151	0.874683	0.877764	0.877135	0.876133	0.880292	0.873178	0.884275	0.868821
0.3	0.869184	0.876495	0.870688	0.875954	0.872539	0.873596	0.875166	0.869306	0.878919	0.862975
0.4	0.864923	0.874003	0.865885	0.873369	0.867102	0.870405	0.869065	0.864971	0.872379	0.856928
0.5	0.859870	0.870687	0.860309	0.870036	0.860870	0.866623	0.862047	0.860286	0.864719	0.850864
0.6	0.854057	0.866560	0.854000	0.865984	0.853892	0.862302	0.854172	0.855343	0.856005	0.844932
0.7	0.847514	0.861637	0.846994	0.861234	0.846217	0.857483	0.845500	0.850213	0.846308	0.839240
0.8	0.840276	0.855935	0.839334	0.855810	0.837896	0.852198	0.836095	0.844940	0.835701	0.833856
0.9	0.832376	0.849472	0.831059	0.849727	0.828980	0.846463	0.826023	0.839545	0.824260	0.828801
1.0	0.823849	0.842268	0.822212	0.842999	0.819522	0.840286	0.815349	0.834021	0.812067	0.824053

proposed method, L_2 and L_∞ error norms for the classical order nonlinear sKdV equation given in Examples 5.3 and 5.4 have been presented in Table 5.16. Again, to examine the accuracy and reliability of the Legendre wavelets for solving a fractional order sKdV equation, we compare the approximate solution of the Legendre wavelet with the fourth-order approximate solution obtained by HAM, taking $\hbar=-1.45$. As pointed out by Liao [33], in general, by means of the so-called $\hbar$-curve, it is straightforward to choose a proper value of $\hbar$ that ensures the convergence of series solution. To investigate the influence of $\hbar$ on the solution series, we plot the $\hbar$-curve of partial derivatives of $u(x,t)$ obtained from the fourth-order HAM solution as shown in Figure 5.1.

The percentage errors of sKdV equation (5.49) for Example 5.4 has been shown in Table 5.17 at various points of x and t. Comparison of approximate solutions obtained by the two-dimensional Legendre wavelet method, HAM, and OHAM for the fractional order nonlinear sKdV equation (5.49) has been demonstrated in Table 5.18, taking $\alpha=0.75$. Agreement between present numerical results and exact solutions appears very satisfactory through illustration in Tables 5.11 through 5.18.

TABLE 5.16

L_2 and L_∞ Error Norms for the Nonlinear Time-Fractional sKdV Equation Using Two-Dimensional Legendre Wavelet Methods at Various Points t, Taking $\alpha = 1$

	Example 5.3		Example 5.4	
t	L_2	L_∞	L_2	L_∞
0.1	3.20987E−5	5.5779E−5	1.81083E−3	2.66655E−3
0.2	4.55341E−4	6.9125E−4	4.03863E−3	5.63322E−3
0.3	2.32241E−3	3.2189E−3	6.80646E−3	9.08072E−3
0.4	7.76660E−3	9.1236E−3	1.02254E−2	1.32161E−2
0.5	1.53778E−2	2.6175E−2	1.43959E−2	1.82721E−2

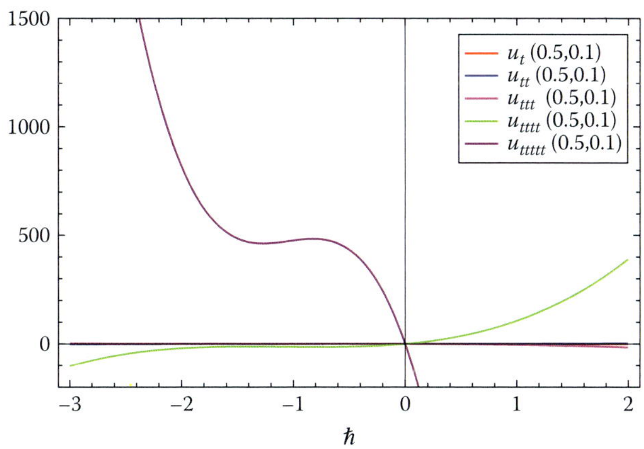

FIGURE 5.1
The $\hbar$-curve for partial derivatives of $u(x,t)$ for the fourth-order HAM solution, taking $x = 0.5$, $t = 0.1$, and $\alpha = 0.75$.

TABLE 5.17

Percentage Errors Obtained by the Two-Dimensional Legendre Wavelet Method for Nonlinear sKdV Equation (5.49) Given in Example 5.4 at Various Points of x and t, Taking $\alpha = 1$

	$\vert u_{Exact} - u_{LW} \vert$			
x	$t = 0$	$t = 0.1$	$t = 0.2$	$t = 0.3$
0.1	3.60870E−4	6.36455E−2	1.88943E−1	3.73007E−1
0.2	4.48452E−4	9.41269E−2	2.50788E−1	4.70472E−1
0.3	5.07751E−4	1.24566E−1	3.11652E−1	5.64999E−1
0.4	5.36868E−4	1.54740E−1	3.71078E−1	6.55872E−1
0.5	5.33951E−4	1.84430E−1	4.28610E−1	7.42358E−1
0.6	4.97215E−4	2.13419E−1	4.83794E−1	8.23722E−1
0.7	4.24966E−4	2.41502E−1	5.36188E−1	8.99238E−1
0.8	3.15643E−4	2.68483E−1	5.85363E−1	9.68192E−1
0.9	1.67864E−4	2.94177E−1	6.30919E−1	1.02990
1.0	1.94903E−4	3.18421E−1	6.72482E−1	1.08372

TABLE 5.18

Comparison of Approximate Solutions Obtained by the Two-Dimensional Legendre Wavelet Method, HAM, and OHAM for the Fractional Order Nonlinear sKdV Equation (5.49) Given in Example 5.4 at Various Points of x and t, Taking $\alpha = 0.75$

	$t=0.1$			$t=0.2$			$t=0.3$			$t=0.4$			$t=0.5$		
x	$u_{Legendre}$	u_{HAM}	$u_{Optimal\ HAM}$	$u_{Legendre}$	u_{HAM}	$u_{Optimal\ HAM}$	$u_{Legendre}$	u_{HAM}	$u_{Optimal\ HAM}$	$u_{Legendre}$	u_{HAM}	$u_{Optimal\ HAM}$	$u_{Legendre}$	u_{HAM}	$u_{Optimal\ HAM}$
0.1	0.875221	0.878962	0.879025	0.877833	0.878771	0.8794	0.880844	0.877951	0.879728	0.884388	0.876460	0.880029	0.888389	0.874258	0.880313
0.2	0.872626	0.878151	0.878288	0.874683	0.877764	0.879035	0.877135	0.876133	0.879689	0.880292	0.873178	0.880291	0.884275	0.868821	0.880855
0.3	0.869184	0.876495	0.876693	0.870688	0.875954	0.877809	0.872539	0.873596	0.878788	0.875166	0.869306	0.879686	0.878919	0.862975	0.88053
0.4	0.864923	0.874003	0.874247	0.865885	0.873369	0.875727	0.867102	0.870405	0.877024	0.869065	0.864971	0.878216	0.872379	0.856928	0.879334
0.5	0.859870	0.870687	0.870956	0.860309	0.870036	0.872793	0.860870	0.866623	0.874403	0.862047	0.860286	0.875882	0.864719	0.850864	0.87727
0.6	0.854057	0.866560	0.86683	0.854000	0.865984	0.869015	0.853892	0.862302	0.870931	0.854172	0.855343	0.872691	0.856005	0.844932	0.874342
0.7	0.847514	0.861637	0.861882	0.846994	0.861234	0.864406	0.846217	0.857483	0.866619	0.845500	0.850213	0.868651	0.846308	0.839240	0.870558
0.8	0.840276	0.855935	0.856127	0.839334	0.855810	0.858978	0.837896	0.852198	0.861478	0.836095	0.844940	0.863774	0.835701	0.833856	0.865929
0.9	0.832376	0.849472	0.849583	0.831059	0.849727	0.852749	0.828980	0.846463	0.855525	0.826023	0.839545	0.858074	0.824260	0.828801	0.860467
1.0	0.823849	0.842268	0.84227	0.822212	0.842999	0.845738	0.819522	0.840286	0.848778	0.815349	0.834021	0.85157	0.812067	0.824053	0.85419

5.9 Convergence of Legendre Wavelet

Theorem 5.1 Convergence Theorem

If a continuous function, $u(x,t) \in L^2(\mathfrak{R} \times \mathfrak{R})$, defined on $[0,1) \times [0,1)$ has bounded mixed fourth partial derivative $|\partial^4 u(x,t)/(\partial x^2 \partial t^2)| \le K$, then the Legendre wavelets expansion of $u(x,t)$ converges uniformly to it.

Proof

Let $u(x,t)$ be a function defined on $[0,1) \times [0,1)$ and $|\partial^4 u(x,t)/(\partial x^2 \partial t^2)| \le K$, K is a positive constant.

The Legendre wavelet coefficients of continuous functions $u(x,t)$ are defined as

$$c_{n_1,m_1,n_2,m_2} = \int_0^1\int_0^1 u(x,t)\psi_{n_1,m_1}(x)\psi_{n_2,m_2}(t)\,dxdt$$
$$= \int_{I_2}\int_{I_1} u(x,t)\sqrt{\left(m_1+\frac{1}{2}\right)\left(m_2+\frac{1}{2}\right)}2^{(k_1+k_2)/2}P_{m_1}\left(2^{k_1}x-\hat{n}_1\right)P_{m_2}\left(2^{k_2}t-\hat{n}_2\right)dxdt$$

where

$$I_1 = \left[\frac{\hat{n}_1-1}{2^{k_1-1}}, \frac{\hat{n}_1+1}{2^{k_1-1}}\right) \quad \text{and} \quad I_2 = \left[\frac{\hat{n}_2-1}{2^{k_2-1}}, \frac{\hat{n}_2+1}{2^{k_2-1}}\right).$$

Now, by change of variable $2^{k_1}x - \hat{n}_1 = y$, we obtain

$$c_{n_1,m_1,n_2,m_2} = \sqrt{\left(m_1+\frac{1}{2}\right)\left(m_2+\frac{1}{2}\right)}\frac{2^{(k_1+k_2)/2}}{2^{k_1}}$$
$$\times \int_{\frac{\hat{n}_2-1}{2^{k_2-1}}}^{\frac{\hat{n}_2+1}{2^{k_2-1}}}\left(\int_{-1}^{1} u\left(\frac{\hat{n}_1+y}{2^{k_1}},t\right)P_{m_1}(y)\,dy\right)P_{m_2}\left(2^{k_2}t-\hat{n}_2\right)dt. \tag{5.59}$$

Now,

$$\int_{-1}^{1} u\left(\frac{\hat{n}_1+y}{2^{k_1}},t\right)P_{m_1}(y)\,dy$$
$$= \frac{1}{(2m_1+1)}\int_{-1}^{1} u\left(\frac{\hat{n}_1+y}{2^{k_1}},t\right)\left(P'_{m_1+1}(y)-P'_{m_1-1}(y)\right)dy, \quad \text{since}\,(2m_1+1)P_{m_1}(y) = P'_{m_1+1}(y)-P'_{m_1-1}(y)$$
$$= \frac{1}{(2m_1+1)}\left[u\left(\frac{\hat{n}_1+y}{2^{k_1}},t\right)\left(P_{m_1+1}(y)-P_{m_1-1}(y)\right)\Big|_{-1}^{1} - \int_{-1}^{1}\frac{\partial u}{\partial y}\left(P_{m_1+1}(y)-P_{m_1-1}(y)\right)dy\right]$$
$$= \frac{-1}{(2m_1+1)}\left[\int_{-1}^{1}\frac{\partial u}{\partial y}\left(P_{m_1+1}(y)-P_{m_1-1}(y)\right)dy\right]$$
$$= \frac{1}{(2m_1+1)}\left[\int_{-1}^{1}\frac{\partial^2 u}{\partial y^2}\left[\frac{P_{m_1+2}(y)-P_{m_1}(y)}{(2m_1+3)} - \frac{P_{m_1}(y)-P_{m_1-2}(y)}{(2m_1-1)}\right]dy\right] \tag{5.60}$$

Substituting Equation 5.60 in Equation 5.59, we have

$$c_{n_1,m_1,n_2,m_2} = \sqrt{\left(m_1+\frac{1}{2}\right)\left(m_2+\frac{1}{2}\right)}\frac{2^{(k_1+k_2)/2}}{2^{k_1}}\frac{1}{(2m_1+1)}\int_{-1}^{1}\left(\int_{\frac{\hat{n}_2-1}{2^{k_2-1}}}^{\frac{\hat{n}_2+1}{2^{k_2-1}}}\frac{\partial^2 u}{\partial y^2}P_{m_2}\left(2^{k_2}t-\hat{n}_2\right)dt\right)$$

$$\times\left[\frac{P_{m_1+2}(y)-P_{m_1}(y)}{(2m_1+3)}-\frac{P_{m_1}(y)-P_{m_1-2}(y)}{(2m_1-1)}\right]dy. \tag{5.61}$$

Similarly,

$$\int_{\frac{\hat{n}_2-1}{2^{k_2-1}}}^{\frac{\hat{n}_2+1}{2^{k_2-1}}}\frac{\partial^2 u}{\partial y^2}P_{m_2}\left(2^{k_2}t-\hat{n}_2\right)dt = \frac{1}{2^{k_2}(2m_2+1)}\int_{-1}^{1}\frac{\partial^4 u}{\partial w^2\partial y^2}\left[\frac{P_{m_2+2}(w)-P_{m_2}(w)}{(2m_2+3)}-\frac{P_{m_2}(w)-P_{m_2-2}(w)}{(2m_2-1)}\right]dw, \tag{5.62}$$

where $w = 2^{k_2}t-\hat{n}_2$.

Substituting Equation 5.62 in Equation 5.61, we have

$$c_{n_1,m_1,n_2,m_2} = \sqrt{\left(m_1+\frac{1}{2}\right)\left(m_2+\frac{1}{2}\right)}\frac{1}{2^{(k_1+k_2)/2}(2m_1+1)(2m_2+1)}\int_{-1}^{1}\int_{-1}^{1}\frac{\partial^4 u}{\partial w^2\partial y^2}$$

$$\times\left[\left(\frac{P_{m_2+2}(w)-P_{m_2}(w)}{(2m_2+3)}-\frac{P_{m_2}(w)-P_{m_2-2}(w)}{(2m_2-1)}\right)\left(\frac{P_{m_1+2}(y)-P_{m_1}(y)}{(2m_1+3)}-\frac{P_{m_1}(y)-P_{m_1-2}(y)}{(2m_1-1)}\right)\right]dwdy.$$

Now suppose

$$\begin{aligned}R_1(y) &= (2m_1-1)P_{m_1+2}(y)-(2m_1-1)P_{m_1}(y)-(2m_1+3)P_{m_1}(y)+(2m_1+3)P_{m_1-2}(y)\\ &= (2m_1-1)P_{m_1+2}(y)-2(2m_1+1)P_{m_1}(y)+(2m_1+3)P_{m_1-2}(y)\end{aligned}$$

and

$$\begin{aligned}R_2(w) &= (2m_2-1)P_{m_2+2}(w)-(2m_2-1)P_{m_2}(w)-(2m_2+3)P_{m_2}(w)+(2m_2+3)P_{m_2-2}(w)\\ &= (2m_2-1)P_{m_2+2}(w)-2(2m_2+1)P_{m_2}(w)+(2m_2+3)P_{m_2-2}(w).\end{aligned}$$

Therefore,

$$\left|c_{n_1,m_1,n_2,m_2}\right| \le \lambda_{m_1,m_2}^{k_1,k_2}\int_{-1}^{1}\int_{-1}^{1}\left|\frac{\partial^4 u}{\partial w^2\partial y^2}\right|\left|R_1(y)\right|\left|R_2(w)\right|dwdy \le \lambda_{m_1,m_2}^{k_1,k_2}\int_{-1}^{1}\int_{-1}^{1}\left|R_1(y)\right|\left|R_2(w)\right|dwdy \tag{5.63}$$

where

$$\lambda_{m_1,m_2}^{k_1,k_2} = \sqrt{\left(m_1+\frac{1}{2}\right)\left(m_2+\frac{1}{2}\right)}\frac{1}{2^{(k_1+k_2)/2}(2m_1+1)^2(2m_1-3)(2m_2+1)^2(2m_2-3)}.$$

Now, using Cauchy–Schwarz inequality

$$\left(\int_{-1}^{1}|R_1(y)|dy\right)^2 \le \left(\int_{-1}^{1}1^2dy\right)\int_{-1}^{1}\left[(2m_1-1)^2+(4m_1+2)^2+(2m_1+3)^2\right]$$
$$\times\left[P_{m_1+2}^2(y)+P_{m_1}^2(y)+P_{m_1-2}^2(y)\right]dy$$
$$\le 2\left[(2m_1-1)^2+(4m_1+2)^2+(2m_1+3)^2\right]\left[\frac{2}{(2m_1+5)}+\frac{2}{(2m_1+1)}+\frac{2}{(2m_1-3)}\right]$$
$$\int_{-1}^{1}|R_1(y)|dy \le 2\sqrt{\left[(2m_1-1)^2+(4m_1+2)^2+(2m_1+3)^2\right]\left[\frac{1}{(2m_1+5)}+\frac{1}{(2m_1+1)}+\frac{1}{(2m_1-3)}\right]}. \tag{5.64}$$

Similarly,

$$\int_{-1}^{1}|R_2(w)|dw \le 2\sqrt{\left[(2m_2-1)^2+(4m_2+2)^2+(2m_2+3)^2\right]\left[\frac{1}{(2m_2+5)}+\frac{1}{(2m_2+1)}+\frac{1}{(2m_2-3)}\right]}. \tag{5.65}$$

Putting Equations 5.64 and 5.65 in Equation 5.63, we have

$$|c_{n_1,m_1,n_2,m_2}| \le 4K\lambda_{m_1,m_2}^{k_1,k_2}\eta_{m_1,m_2}, \tag{5.66}$$

where

$$\eta_{m_1,m_2} = \sqrt{\left[(2m_1-1)^2+(4m_1+2)^2+(2m_1+3)^2\right]\left[\frac{1}{(2m_1+5)}+\frac{1}{(2m_1+1)}+\frac{1}{(2m_1-3)}\right]}$$
$$\times\sqrt{\left[(2m_2-1)^2+(4m_2+2)^2+(2m_2+3)^2\right]\left[\frac{1}{(2m_2+5)}+\frac{1}{(2m_2+1)}+\frac{1}{(2m_2-3)}\right]}.$$

Therefore, $\sum_{n_1=0}^{\infty}\sum_{m_1=0}^{\infty}\sum_{n_2=0}^{\infty}\sum_{m_2=0}^{\infty} c_{n_1,m_1,n_2,m_2}$ is absolutely convergent.

Hence, according to Reference [44], the Legendre series expansion of $u(x,t)$ converges uniformly. □

Theorem 5.2 **Error Estimate**

If a continuous function, $u(x,t) \in L^2(\mathfrak{R}\times\mathfrak{R})$, defined on $[0,1)\times[0,1)$ is bounded, namely, $|u(x,t)| \le K$, then

$$\|\varepsilon_{u,k_1,M_1,k_2,M_2}\|_{L^2_\omega[[0,1)\times[0,1)]} \le 4K\left[\sum_{n_1=2^{k_1-1}+1}^{\infty}\sum_{m_1=M_1}^{\infty}\sum_{n_2=2^{k_2-1}+1}^{\infty}\sum_{m_2=M_2}^{\infty}\left(\lambda_{m_1,m_2}^{k_1,k_2}\eta_{m_1,m_2}\right)^2\right]^{1/2},$$

where

$$\lambda_{m_1,m_2}^{k_1,k_2} = \sqrt{\left(m_1+\frac{1}{2}\right)\left(m_2+\frac{1}{2}\right)}\frac{1}{2^{(k_1+k_2)/2}(2m_1+1)^2(2m_1-3)(2m_2+1)^2(2m_2-3)}$$

$$\eta_{m_1,m_2} = \sqrt{\left[(2m_1-1)^2+(4m_1+2)^2+(2m_1+3)^2\right]\left[\frac{1}{(2m_1+5)}+\frac{1}{(2m_1+1)}+\frac{1}{(2m_1-3)}\right]}$$
$$\times\sqrt{\left[(2m_2-1)^2+(4m_2+2)^2+(2m_2+3)^2\right]\left[\frac{1}{(2m_2+5)}+\frac{1}{(2m_2+1)}+\frac{1}{(2m_2-3)}\right]}.$$

Proof

$$\begin{aligned}\left\|\varepsilon_{u,k_1,M_1,k_2,M_2}\right\|^2_{L^2_\omega[[0,1)\times[0,1)]} &= \int_0^1\int_0^1\left|u(x,t)-\sum_{n_1=1}^{2^{k_1-1}}\sum_{m_1=0}^{M_1-1}\sum_{n_2=1}^{2^{k_2-1}}\sum_{m_2=0}^{M_2-1}c_{n_1,m_1,n_2,m_2}\psi_{n_1,m_1}(x)\psi_{n_2,m_2}(t)\right|^2\omega(x)\omega(t)\,dxdt\\ &= \int_0^1\int_0^1\left|\sum_{n_1=2^{k_1-1}+1}^{\infty}\sum_{m_1=M_1}^{\infty}\sum_{n_2=2^{k_2-1}+1}^{\infty}\sum_{m_2=M_2}^{\infty}c_{n_1,m_1,n_2,m_2}\psi_{n_1,m_1}(x)\psi_{n_2,m_2}(t)\right|^2 dxdt\\ &= \sum_{n_1=2^{k_1-1}+1}^{\infty}\sum_{m_1=M_1}^{\infty}\sum_{n_2=2^{k_2-1}+1}^{\infty}\sum_{m_2=M_2}^{\infty}\left|c_{n_1,m_1,n_2,m_2}\right|^2 \end{aligned}\tag{5.67}$$

Substituting Equation 5.66 of Theorem 5.1 in Equation 5.67, we obtain

$$\left\|\varepsilon_{u,k_1,M_1,k_2,M_2}\right\|^2_{L^2_\omega[[0,1)\times[0,1)]} \le \sum_{n_1=2^{k_1-1}+1}^{\infty}\sum_{m_1=M_1}^{\infty}\sum_{n_2=2^{k_2-1}+1}^{\infty}\sum_{m_2=M_2}^{\infty}\left(4K\lambda_{m_1,m_2}^{k_1,k_2}\eta_{m_1,m_2}\right)^2 \tag{5.68}$$

where

$$\lambda_{m_1,m_2}^{k_1,k_2} = \sqrt{\left(m_1+\frac{1}{2}\right)\left(m_2+\frac{1}{2}\right)}\frac{1}{2^{(k_1+k_2)/2}(2m_1+1)^2(2m_1-3)(2m_2+1)^2(2m_2-3)}$$

$$\eta_{m_1,m_2} = \sqrt{\left[(2m_1-1)^2+(4m_1+2)^2+(2m_1+3)^2\right]\left[\frac{1}{(2m_1+5)}+\frac{1}{(2m_1+1)}+\frac{1}{(2m_1-3)}\right]}$$
$$\times\sqrt{\left[(2m_2-1)^2+(4m_2+2)^2+(2m_2+3)^2\right]\left[\frac{1}{(2m_2+5)}+\frac{1}{(2m_2+1)}+\frac{1}{(2m_2-3)}\right]}.$$

Equation 5.68 implies

$$\left\|\varepsilon_{u,k_1,M_1,k_2,M_2}\right\|_{L^2_\omega[[0,1)\times[0,1)]} \le \left[\sum_{n_1=2^{k_1-1}+1}^{\infty}\sum_{m_1=M_1}^{\infty}\sum_{n_2=2^{k_2-1}+1}^{\infty}\sum_{m_2=M_2}^{\infty}\left(4K\lambda_{m_1,m_2}^{k_1,k_2}\eta_{m_1,m_2}\right)^2\right]^{1/2}. \qquad \square$$

5.10 Solution of Fractional KK Equation Using Legendre Multiwavelets

5.10.1 Introduction of Legendre Multiwavelets

Legendre multiwavelets $\psi_{n,m}(x)=\psi(k,n,m,x)$ have four arguments; $n=0,1,2,\ldots,2^k-1$, $k\in Z^+$, m is the order of Legendre polynomials and x is normalized time. They are defined on the interval $[0,1)$ as [126]

$$\psi_{n,m}(x)=\begin{cases}\sqrt{2m+1}2^{k/2}L_m\left(2^kx-n\right), & \frac{n}{2^k}\le x<\frac{n+1}{2^k},\\ 0, & \text{elsewhere}\end{cases}$$

where $m=0,1,\ldots,M-1$, $n=0,1,2,\ldots,2^k-1$.

Here, $L_m(x)$ are the well-known shifted Legendre polynomials of order m, which are defined on the interval [0,1], and can be determined with the aid of the following recurrence formulae:

$$\begin{aligned}&L_0(x)=1,\\&L_1(x)=2x-1,\\&L_{m+1}(x)=\left(\frac{2m+1}{m+1}\right)(2x-1)L_m(x)-\left(\frac{m}{m+1}\right)L_{m-1}(x),\quad m=1,2,3,\ldots.\end{aligned}$$

The two-dimensional Legendre multiwavelets are defined as

$$\psi_{n_1,m_1,n_2,m_2}(x,t)=\begin{cases}AL_{m_1}\left(2^{k_1}x-n_1\right)L_{m_2}\left(2^{k_2}t-n_2\right), & \frac{n_1}{2^{k_1}}\le x<\frac{n_1+1}{2^{k_1}},\ \frac{n_2}{2^{k_2}}\le t<\frac{n_2+1}{2^{k_2}}\\ 0, & \text{elsewhere}\end{cases}$$

where

$A=\sqrt{(2m_1+1)(2m_2+1)}2^{(k_1+k_2)/2}$

n_1 and n_2 are defined similarly to n

k_1 and k_2 are any positive integers

m_1 and m_2 are the orders for shifted Legendre polynomials

$\psi_{n_1,m_1,n_2,m_2}(x,t)$ forms a basis for $L^2([0,1)\times[0,1))$

5.10.2 Function Approximation

A function, $f(x,t)$, defined over $[0,1)\times[0,1)$ can be expanded in terms of Legendre multiwavelet as [126]

$$f(x,t)=\sum_{n=0}^{\infty}\sum_{i=0}^{\infty}\sum_{l=0}^{\infty}\sum_{j=0}^{\infty}c_{n,i,l,j}\psi_{n,i,l,j}(x,t). \tag{5.69}$$

If the infinite series in Equation 5.69 is truncated, then it can be written as

$$f(x,t)\cong\sum_{n=0}^{2^{k_1}-1}\sum_{i=0}^{M_1}\sum_{l=0}^{2^{k_2}-1}\sum_{j=0}^{M_2}c_{n,i,l,j}\psi_{n,i,l,j}(x,t)=\Psi^T(x)C\Psi(t), \tag{5.70}$$

where $\Psi(x)$ and $\Psi(t)$ are $2^{k_1}(M_1+1)\times 1$ and $2^{k_2}(M_2+1)\times 1$ matrices, respectively:

$$\Psi(x)\equiv\left[\psi_{0,0}(x),\psi_{0,1}(x),\ldots,\psi_{0,M_1}(x),\psi_{1,0}(x),\ldots,\psi_{1,M_1}(x),\ldots,\psi_{(2^{k1}-1),0}(x),\ldots,\psi_{(2^{k1}-1),M_1}(x)\right]^T,$$

$$\Psi(t)\equiv\left[\psi_{0,0}(t),\psi_{0,1}(t),\ldots,\psi_{0,M_2}(t),\psi_{1,0}(t),\ldots,\psi_{1,M_2}(t),\ldots,\psi_{(2^{k2}-1),0}(t),\ldots,\psi_{(2^{k2}-1),M_2}(t)\right]^T.$$

Also, C is a $2^{k_1}(M_1+1)\times 2^{k_2}(M_2+1)$ matrix whose elements can be calculated from the formula

$$c_{n,i,l,j}=\int_0^1\int_0^1\psi_{n,i}(x)\psi_{l,j}(t)f(x,t)\,dt\,dx, \tag{5.71}$$

with $n=0,1,\ldots,2^{k_1}-1$, $i=0,\ldots,M_1$, $l=0,1,\ldots,2^{k_2}-1$, $j=0,\ldots,M_2$.

5.10.3 Operational Matrix of the General Order Integration

The integration of

$$\Psi(t)\equiv\left[\psi_{0,0}(t),\ldots,\psi_{0,M}(t),\psi_{1,0}(t),\ldots,\psi_{1,M}(t),\ldots,\psi_{(2^k-1),0}(t),\ldots,\psi_{(2^k-1),M}(t)\right]^T$$

can be approximated by [127]

$$\int_0^t\Psi(\tau)\,d\tau\cong Q\Psi(t), \tag{5.72}$$

where Q is called the Legendre multiwavelet operational matrix of integration. To derive the Legendre multiwavelet operational matrix of the general order of integration, let us recall the fractional integral of order $\alpha(>0)$, defined by Podlubny [41]:

$$J^{\alpha}f(t)=\frac{1}{\Gamma(\alpha)}\int_0^t(t-\tau)^{\alpha-1}f(\tau)\,d\tau,\quad \alpha>0,\quad \alpha\in\Re^+ \tag{5.73}$$

where $\Re^+$ is the set of positive real numbers.

The Legendre multiwavelet operational matrix Q^{α} for integration of the general order α is given by

$$\begin{aligned}Q^{\alpha}\Psi(t)&=J^{\alpha}\Psi(t)\\&=\left[J^{\alpha}\psi_{0,0}(t),\ldots,J^{\alpha}\psi_{0,M}(t),J^{\alpha}\psi_{1,0}(t),\ldots,J^{\alpha}\psi_{1,M}(t),J^{\alpha}\psi_{(2^k-1),0}(t),\ldots,J^{\alpha}\psi_{(2^k-1),M}(t)\right]^T,\end{aligned}$$

where

$$J^{\alpha}\psi_{n,m}(t)=\begin{cases}(2m+1)^{1/2}\,2^{k/2}J^{\alpha}L_m\left(2^k t-n\right), & \dfrac{n}{2^k}\le t<\dfrac{n+1}{2^k},\\ 0, & \text{elsewhere}\end{cases}$$

for $n=0,1,2,\ldots,2^k-1$, $m=0,1,2,\ldots,M$ is the order of the Legendre polynomials, and M is a fixed positive integer.

5.11 Application of Analytical and Numerical Methods for Solving the Time-Fractional KK Equation

5.11.1 Solution of the Fractional KK Equation Using Legendre Multiwavelets

To exhibit the effectiveness and accuracy of the proposed numerical scheme, we consider the time-fractional KK equation. The numerical solutions thus obtained are compared with the exact solutions as well as with the solutions obtained by OHAM.

Consider the nonlinear time-fractional generalized KK equation [121]

$$\frac{\partial^{\alpha}u}{\partial t^{\alpha}}+45u^2\frac{\partial u}{\partial x}\quad 15p\frac{\partial u}{\partial x}\frac{\partial^2 u}{\partial x^2}-15u\frac{\partial^3 u}{\partial x^3}+\frac{\partial^5 u}{\partial x^5}=0, \tag{5.74}$$

with initial condition

$$u(x,0)=\frac{1}{4}w^2\lambda^2\text{sech}^2\left(\frac{\lambda w x}{2}\right)+\frac{w^2\lambda^2}{12}. \tag{5.75}$$

The exact solution of Equation 5.74 is given by [121]

$$u(x,t)=\frac{1}{4}w^2\lambda^2\text{sech}^2\left\{\frac{\lambda}{2}\left(\frac{-w^5\left(-8\lambda^2\mu+16\mu^2+\lambda^4\right)}{16\Gamma(1+\alpha)}t^{\alpha}+wx\right)\right\}+\frac{w^2\lambda^2}{12}, \tag{5.76}$$

where λ, μ, and w are constant with $w\neq 0$.

The Legendre multiwavelet solution of $u(x,t)$ is sought by assuming that $u(x,t)$ can be expanded in terms of the Legendre multiwavelet as

$$u(x,t)=\sum_{n=0}^{2^{k_1}-1}\sum_{i=0}^{M_1}\sum_{l=0}^{2^{k_2}-1}\sum_{j=0}^{M_2}d_{n,i,l,j}\psi_{n,i,l,j}(x,t), \tag{5.77}$$

where $n=0,\ldots,2^{k_1}-1$, $i=0,\ldots,M_1$, $l=0,\ldots,2^{k_2}-1$, $j=0,\ldots,M_2$.

The nonlinear terms presented in Equation 5.74 can be approximated by using the Legendre multiwavelet function as

$$u^2\frac{\partial u}{\partial x}=\sum_{n=0}^{2^{k1}-1}\sum_{i=0}^{M_1}\sum_{l=0}^{2^{k2}-1}\sum_{j=0}^{M_2}a_{n,i,l,j}\psi_{n,i,l,j}(x,t), \tag{5.78}$$

$$\frac{\partial u}{\partial x}\frac{\partial^2 u}{\partial x^2}=\sum_{n=0}^{2^{k1}-1}\sum_{i=0}^{M_1}\sum_{l=0}^{2^{k2}-1}\sum_{j=0}^{M_2}b_{n,i,l,j}\psi_{n,i,l,j}(x,t), \tag{5.79}$$

and

$$u\frac{\partial^3 u}{\partial x^3}=\sum_{n=0}^{2^{k1}-1}\sum_{i=0}^{M_1}\sum_{l=0}^{2^{k2}-1}\sum_{j=0}^{M_2}c_{n,i,l,j}\psi_{n,i,l,j}(x,t). \tag{5.80}$$

This implies

$$\left(\sum_{n=0}^{2^{k1}-1}\sum_{i=0}^{M_1}\sum_{l=0}^{2^{k2}-1}\sum_{j=0}^{M_2}d_{n,i,l,j}\psi_{n,i,l,j}(x,t)\right)^2\left[\sum_{n=0}^{2^{k1}-1}\sum_{i=0}^{M_1}\sum_{l=0}^{2^{k2}-1}\sum_{j=0}^{M_2}d_{n,i,l,j}\frac{\partial\psi_{n,i,l,j}(x,t)}{\partial x}\right]$$
$$=\sum_{n=0}^{2^{k1}-1}\sum_{i=0}^{M_1}\sum_{l=0}^{2^{k2}-1}\sum_{j=0}^{M_2}a_{n,i,l,j}\psi_{n,i,l,j}(x,t), \tag{5.81}$$

$$\left(\sum_{n=0}^{2^{k1}-1}\sum_{i=0}^{M_1}\sum_{l=0}^{2^{k2}-1}\sum_{j=0}^{M_2}d_{n,i,l,j}\frac{\partial\psi_{n,i,l,j}(x,t)}{\partial x}\right)\left[\sum_{n=0}^{2^{k1}-1}\sum_{i=0}^{M_1}\sum_{l=0}^{2^{k2}-1}\sum_{j=0}^{M_2}d_{n,i,l,j}\frac{\partial^2\psi_{n,i,l,j}(x,t)}{\partial x^2}\right]$$
$$=\sum_{n=0}^{2^{k1}-1}\sum_{i=0}^{M_1}\sum_{l=0}^{2^{k2}-1}\sum_{j=0}^{M_2}b_{n,i,l,j}\psi_{n,i,l,j}(x,t), \tag{5.82}$$

and

$$\left(\sum_{n=0}^{2^{k1}-1}\sum_{i=0}^{M_1}\sum_{l=0}^{2^{k2}-1}\sum_{j=0}^{M_2}d_{n,i,l,j}\psi_{n,i,l,j}(x,t)\right)\left[\sum_{n=0}^{2^{k1}-1}\sum_{i=0}^{M_1}\sum_{l=0}^{2^{k2}-1}\sum_{j=0}^{M_2}d_{n,i,l,j}\frac{\partial^3\psi_{n,i,l,j}(x,t)}{\partial x^3}\right]$$
$$=\sum_{n=0}^{2^{k1}-1}\sum_{i=0}^{M_1}\sum_{l=0}^{2^{k2}-1}\sum_{j=0}^{M_2}c_{n,i,l,j}\psi_{n,i,l,j}(x,t). \tag{5.83}$$

Again applying J_t^α on both sides of Equation 5.74, we have

$$u(x,t)-u(x,0)=J_t^\alpha\left[-45u^2\frac{\partial u}{\partial x}+15p\frac{\partial u}{\partial x}\frac{\partial^2 u}{\partial x^2}+15u\frac{\partial^3 u}{\partial x^3}-\frac{\partial^5 u}{\partial x^5}\right]. \tag{5.84}$$

Putting Equations 5.77 through 5.80 in Equation 5.84, we have

$$\sum_{n=0}^{2^{k_1}-1}\sum_{i=0}^{M_1}\sum_{l=0}^{2^{k_2}-1}\sum_{j=0}^{M_2} d_{n,i,l,j}\psi_{n,i,l,j}(x,t)-u(x,0)=J_t^{\alpha}\left[-45\left(\sum_{n=0}^{2^{k_1}-1}\sum_{i=0}^{M_1}\sum_{l=0}^{2^{k_2}-1}\sum_{j=0}^{M_2} a_{n,i,l,j}\psi_{n,i,l,j}(x,t)\right)\right.$$

$$+15p\left(\sum_{n=0}^{2^{k_1}-1}\sum_{i=0}^{M_1}\sum_{l=0}^{2^{k_2}-1}\sum_{j=0}^{M_2} b_{n,i,l,j}\psi_{n,i,l,j}(x,t)\right)+15\left(\sum_{n=0}^{2^{k_1}-1}\sum_{i=0}^{M_1}\sum_{l=0}^{2^{k_2}-1}\sum_{j=0}^{M_2} c_{n,i,l,j}\psi_{n,i,l,j}(x,t)\right)$$

$$\left.-\left(\sum_{n=0}^{2^{k_1}-1}\sum_{i=0}^{M_1}\sum_{l=0}^{2^{k_2}-1}\sum_{j=0}^{M_2} d_{n,i,l,j}\frac{\partial^5}{\partial x^5}\psi_{n,i,l,j}(x,t)\right)\right]. \tag{5.85}$$

Now substituting the collocation points $x_l=(l-0.5)/(2^{k_1}(M_1+1))$ and $t_r=(r-0.5)/(2^{k_2}(M_2+1))$ for $l=1,2,\ldots,2^{k_1}(M_1+1)$ and $r=1,2,\ldots,2^{k_2}(M_2+1)$ in Equations 5.81 through 5.83 and 5.85, we have $4(2^{k_1}(M_1+1))(2^{k_2}(M_2+1))$ equations in $4(2^{k_1}(M_1+1))(2^{k_2}(M_2+1))$ unknowns in $a_{n,i,l,j}$, $b_{n,i,l,j}$, $c_{n,i,l,j}$, and $d_{n,i,l,j}$. By solving this system of equations using mathematical software, the Legendre multiwavelet coefficients $a_{n,i,l,j}$, $b_{n,i,l,j}$, $c_{n,i,l,j}$, and $d_{n,i,l,j}$ can be obtained.

5.11.2 Comparison with OHAM for a Solution of the Time-Fractional KK Equation

Using OHAM [128], the homotopy for Equation 5.74 can be written as

$$(1-p)L(\varphi(x,t;p))=H(p)\left[\frac{\partial^{\alpha}\varphi(x,t;p)}{\partial t^{\alpha}}+45\varphi(x,t;p)^2\frac{\partial\varphi(x,t;p)}{\partial x}\right.$$

$$\left.-15p\frac{\partial\varphi(x,t;p)}{\partial x}\frac{\partial^2\varphi(x,t;p)}{\partial x^2}-15\varphi(x,t;p)\frac{\partial^3\varphi(x,t;p)}{\partial x^3}+\frac{\partial^5\varphi(x,t;p)}{\partial x^5}\right], \tag{5.86}$$

where

$$\varphi(x,t;p)=u_0(x,t)+\sum_{i=1}^{\infty}u_i(x,t)p^i \tag{5.87}$$

$$H(p)=C_1p+C_2p^2+C_3p^3+\cdots \tag{5.88}$$

$$N(\varphi(x,t,p))=N_0(u_0(x,t))+\sum_{k=1}^{\infty}N_k(u_0,u_1,\ldots,u_k)p^k. \tag{5.89}$$

Substituting Equations 5.87 through 5.89 in Equation 5.86 and equating the coefficients of different powers in p, we have the following system of PDEs:

$$\text{Coefficients of } p^0:\ \frac{\partial^{\alpha}u_0(x,t)}{\partial t^{\alpha}}=0 \tag{5.90}$$

$$\text{Coefficients of } p^1: \frac{\partial^\alpha u_1(x,t)}{\partial t^\alpha} - \frac{\partial^\alpha u_0(x,t)}{\partial t^\alpha} = C_1\left[\frac{\partial^\alpha u_0(x,t)}{\partial t^\alpha} + 45(u_0(x,t))^2\frac{\partial u_0(x,t)}{\partial x}\right.$$

$$\left. -15p\frac{\partial u_0(x,t)}{\partial x}\frac{\partial^2 u_0(x,t)}{\partial x^2} - 15u_0(x,t)\frac{\partial^3 u_0(x,t)}{\partial x^3} + \frac{\partial^5 u_0(x,t)}{\partial x^5}\right] \tag{5.91}$$

$$\text{Coefficients of } p^2: \frac{\partial^\alpha u_2(x,t)}{\partial t^\alpha} - \frac{\partial^\alpha u_1(x,t)}{\partial t^\alpha}$$

$$= C_1\left[\frac{\partial^\alpha u_1(x,t)}{\partial t^\alpha} + 45\left(2u_0(x,t)u_1(x,t)\frac{\partial u_0(x,t)}{\partial x} + (u_0(x,t))^2\frac{\partial u_1(x,t)}{\partial x}\right)\right.$$

$$-15p\left(\frac{\partial u_0(x,t)}{\partial x}\frac{\partial^2 u_1(x,t)}{\partial x^2} + \frac{\partial u_1(x,t)}{\partial x}\frac{\partial^2 u_0(x,t)}{\partial x^2}\right)$$

$$\left.-15\left(u_0(x,t)\frac{\partial^3 u_1(x,t)}{\partial x^3} + u_1(x,t)\frac{\partial^3 u_0(x,t)}{\partial x^3}\right) + \frac{\partial^5 u_1(x,t)}{\partial x^5}\right]$$

$$+C_2\left[\frac{\partial^\alpha u_0(x,t)}{\partial t^\alpha} + 45(u_0(x,t))^2\frac{\partial u_0(x,t)}{\partial x}\right.$$

$$-15p\frac{\partial u_0(x,t)}{\partial x}\frac{\partial^2 u_0(x,t)}{\partial x^2}$$

$$\left.-15u_0(x,t)\frac{\partial^3 u_0(x,t)}{\partial x^3} + \frac{\partial^5 u_0(x,t)}{\partial x^5}\right] \tag{5.92}$$

and so on.

For solving a fractional KK equation using OHAM, we consider the following initial condition for Equation 5.74:

$$u(x,0) = \frac{1}{4}w^2\lambda^2\text{sech}^2\left(\frac{\lambda w x}{2}\right) + \frac{w^2\lambda^2}{12}.$$

Using the initial condition $u_0 = u(x,0)$ and solving Equations 5.90 through 5.92, we obtain the expressions for u_0, u_1, and u_2.

Finally, the second-order approximate solution for time-fractional KK equation is given by

$$u = u_0(x,t) + u_1(x,t) + u_2(x,t). \tag{5.93}$$

The optimal values of the convergence control parameters C_1 and C_2 can be obtained by using the collocation method given in Equation 1.33.

5.12 Numerical Results of the Fractional KK Equation

The comparison of the absolute errors for the time-fractional KK equation (5.74) has been exhibited in Tables 5.19 and 5.22, which are constructed using the results obtained by two-dimensional Legendre multiwavelet method and OHAM at different values of x and t,

TABLE 5.19

Comparison of Absolute Errors Obtained by the Two-Dimensional Legendre Multiwavelet Method for the Nonlinear KK Equation Given in Equation 5.74 at Various Points of x and t, Taking $\lambda=0.1$, $\mu=0$, $w=1$, and $\alpha=1$

	$\lvert u_{Exact}-u_{Legendre\ Multiwavelet}\rvert$								
x	$t=0.1$	$t=0.2$	$t=0.3$	$t=0.4$	$t=0.5$	$t=0.6$	$t=0.7$	$t=0.8$	$t=0.9$
0.1	3.5268E−10	7.0333E−10	1.0519E−9	1.3985E−9	1.7430E-9	2.0855E−9	2.4259E−9	2.7643E−9	3.1007E−9
0.2	7.0308E−10	1.4041E−9	2.1031E−9	2.8001E−9	3.4950E−9	4.1879E−9	4.8788E−9	5.5676E−9	6.2544E−9
0.3	1.0532E−9	2.1043E−9	3.1535E−9	4.2006E−9	5.2456E−9	6.2887E−9	7.3297E−9	8.3687E−9	9.4057E−9
0.4	1.4028E−9	2.8037E−9	4.2025E−9	5.5994E−9	6.9942E−9	8.3869E−9	9.7777E−9	1.1166E−8	1.2553E−8
0.5	1.7520E−9	3.5020E−9	5.2500E−9	6.9959E−9	8.7399E−9	1.0482E−8	1.2222E−8	1.3959E−8	1.5695E−8
0.6	2.1004E−9	4.1988E−9	6.2953E−9	8.3897E−9	1.0482E−8	1.2572E−8	1.4661E−8	1.6747E−8	1.8832E−8
0.7	2.4480E−9	4.8941E−9	7.3381E−9	9.7802E−9	1.2220E−8	1.4658E−8	1.7094E−8	1.9528E−8	2.1960E−8
0.8	2.7946E−9	5.5873E−9	8.3780E−9	1.1166E−8	1.3953E−8	1.6738E−8	1.9521E−8	2.2302E−8	2.5081E−8
0.9	3.1402E−9	6.2783E−9	9.4146E−9	1.2548E−8	1.5881E−8	1.8811E−8	2.1939E−8	2.5066E−8	2.8191E−8

TABLE 5.20

Comparison of Absolute Errors Obtained by Two-Dimensional Legendre Multiwavelet Method for Fractional Order Nonlinear KK Equation Given in Equation 5.74 at Various Points of x and t Taking $\lambda=0.1$, $\mu=0$, $w=1$ and $\alpha=0.75$

	$\|u_{Exact}-u_{Legendre\ Multiwavelet}\|$								
x	$t=0.1$	$t=0.2$	$t=0.3$	$t=0.4$	$t=0.5$	$t=0.6$	$t=0.7$	$t=0.8$	$t=0.9$
0.1	6.7734E−10	1.1369E−9	1.5349E−9	1.8967E−9	2.2340E−9	2.5523E−9	2.8551E−9	3.1456E−9	3.4253E−9
0.2	1.3533E−9	2.2769E−9	3.0806E−9	3.8142E−9	4.5010E−9	5.1516E−9	5.7729E−9	6.3709E−9	6.9489E−9
0.3	2.0287E−9	3.4161E−9	4.6251E−9	5.7303E−9	6.7663E−9	7.7489E−9	8.6884E−9	9.5937E−9	1.0469E−8
0.4	2.7033E−9	4.5538E−9	6.1677E−9	7.6441E−9	9.0289E−9	1.0343E−8	1.1600E−8	1.2812E−8	1.3986E−8
0.5	3.3768E−9	5.6898E−9	7.7079E−9	9.5548E−9	1.1287E−8	1.2933E−8	1.4507E−8	1.6026E−8	1.7497E−8
0.6	4.0490E−9	6.8234E−9	9.2450E−9	1.1462E−8	1.3542E−8	1.5518E−8	1.7409E−8	1.9234E−8	2.1001E−8
0.7	4.7196E−9	7.9544E−9	1.0778E−8	1.3364E−8	1.5791E−8	1.8096E−8	2.0304E−8	2.2434E−8	2.4497E−8
0.8	5.3883E−9	9.0822E−9	1.2307E−8	1.5261E−8	1.8034E−8	2.0668E−8	2.3191E−8	2.5624E−8	2.7983E−8
0.9	6.0548E−9	1.0206E−8	1.3832E−8	1.7152E−8	2.0269E−8	2.3232E−8	2.6068E−8	2.8805E−8	3.1458E−8

TABLE 5.21

Comparison of Absolute Errors Cbtained by the Two-Dimensional Legendre Multiwavelet Method for the Fractional Order Nonlinear KK Equation Given in Equation 5.74 at Various Points of x and t, Taking $\lambda = 0.1$, $\mu = 0$, $w = 1$, and $\alpha = 0.5$

	$\lvert u_{Exact} - u_{Legendre\ Multiwavelet} \rvert$								
x	$t=0.1$	$t=0.2$	$t=0.3$	$t=0.4$	$t=0.5$	$t=0.6$	$t=0.7$	$t=0.8$	$t=0.9$
0.1	1.2348E−9	1.7431E−9	2.1251E−9	2.4420E−9	2.7198E−9	2.9690E−9	3.1958E−9	3.4065E−9	3.6035E−9
0.2	2.4789E−9	3.5107E−9	4.2915E−9	4.9425E−9	5.5158E−9	6.0321E−9	6.5038E−9	6.9435E−9	7.3559E−9
0.3	3.7221E−9	5.2770E−9	6.4561E−9	7.4411E−9	8.3096E−9	9.0928E−9	9.8093E−9	1.0477E−8	1.1105E−8
0.4	4.9638E−9	7.0412E−9	8.6182E−9	9.9366E−9	1.1100E−8	1.2149E−8	1.3111E−8	1.4007E−8	1.4850E−8
0.5	6.2035E−9	8.8026E−9	1.0776E−8	1.2428E−8	1.3886E−8	1.5202E−8	1.6407E−8	1.7532E−8	1.8589E−8
0.6	7.4407E−9	1.0560E−8	1.2931E−8	1.4915E−8	1.6667E−8	1.8248E−8	1.9697E−8	2.1049E−8	2.2321E−8
0.7	8.6750E−9	1.2314E−8	1.5080E−8	1.7395E−8	1.9440E−8	2.1287E−8	2.2979E−8	2.4558E−8	2.6044E−8
0.8	9.9058E−9	1.4063E−8	1.7223E−8	1.9869E−8	2.2209E−8	2.4317E−8	2.6251E−8	2.8058E−8	2.9756E−8
0.9	1.1132E−8	1.5806E−8	1.9359E−8	2.2335E−8	2.4964E−8	2.7338E−8	2.9514E−8	3.1546E−8	3.3457E−8

TABLE 5.22

Comparison of Absolute Errors Obtained by OHAM for the Nonlinear KK Equation Given in Equation 5.74 at Various Points of x and t, Taking $\lambda = 0.1$, $\mu = 0$, $w = 1$, and $\alpha = 1$

	$\lvert u_{Exact} - u_{OHAM} \rvert$								
x	$t = 0.1$	$t = 0.2$	$t = 0.3$	$t = 0.4$	$t = 0.5$	$t = 0.6$	$t = 0.7$	$t = 0.8$	$t = 0.9$
0.1	3.4968E−10	3.6511E−9	6.5846E−9	9.1501E−9	1.1377E−8	1.3177E−8	1.4638E−8	1.5732E−8	1.6457E−8
0.2	7.2934E−6	7.2553E−6	7.2176E−6	7.1802E−6	7.1432E−6	7.1065E−6	7.0701E−6	7.0341E−6	6.9983E−6
0.3	2.6793E−5	2.6721E−5	2.6651E−5	2.6581E−5	2.6511E−5	2.6442E−5	2.6372E−5	2.6303E−5	2.6235E−5
0.4	5.8103E−5	5.8005E−5	5.7906E−5	5.7807E−5	5.7709E−5	5.7611E−5	5.7513E−5	5.7426E−5	5.7330E−5
0.5	1.0061E−4	1.0049E−4	1.0037E−4	1.0024E−4	1.0013E−4	1.0001E−4	9.9885E−5	9.9976E−5	9.9643E−5
0.6	1.5350E−4	1.5336E−4	1.5323E−4	1.5309E−4	1.5295E−4	1.5281E−4	1.5268E−4	1.5254E−4	1.5240E−4
0.7	2.1579E−4	2.1564E−4	2.1549E−4	2.1535E−4	2.1520E−4	2.1506E−4	2.1491E−4	2.1476E−4	2.1461E−4
0.8	2.8635E−4	2.8621E−4	2.8606E−4	2.8591E−4	2.8576E−4	2.8561E−4	2.8546E−4	2.8531E−4	2.8516E−4
0.9	3.6399E−4	3.6384E−4	3.6370E−4	3.6355E−4	3.6341E−4	3.6326E−4	3.6312E−4	3.6297E−4	3.6282E−4

taking $\alpha = 1$. In the present analysis, to examine the accuracy and reliability of the Legendre multiwavelets for solving a fractional order KK equation, we compare the approximate solution of Legendre multiwavelets with the exact solution as well as with second-order approximate solution obtained by OHAM. Tables 5.20, 5.21, 5.23, and 5.24 show the comparison of absolute errors of the fractional order KK equation (5.74) at various points of x and t, taking $\alpha = 0.5$ and 0.75. Agreement between present numerical results obtained by Legendre multiwavelets and exact solutions appears very satisfactory through illustrations in Tables 5.19 through 5.21.

It may be observed from Tables 5.19 through 5.24 that absolute errors for the Legendre multiwavelet method are lesser than those of OHAM. Moreover, L_2 and L_∞ error norms confirm that the Legendre multiwavelet method provides more accurate and better solution than OHAM.

5.13 Conclusion

In this chapter, the two-dimensional Legendre wavelet method has been successfully implemented to obtain the numerical solution of a fractional order parabolic PDE subject to Dirichlet boundary conditions, fractional KBK equation, and fractional sKdV equation. In the case of a fractional order PDE with Dirichlet boundary conditions, the acquired numerical results of Legendre wavelet methods are compared with exact solutions obtained by HPM as well as with the numerical solution of the Haar wavelet method. These results have been cited in the tables in order to justify the accuracy and efficiency of the proposed schemes. For a fractional order parabolic PDE, the Legendre wavelet method provides more accurate results than the Haar wavelet method as shown in Tables 5.1 through 5.5. In the case of a fractional KBK equation, the obtained results are compared with exact solutions. Agreement between present numerical results and exact solutions appear very satisfactory through illustrations in Tables 5.6 through 5.10. The obtained results demonstrate the efficiency, accuracy, and reliability of the proposed algorithm based on the two-dimensional Legendre wavelet method and its applicability to a nonlinear time-fractional KBK equation.

The fractional order sKdV equation has been solved by using the two-dimensional Legendre wavelet method. The results are compared with exact solutions and also with HAM and OHAM solutions. Tables 5.11 through 5.18 illustrate a pretty good agreement between present numerical results obtained by the Legendre wavelet method with HAM and exact solutions. The present scheme is very simple, effective, and convenient for obtaining numerical solutions of a nonlinear time-fractional sKdV equation.

Next, the fractional KK equation has been solved numerically by using two-dimensional Legendre multiwavelet method and OHAM. The results obtained by the Legendre multiwavelet method are then compared with exact solutions as well as with OHAM. One can observe a pretty good agreement between present numerical results obtained by Legendre multiwavelet method with OHAM and exact solutions through illustrated results in Tables 5.19 through 5.25. The obtained results demonstrate the accuracy, efficiency, and reliability of the proposed algorithm based on the two-dimensional Legendre multiwavelet method and its applicability to a nonlinear time-fractional KK equation. It may be observed from Tables 5.19 through 5.25 that absolute errors for the Legendre multiwavelet method are lesser than those of OHAM. Moreover, L_2 and L_∞ error norms confirm that the Legendre

TABLE 5.23

Comparison of Absolute Errors Obtained by OHAM for the Fractional Order Nonlinear KK Equation Given in Equation 5.74 at Various Points of x and t, Taking $\lambda=0.1$, $\mu=0$, $w=1$, and $\alpha=0.75$

	$\lvert u_{Exact}-u_{OHAM}\rvert$								
x	$t=0.1$	$t=0.2$	$t=0.3$	$t=0.4$	$t=0.5$	$t=0.6$	$t=0.7$	$t=0.8$	$t=0.9$
0.1	6.7141E−10	3.7954E−9	5.8412E−9	7.4404E−9	8.7416E−9	9.8151E−9	1.0702E−8	1.1428E−8	1.2015E−8
0.2	7.2899E−6	7.2528E−6	7.2254E−6	7.2014E−6	7.1794E−6	7.1589E−6	7.1395E−6	7.1209E−6	7.1032E−6
0.3	2.6785E−5	2.6716E−5	2.6665E−5	2.6620E−5	2.6578E−5	2.6540E−5	2.6503E−5	2.6467E−5	2.6434E−5
0.4	5.8094E−5	5.7998E−5	5.7926E−5	5.7862E−5	5.7804E−5	5.7749E−5	5.7697E−5	5.7647E−5	5.7599E−5
0.5	1.0060E−4	1.0048E−4	1.0039E−4	1.0031E−4	1.0024E−4	1.0017E−4	1.0011E−4	1.0005E−4	9.9992E−5
0.6	1.5349E−4	1.5335E−4	1.5326E−4	1.5317E−4	1.5308E−4	1.5301E−4	1.5293E−4	1.5286E−4	1.5279E−4
0.7	2.1577E−4	2.1563E−4	2.1552E−4	2.1543E−4	2.1534E−4	2.1526E−4	2.1518E−4	2.1511E−4	2.1503E−4
0.8	2.8634E−4	2.8619E−4	2.8608E−4	2.8599E−4	2.8590E−4	2.8582E−4	2.8574E−4	2.8566E−4	2.8559E−4
0.9	3.6397E−4	3.6384E−4	3.6373E−4	3.6364E−4	3.6355E−4	3.6347E−4	3.6339E−4	3.6331E−4	3.6324E−4

TABLE 5.24

Comparison of Absolute Errors Obtained by OHAM for the Fractional Order Nonlinear KK Equation Given in Equation 5.74 at Various Points of x and t, Taking $\lambda=0.1$, $\mu=0$, $w=1$, and $\alpha=0.5$

	$\lvert u_{Exact}-u_{OHAM}\rvert$								
x	$t=0.1$	$t=0.2$	$t=0.3$	$t=0.4$	$t=0.5$	$t=0.6$	$t=0.7$	$t=0.8$	$t=0.9$
0.1	1.2175E−9	3.7606E−9	4.9831E−9	5.8529E−9	6.5343E−9	7.0939E−9	7.5661E−8	7.9718E−9	8.3247E−9
0.2	7.2836E−6	7.2523E−6	7.2354E−6	7.2223E−6	7.2112E−6	7.2015E−6	7.1927E−6	7.1846E−6	7.1771E−6
0.3	2.6773E−5	2.6715E−5	2.6683E−5	2.6659E−5	2.6638E−5	2.6620E−5	2.6603E−5	2.6588E−5	2.6573E−5
0.4	5.8078E−5	5.7996E−5	5.7952E−5	5.7917E−5	5.7888E−5	5.7862E−5	5.7839E−5	5.7817E−5	5.7797E−5
0.5	1.0058E−4	1.0048E−4	1.0042E−4	1.0038E−4	1.0034E−4	1.0031E−4	1.0028E−4	1.0026E−4	1.0023E−4
0.6	1.5346E−4	1.5335E−4	1.5329E−4	1.5324E−4	1.5320E−4	1.5316E−4	1.5313E−4	1.5310E−4	1.5307E−4
0.7	2.1575E−4	2.1563E−4	2.1556E−4	2.1551E−4	2.1547E−4	2.1543E−4	2.1539E−4	2.1536E−4	2.1533E−4
0.8	2.8631E−4	2.8619E−4	2.8612E−4	2.8607E−4	2.8603E−4	2.8599E−4	2.8595E−4	2.8592E−4	2.8589E−4
0.9	3.6395E−4	3.6383E−4	3.6376E−4	3.6371E−4	3.6367E−4	3.6363E−4	3.6360E−4	3.6356E−4	3.6354E−4

TABLE 5.25

L_2 and L_∞ Error Norms for the Nonlinear Time-Fractional KK Equation Using the Two-Dimensional Legendre Multiwavelet Method and OHAM at Various Points x, Taking $\alpha = 0.5$, 0.75, and 1

	Error Analysis with Regard to Legendre Multiwavelet						Error Analysis with Regard to OHAM					
	$\alpha=1$		$\alpha=0.75$		$\alpha=0.5$		$\alpha=1$		$\alpha=0.75$		$\alpha=0.5$	
x	L_2	L_∞	L_2	L_∞	L_2	L_∞	L_2	L_∞	L_2	L_∞	L_2	L_∞
0.1	5.8488E−9	3.1007E−9	6.9992E−9	3.4253E−9	8.1275E−9	3.6035E−9	3.4295E−8	1.6457E−8	2.5818E−8	7.2899E−6	7.7622E−8	8.3247E−9
0.2	1.1764E−8	6.2544E−9	1.4079E−8	6.9489E−9	1.6515E−8	7.3559E−9	2.1434E−5	7.2934E−6	2.1557E−5	1.2015E−8	2.1653E−5	7.2836E−6
0.3	1.7675E−8	9.4057E−9	2.1283E−8	1.0469E−8	2.4895E−8	1.1105E−8	7.9538E−5	2.6793E−5	7.9770E−5	2.6785E−5	7.9951E−5	2.6773E−5
0.4	2.3578E−8	1.2553E−8	2.8413E−8	1.3986E−8	3.3266E−8	1.4850E−8	1.7313E−4	5.8103E−5	1.7345E−4	5.8094E−5	1.7372E−4	5.8078E−5
0.5	2.9473E−8	1.5695E−8	3.5532E−8	1.7497E−8	4.1624E−8	1.8589E−8	3.0045E−4	1.0061E−4	3.0045E−4	1.0060E−4	3.0109E−4	1.0058E−4
0.6	3.2498E−8	1.8832E−8	4.2638E−8	2.1001E−8	4.9965E−8	2.2321E−8	4.5885E−4	1.5350E−4	4.5931E−4	1.5349E−4	4.5967E−4	1.5346E−4
0.7	4.1224E−8	2.1960E−8	4.9727E−8	2.4497E−8	5.8286E−8	2.6044E−8	6.4560E−4	2.1579E−4	6.4609E−4	2.1577E−4	6.4647E−4	2.1575E−4
0.8	4.7077E−8	2.5081E−8	5.6795E−8	2.7983E−8	6.6585E−8	2.9756E−8	8.5727E−4	2.8635E−4	8.5777E−4	2.8634E−4	8.5815E−4	2.8631E−4
0.9	5.2969E−8	2.8191E−8	6.3842E−8	3.1458E−8	7.6494E−8	3.3457E−8	1.0902E−3	3.6399E−4	1.0907E−3	3.6397E−4	1.0911E−3	3.6395E−4

multiwavelet method provides more accurate and better solution than OHAM. Thus, the Legendre multiwavelet method provides more accurate and better solution in comparison to OHAM. The application of the proposed numerical method based on the two-dimensional Legendre multiwavelet method for the solutions of a time-fractional KK equation satisfactorily justifies its simplicity, efficiency, and applicability. The present numerical scheme is quite simple, effective, and expedient for obtaining the numerical solution of a fractional KK equation in comparison to the analytical approach of OHAM.

6

Application of Chebyshev Wavelet Methods for Numerical Simulation of Fractional Differential Equations

6.1 Introduction

Nowadays, Chebyshev polynomials have become more significant in numerical evaluation. Among the four forms of Chebyshev polynomials, the first and second kinds are certain cases of the symmetric Jacobi polynomials, whereas the third and fourth kinds are unique instances of the nonsymmetric Jacobi polynomials. Much attention has been focused on the first and second kinds of Chebyshev polynomials $T_n(x)$ and $U_n(x)$ and their various uses in numerous applications. Nevertheless, there are very few articles that concentrate on the wavelets shaped through these two types of Chebyshev polynomials for application in fractional partial differential equations. This motivates our curiosity in such wavelets. In this chapter, our aim is to study the application of Chebyshev wavelets for the solution of fractional order differential equations. Moreover, the Chebyshev wavelets are competent for solving some fractional and integral equations [45,46].

There are several advantages of using Chebyshev wavelets approximations based on the collocation spectral method. First, unlike most numerical methods, it is now conventional that they are characterized by the use of exponentially decaying errors. Second, various numerical methods do not perform well near singularities, whereas approximations through wavelets effectively handle singularities in the problem. Also, due to their fast convergence, the Chebyshev wavelets method does not face the instability problems related with other numerical methods.

In this chapter, fractional order partial differential equations comprising the Caputo fractional derivative and the Riesz fractional derivative are considered. This chapter is devoted to the study of the application of Chebyshev wavelets for the numerical solution of fractional differential equations involving Caputo and Riesz fractional derivatives. The prime focus of the present chapter is to implement the two-dimensional Chebyshev wavelet technique for solving nonlinear fractional differential equations like the time-fractional Sawada-Kotera (SK) equation, Riesz fractional Camassa–Holm (CH) equation, and Riesz fractional sine-Gordon equation (SGE) in order to demonstrate the efficiency and accuracy of the proposed method.

6.2 Outline of the Present Study

Consider the time-fractional generalized fifth-order SK equation [129]:

$$\frac{\partial^\alpha u}{\partial t^\alpha}+45u^2\frac{\partial u}{\partial x}+15\frac{\partial u}{\partial x}\frac{\partial^2 u}{\partial x^2}+15\frac{\partial^3 u}{\partial x^3}u+\frac{\partial^5 u}{\partial x^5}=0,\quad t>0,\ x>0, \tag{6.1}$$

which is the variation of the fifth-order SK equation [49,130–132]. Here, $0<\alpha\leq 1$ is the parameter describing the order of the fractional time derivative. The fractional derivative is considered in the Caputo sense.

The classical SK equation is an important mathematical model arising in many different physical contexts to describe the motion of long waves in shallow water under gravity and in a one-dimensional nonlinear lattice and has wide applications in quantum mechanics and nonlinear optics. It is well known that wave phenomena of plasma media and fluid dynamics are modeled by kink-shaped tanh solution or by bell-shaped sech solutions. This equation is also used in modeling waves that propagate in opposite directions. There are a lot of studies for the classical SK equation, and some profound results have been established. But according to the best possible information of the authors, the detailed study of the nonlinear fractional order SK equation is only beginning.

Consider the following CH equation with a Riesz time-fractional derivative [133]:

$${}_0^R D_t^\alpha u(x,t)+2ku_x(x,t)-u_{xxt}(x,t)+3u(x,t)u_x(x,t)-2u_x(x,t)u_{xx}(x,t)-u(x,t)u_{xxx}(x,t)=0 \tag{6.2}$$

where

$k\neq 0$ is a constant

$u(x,t)$ is the unknown function depending on the temporal variable t and spatial variable x

${}_0^R D_t^\alpha$ is the Riesz fractional derivative

The CH equation is used to describe the physical model for the unidirectional propagation of waves in shallow water [134,135]. This equation is widely used in fluid dynamics, continuum mechanics, aerodynamics, and models for shock wave formation, solitons, turbulence, mass transport, and the solution representing the water's free surface above a flat bottom [136,137]. The CH equation has been obtained by Fokas and Fuchssteiner [138] and Lenells [139]. Camassa and Holm [140] put forward the derivation of the solution as a model for dispersive shallow water waves and revealed that it is a formally integrable finite-dimensional Hamiltonian system and its solitary waves are solitons. The classical CH equation has attracted much research interest in recent years both from analytical and numerical point of view, and some exhaustive results have been established. The intention of the present work is to perform the two-dimensional Chebyshev wavelet technique in order to exhibit the competency of this method for the numerical solution of nonlinear CH equations with the Riesz time-fractional derivative.

Next, a numerical process involving the Chebyshev wavelet method has been implemented for computing the approximate solution of a Riesz space-fractional SGE. The fractional SGE is considered as an interpolation between the classical SGE (corresponding to $\alpha=2$) and nonlocal SGE (corresponding to $\alpha=1$).

Consider the Riesz space-fractional SGE proposed in [141,142] as follows:

$$u_{tt}-{}^R D_x^\alpha u+\sin u=0,\quad 1<\alpha\leq 2, \tag{6.3}$$

where ${}^{R}D_x^{\alpha}$ denotes the Riesz fractional derivative. Equation 6.3 is the variation of the SGE. The classical SGE, an elementary equation of modern nonlinear wave theory, has accomplished great repute primarily due to its substantiated applications in diverse fields of science and engineering. It arises in various disciplines of physics, such as propagation of magnetic flux on Josephson junctions, propagation of optical pulses in resonant laser media, field theory, sound propagation in a crystal lattice, in ferromagnetism and ferroelectric substances, and in nonlinear optics [132,143]. In these applications, the SGE provides the simplest nonlinear description of physical phenomena in different configurations. The more adequate modeling can be prevailed corresponding to the generalization of classical SGE. In particular, taking into account of nonlinear effects corresponding to long-range interactions of particles, the complex law of medium dispersion or curvilinear geometry of the initial boundary problem, classical SGE results in nonlocal generalization of SGE.

Various methods such as the homotopy analysis method (HAM) [144], modified decomposition method [145], variational iteration method (VIM) [146], and tanh method [147] have been implemented to evaluate an approximate analytical solution of the classical SGE. Using numerical experiments, Ablowitz et al. [148] examined the numerical behavior of a double-discrete, completely integrable discretization of the SGE. Herbst and Ablowitz [149] provided the numerical results of the SGE acquired by the use of the explicit symplectic method. However, the comprehensive study of the fractional SGE is only the beginning. In this work, we will consider fractional SGE with a number of initial values. The motivation of the present work is to establish the Chebyshev wavelet method as a powerful tool for solving the Riesz fractional SGE.

6.3 Formulation of a Time-Fractional SK Equation

Consider the following generalized SK equation:

$$\frac{\partial u}{\partial t}+au^2\frac{\partial u}{\partial x}+b\frac{\partial u}{\partial x}\frac{\partial^2 u}{\partial x^2}+c\frac{\partial^3 u}{\partial x^3}u+d\frac{\partial^5 u}{\partial x^5}=0, \tag{6.4}$$

where

a, b, c, and d are constants

$u(x,t)$ is a field variable

$x\in\Omega$ is a space coordinate in the propagation direction of the field

$t\in\Gamma$ is the time

Employing a potential function, $v(x,t)$, on the field variable and setting $u(x,t)=v_x(x,t)$ yields the potential equation of the SK equation (6.4) in the form

$$v_{xt}(x,t)+av_x^2(x,t)v_{xx}(x,t)+bv_{xx}(x,t)v_{xxx}(x,t)+cv_{xxxx}(x,t)v_x(x,t)+dv_{6x}(x,t)=0. \tag{6.5}$$

The functional of the potential equation (6.5) can be represented as

$$\begin{aligned}J(v)=\int_{\Omega}dx\int_{\Gamma}\Big[v(x,t)\big[&c_1v_{xt}(x,t)+c_2av_x^2(x,t)v_{xx}(x,t)+c_3bv_{xx}(x,t)v_{xxx}(x,t)\\&+c_4cv_{xxxx}(x,t)v_x(x,t)+c_5dv_{6x}(x,t)\big]\Big]dt\end{aligned} \tag{6.6}$$

where c_1, c_2, c_3, c_4, and c_5 are unknown constants to be determined. Integrating Equation 6.6 by parts and taking $v_t|_\Omega = v_x|_\Omega = v_{xx}|_\Omega = v_{xxx}|_\Omega = v_{5x}|_\Omega = 0$ yields

$$J(v) = \int_\Omega dx \int_\Gamma \Big[-c_1 v_t(x,t) v_x(x,t) - \frac{c_2 a}{3} v_x^4(x,t) + \left(\frac{c_4 c}{2} - \frac{b c_3}{2} \right) v_x(x,t) v_{xx}^2(x,t) \\ - c_4 c v_x^2(x,t) v_{xxx}(x,t) + c_5 d v_x(x,t) v_{5x}(x,t) \Big] dt. \tag{6.7}$$

The constants c_1, c_2, c_3, c_4, and c_5 can be determined, taking the variation of the functional (6.7) to make it optimal. Integrating each term by parts and using the variation optimum condition of the functional $J(v)$, the following expression can be obtained:

$$2c_1 v_{xt}(x,t) + 4c_2 a v_x^2(x,t) v_{xx}(x,t) + c_3 b v_{xx}(x,t) v_{xxx}(x,t) \\ - 2c_4 c v_{xxxx}(x,t) v_x(x,t) - 2c_5 d v_{6x}(x,t) = 0. \tag{6.8}$$

Comparing the obtained result in Equation 6.8 with the equivalent Equation 6.5, we get

$$c_1 = \frac{1}{2}, \quad c_2 = \frac{1}{4}, \quad c_3 = 1, \quad c_4 = -\frac{1}{2}, \quad \text{and} \quad c_5 = -\frac{1}{2}.$$

The functional expression given by Equation 6.7 obtains directly the Lagrangian form of the SK equation:

$$L(v_t, v_x, v_{xx}, v_{xxx}, v_{5x}) = -\frac{1}{2} v_t(x,t) v_x(x,t) - \frac{a}{12} v_x^4(x,t) - \left(\frac{2b+c}{4} \right) v_x(x,t) v_{xx}^2(x,t) \\ + \frac{c}{2} v_x^2(x,t) v_{xxx}(x,t) - \frac{d}{2} v_x(x,t) v_{5x}(x,t). \tag{6.9}$$

Similarly, the Lagrangian of the time-fractional SK equation can be written as

$$F\left({}_0D_t^\alpha v, v_x, v_{xx}, v_{xxx}, v_{5x}\right) = -\frac{1}{2} {}_0D_t^\alpha v(x,t) v_x(x,t) - \frac{a}{12} v_x^4(x,t) \\ - \left(\frac{2b+c}{4} \right) v_x(x,t) v_{xx}^2(x,t) + \frac{c}{2} v_x^2(x,t) v_{xxx}(x,t) - \frac{d}{2} v_x(x,t) v_{5x}(x,t). \tag{6.10}$$

Then the functional of the time-fractional SK equation will be

$$J(v) = \int_\Omega dx \int_\Gamma F\left({}_0D_t^\alpha v, v_x, v_{xx}, v_{xxx}, v_{5x}\right) dt.$$

The variation of above functional with respect to $v(x,t)$ leads to

$$\delta J(v) = \int_\Omega dx \int_\Gamma \Big[\frac{\partial F}{\partial {}_0D_t^\alpha v} \delta\left({}_0D_t^\alpha v(x,t)\right) + \frac{\partial F}{\partial v_x} \delta v_x(x,t) + \frac{\partial F}{\partial v_{xx}} \delta v_{xx}(x,t) \\ + \frac{\partial F}{\partial v_{xxx}} \delta v_{xxx}(x,t) + \frac{\partial F}{\partial v_{5x}} \delta v_{5x}(x,t) \Big] dt. \tag{6.11}$$

By Lemma 4.1, upon integrating the right-hand side of Equation 6.11, we have

$$\delta J(v) = \int_{\Omega} dx \int_{\Gamma} \left[{}_tD_T^{\alpha}\left(\frac{\partial F}{\partial_0 D_t^{\alpha} v}\right) - \frac{\partial}{\partial x}\left(\frac{\partial F}{\partial v_x}\right) + \frac{\partial^2}{\partial x^2}\left(\frac{\partial F}{\partial v_{xx}}\right) - \frac{\partial^3}{\partial x^3}\left(\frac{\partial F}{\partial v_{xxx}}\right) - \frac{\partial^5}{\partial x^5}\left(\frac{\partial F}{\partial v_{5x}}\right)\right] \delta v(x,t)\,dt. \tag{6.12}$$

Optimizing the variation of the functional $J(v)$, that is, $\delta J(v) = 0$, yields the Euler–Lagrange equation for the time-fractional SK equation in the following expression:

$${}_tD_T^{\alpha}\left(\frac{\partial F}{\partial_0 D_t^{\alpha} v}\right) - \frac{\partial}{\partial x}\left(\frac{\partial F}{\partial v_x}\right) + \frac{\partial^2}{\partial x^2}\left(\frac{\partial F}{\partial v_{xx}}\right) - \frac{\partial^3}{\partial x^3}\left(\frac{\partial F}{\partial v_{xxx}}\right) - \frac{\partial^5}{\partial x^5}\left(\frac{\partial F}{\partial v_{5x}}\right) = 0. \tag{6.13}$$

Substituting the Lagrangian of the time-fractional SK equation (6.10) into the Euler–Lagrange formula (6.13) yields

$$-{}_tD_T^{\alpha} v_x(x,t) + {}_0D_t^{\alpha} v_x(x,t) + 2a v_x^2(x,t) v_{xx}(x,t) - (4b + 10c) v_{xx}(x,t) v_{xxx}(x,t)$$
$$-(2b + 3c) v_{xxxx}(x,t) v_x(x,t) + 2d v_{6x}(x,t) = 0.$$

Substituting the potential function $v_x(x,t) = u(x,t)$ and taking $a = 45/2$, $b = -105/8$, $c = 15/4$, and $d = 1/2$ yields

$$-{}_tD_T^{\alpha} u(x,t) + {}_0D_t^{\alpha} u(x,t) + 45u^2(x,t) u_x(x,t) + 15u_x(x,t) u_{xx}(x,t)$$
$$+ 15u_{xxx}(x,t) u(x,t) + u_{5x}(x,t) = 0. \tag{6.14}$$

The right-hand side Riemann–Liouville fractional derivative is interpreted as a future state of the process in physics. So the right derivative is generally neglected in applications, when the present state of the process does not depend on the results of the future development, and so the right derivative in Equation 6.14 can be neglected. Hence, the time-fractional SK equation can be represented by

$${}_0D_t^{\alpha} u(x,t) + 45u^2(x,t) u_x(x,t) + 15u_x(x,t) u_{xx}(x,t) + 15u_{xxx}(x,t) u(x,t) + u_{5x}(x,t) = 0. \tag{6.15}$$

6.4 Application of Analytical and Numerical Methods for Solving a Fractional SK Equation

6.4.1 Implementation of Chebyshev Wavelet on a Time-Fractional SK Equation

To show the effectiveness and accuracy of the proposed scheme, we consider the time-fractional SK equation. The numerical solutions thus obtained are compared with the exact solutions for a classical case and also solutions obtained by HAM in a fractional order case.

Consider the nonlinear time-fractional generalized SK equation [150]

$$\frac{\partial^\alpha u}{\partial t^\alpha}+45u^2\frac{\partial u}{\partial x}+15\frac{\partial u}{\partial x}\frac{\partial^2 u}{\partial x^2}+15\frac{\partial^3 u}{\partial x^3}u+\frac{\partial^5 u}{\partial x^5}=0, \tag{6.16}$$

with initial condition $u(x,t_0)=g(x)$.

The Chebyshev wavelet solution of $u(x,t)$ is sought by assuming that $u(x,t)$ can be expanded in terms of a Chebyshev wavelet as

$$u(x,t)=\sum_{n=1}^{2^{k1-1}}\sum_{i=0}^{M_1-1}\sum_{l=1}^{2^{k2-1}}\sum_{j=0}^{M_2-1}c_{n,i,l,j}\psi_{n,i,l,j}(x,t), \tag{6.17}$$

where $n=1,\ldots,2^{k_1-1}$, $i=0,\ldots,M_1-1$, $l=1,\ldots,2^{k_2-1}$, $j=0,\ldots,M_2-1$.

The nonlinear terms presented in Equation 6.16 can be approximated using the Chebyshev wavelet function as follows:

$$u^2\frac{\partial u}{\partial x}=\sum_{n=1}^{2^{k1-1}}\sum_{i=0}^{M_1-1}\sum_{l=1}^{2^{k2-1}}\sum_{j=0}^{M_2-1}a_{n,i,l,j}\psi_{n,i,l,j}(x,t) \tag{6.18}$$

and

$$\frac{\partial u}{\partial x}\frac{\partial^2 u}{\partial x^2}+\frac{\partial^3 u}{\partial x^3}u=\sum_{n=1}^{2^{k1-1}}\sum_{i=0}^{M_1-1}\sum_{l=1}^{2^{k2-1}}\sum_{j=0}^{M_2-1}b_{n,i,l,j}\psi_{n,i,l,j}(x,t). \tag{6.19}$$

This implies

$$\left(\sum_{n=1}^{2^{k1-1}}\sum_{i=0}^{M_1-1}\sum_{l=1}^{2^{k2-1}}\sum_{j=0}^{M_2-1}c_{n,i,l,j}\psi_{n,i,l,j}(x,t)\right)^2\left[\sum_{n=1}^{2^{k1-1}}\sum_{i=0}^{M_1-1}\sum_{l=1}^{2^{k2-1}}\sum_{j=0}^{M_2-1}c_{n,i,l,j}\frac{\partial\psi_{n,i,l,j}(x,t)}{\partial x}\right]$$
$$=\sum_{n=1}^{2^{k1-1}}\sum_{i=0}^{M_1-1}\sum_{l=1}^{2^{k2-1}}\sum_{j=0}^{M_2-1}a_{n,i,l,j}\psi_{n,i,l,j}(x,t) \tag{6.20}$$

and

$$\left(\sum_{n=1}^{2^{k1-1}}\sum_{i=0}^{M_1-1}\sum_{l=1}^{2^{k2-1}}\sum_{j=0}^{M_2-1}c_{n,i,l,j}\frac{\partial\psi_{n,i,l,j}(x,t)}{\partial x}\right)\left[\sum_{n=1}^{2^{k1-1}}\sum_{i=0}^{M_1-1}\sum_{l=1}^{2^{k2-1}}\sum_{j=0}^{M_2-1}c_{n,i,l,j}\frac{\partial^2\psi_{n,i,l,j}(x,t)}{\partial x^2}\right]$$
$$+\left(\sum_{n=1}^{2^{k1-1}}\sum_{i=0}^{M_1-1}\sum_{l=1}^{2^{k2-1}}\sum_{j=0}^{M_2-1}c_{n,i,l,j}\frac{\partial^3\psi_{n,i,l,j}(x,t)}{\partial x^3}\right)\left[\sum_{n=1}^{2^{k1-1}}\sum_{i=0}^{M_1-1}\sum_{l=1}^{2^{k2-1}}\sum_{j=0}^{M_2-1}c_{n,i,l,j}\psi_{n,i,l,j}(x,t)\right]$$
$$=\sum_{n=1}^{2^{k1-1}}\sum_{i=0}^{M_1-1}\sum_{l=1}^{2^{k2-1}}\sum_{j=0}^{M_2-1}b_{n,i,l,j}\psi_{n,i,l,j}(x,t). \tag{6.21}$$

Again, applying J_t^α on both sides of Equation 6.16, we have

$$u(x,t)-u(x,0)=J_t^\alpha\left[-45u^2\frac{\partial u}{\partial x}-15\left(\frac{\partial u}{\partial x}\frac{\partial^2 u}{\partial x^2}+\frac{\partial^3 u}{\partial x^3}u\right)-\frac{\partial^5 u}{\partial x^5}\right]. \tag{6.22}$$

Putting Equations 6.17 through 6.19 in Equation 6.22, we have

$$\sum_{n=1}^{2^{k_1-1}}\sum_{i=0}^{M_1-1}\sum_{l=1}^{2^{k_2-1}}\sum_{j=0}^{M_2-1}c_{n,i,l,j}\psi_{n,i,l,j}(x,t)-u(x,0)=J_t^\alpha\left[-45\left(\sum_{n=1}^{2^{k_1-1}}\sum_{i=0}^{M_1-1}\sum_{l=1}^{2^{k_2-1}}\sum_{j=0}^{M_2-1}a_{n,i,l,j}\psi_{n,i,l,j}(x,t)\right)\right.$$

$$-15\left(\sum_{n=1}^{2^{k_1-1}}\sum_{i=0}^{M_1-1}\sum_{l=1}^{2^{k_2-1}}\sum_{j=0}^{M_2-1}b_{n,i,l,j}\psi_{n,i,l,j}(x,t)\right)$$

$$\left.-\sum_{n=1}^{2^{k_1-1}}\sum_{i=0}^{M_1-1}\sum_{l=1}^{2^{k_2-1}}\sum_{j=0}^{M_2-1}c_{n,i,l,j}\frac{\partial^5\psi_{n,i,l,j}(x,t)}{\partial x^5}\right]. \tag{6.23}$$

Now, substituting the collocation points $x_l=(l-0.5)/(2^{k_1-1}M_1)$ and $t_r=(r-0.5)/(2^{k_2-1}M_2)$ for $l=1,2,\ldots,2^{k_1-1}M_1$ and $r=1,2,\ldots,2^{k_2-1}M_2$ in Equations 6.20, 6.21, and 6.23, we have $3(2^{k_1-1}M_1)(2^{k_2-1}M_2)$ equations in $3(2^{k_1-1}M_1)(2^{k_2-1}M_2)$ unknowns in $a_{n,i,l,j}$, $b_{n,i,l,j}$, and $c_{n,i,l,j}$. By solving this system of equations using mathematical software, the Chebyshev wavelet coefficients $a_{n,i,l,j}$, $b_{n,i,l,j}$, and $c_{n,i,l,j}$ can be obtained.

6.4.2 Comparison with HAM for the Solution of a Time-Fractional SK Equation

Consider the nonlinear time-fractional generalized SK equation [150]

$$\frac{\partial^\alpha u}{\partial t^\alpha}+45u^2\frac{\partial u}{\partial x}+15u\frac{\partial^2 u}{\partial x^2}+15\frac{\partial^3 u}{\partial x^3}u+\frac{\partial^5 u}{\partial x^5}=0, \tag{6.24}$$

subject to the initial condition [131]

$$u(x,0)=2k^2\operatorname{sech}^2\left(k(x-\lambda)\right),$$

where $k(\neq 0)$ and λ are arbitrary constants and $0<\alpha\leq 1$.

To obtain the approximate solution of the time-fractional SK equation (6.24), we choose the linear operator

$$L\left[\phi(x,t;p)\right]=D_t^\alpha\phi(x,t;p). \tag{6.25}$$

Now, we construct the mth order deformation equation for Equation 6.24 as follows [34,151–153]:

$$L\left[u_m(x,t)-\chi_m u_{m-1}(x,t)\right]=\hbar\Re_m\left(u_0,u_1,\ldots,u_{m-1}\right), \tag{6.26}$$

where

$$\Re_m\left(u_0,u_1,\ldots,u_{m-1}\right)=\frac{1}{(m-1)!}\frac{\partial^{m-1}N\left[\phi(x,t;p)\right]}{\partial p^{m-1}}\Bigg|_{p=0}. \tag{6.27}$$

$$N\left[\phi(x,t;p)\right]=D_t^{\alpha}\phi(x,t;p)+45\left(\phi(x,t;p)\right)^2\frac{\partial\phi(x,t;p)}{\partial x}+15\frac{\partial\phi(x,t;p)}{\partial x}\frac{\partial^2\phi(x,t;p)}{\partial x^2}$$
$$+15\frac{\partial^3\phi(x,t;p)}{\partial x^3}\phi(x,t;p)+\frac{\partial^5\phi(x,t;p)}{\partial x^5},$$

and

$$\phi(x,t;p)=u_0(x,t)+\sum_{m=1}^{\infty}p^m u_m(x,t),$$
$$u_m(x,t)=\frac{1}{m!}\frac{\partial^m\phi(x,t;p)}{\partial p^m}\Bigg|_{p=0}.$$

Now, the solution of the first deformation equation in Equation 6.26 is given by

$$u_1(x,t)=\hbar J_t^{\alpha}\left[\frac{\partial^{\alpha}u_0}{\partial t^{\alpha}}+45{u_0}^2\frac{\partial u_0}{\partial x}+15\frac{\partial u_0}{\partial x}\frac{\partial^2u_0}{\partial x^2}+15\frac{\partial^3u_0}{\partial x^3}u_0+\frac{\partial^5u_0}{\partial x^5}\right]. \tag{6.28}$$

Similarly, the solutions of second-, third-, and fourth-order deformation equations are

$$u_2(x,t)=u_1(x,t)+\hbar J_t^{\alpha}\left[\frac{\partial^{\alpha}u_1}{\partial t^{\alpha}}+90u_0u_1\frac{\partial u_0}{\partial x}+45{u_0}^2\frac{\partial u_1}{\partial x}+15\left(\frac{\partial u_0}{\partial x}\frac{\partial^2u_1}{\partial x^2}+\frac{\partial u_1}{\partial x}\frac{\partial^2u_0}{\partial x^2}\right)\right.$$
$$\left.+15\left(\frac{\partial^3u_1}{\partial x^3}u_0+\frac{\partial^3u_0}{\partial x^3}u_1\right)+\frac{\partial^5u_1}{\partial x^5}\right], \tag{6.29}$$

$$u_3(x,t)=u_2(x,t)+\hbar J_t^{\alpha}\left[\frac{\partial^{\alpha}u_2}{\partial t^{\alpha}}+45\left(u_1^2\frac{\partial u_0}{\partial x}+2u_0u_2\frac{\partial u_0}{\partial x}+2u_0u_1\frac{\partial u_1}{\partial x}+u_0^2\frac{\partial u_2}{\partial x}\right)\right.$$
$$\left.+15\left(\frac{\partial u_0}{\partial x}\frac{\partial^2u_2}{\partial x^2}+\frac{\partial u_1}{\partial x}\frac{\partial^2u_1}{\partial x^2}+\frac{\partial u_2}{\partial x}\frac{\partial^2u_0}{\partial x^2}\right)+15\left(\frac{\partial^3u_2}{\partial x^3}u_0+\frac{\partial^3u_1}{\partial x^3}u_1+\frac{\partial^3u_0}{\partial x^3}u_2\right)+\frac{\partial^5u_2}{\partial x^5}\right], \tag{6.30}$$

$$u_4(x,t) = u_3(x,t) + \hbar J_t^{\alpha}\left[\frac{\partial^{\alpha} u_3}{\partial t^{\alpha}} + 45\left(2u_1u_2\frac{\partial u_0}{\partial x} + 2u_0u_3\frac{\partial u_0}{\partial x} + u_1^2\frac{\partial u_1}{\partial x} + 2u_0u_2\frac{\partial u_1}{\partial x}\right.\right.$$

$$\left. + 2u_0u_1\frac{\partial u_2}{\partial x} + u_0^2\frac{\partial u_3}{\partial x}\right) + 15\left(\frac{\partial u_0}{\partial x}\frac{\partial^2 u_3}{\partial x^2} + \frac{\partial u_1}{\partial x}\frac{\partial^2 u_2}{\partial x^2} + \frac{\partial u_2}{\partial x}\frac{\partial^2 u_1}{\partial x^2} + \frac{\partial u_3}{\partial x}\frac{\partial^2 u_0}{\partial x^2}\right)$$

$$\left. + 15\left(\frac{\partial^3 u_3}{\partial x^3}u_0 + \frac{\partial^3 u_2}{\partial x^3}u_1 + \frac{\partial^3 u_1}{\partial x^3}u_2 + \frac{\partial^3 u_0}{\partial x^3}u_3\right) + \frac{\partial^5 u_3}{\partial x^5}\right], \tag{6.31}$$

and so on.

By putting the initial condition $u_0 = u(x,0)$ in Equations 6.28 through 6.31 and solving them, we obtain the expressions for u_1, u_2, u_3, u_4, and so on.

Finally, the approximate solution for a time-fractional fifth-order SK equation is given by

$$u = u_1(x,t) + u_2(x,t) + u_3(x,t) + u_4(x,t) + \cdots. \tag{6.32}$$

6.5 Numerical Results of a Fractional SK Equation

The comparison of the absolute errors for time-fractional fifth-order SK equation (6.16) has been exhibited in Table 6.1, which is constructed using the results obtained by the Chebyshev wavelet method at different values of x and t, taking $\alpha=1$. Similarly, Table 6.2 shows the comparison of approximate solutions of the fractional order SK equation (6.16) at various points of x and t, taking $\alpha=0.5$. Agreement between present numerical results and exact solutions appears very satisfactory through illustrations in Tables 6.1 and 6.2. To show the accuracy of the proposed method, L_2 and L_∞ error norms for a classical order nonlinear SK equation have been presented in Table 6.3. In the present analysis, to examine the accuracy and reliability of the Chebyshev wavelets for solving a fractional order SK equation, we compare the approximate solution of a Chebyshev wavelet with the fifth-order approximate solution obtained by HAM, taking $\hbar =$ 1. As pointed out by Liao [33], in general, by means of the so-called $\hbar$-curve, it is straightforward to choose a proper value of $\hbar$ that ensures the convergence of series solution. To investigate the influence of $\hbar$ on the solution series, we plot $\hbar$-curve of partial derivatives of $u(x,t)$ obtained from the fifth-order HAM solution as shown in Figure 6.1.

6.6 Application of the Two-Dimensional Chebyshev Wavelet Method on a Time-Fractional CH Equation

Consider the nonlinear CH equations with the Riesz time-fractional derivative [133]

$${}_0^R D_t^{\alpha} u(x,t) + 2ku_x(x,t) - \frac{2}{3}u_{xxt}(x,t) + 3u(x,t)u_x(x,t) + u_x(x,t)u_{xx}(x,t) + \frac{1}{2}u(x,t)u_{xxx}(x,t) = 0, \tag{6.33}$$

TABLE 6.1

Comparison of Absolute Errors Obtained by the Two-Dimensional Chebyshev Wavelet Method for the Classical Nonlinear SK Equation Given in Equation 6.16 at Various Points of x and t, Taking $\alpha = 1$

	$\lvert u_{Exact} - u_{CWM} \rvert$									
x	$t=0$	$t=0.1$	$t=0.2$	$t=0.3$	$t=0.4$	$t=0.5$	$t=0.6$	$t=0.7$	$t=0.8$	$t=0.9$
0.1	2.3394E−3	1.3646E−3	2.3696E−3	2.9815E−3	3.8367E−3	4.6925E−3	5.2186E−3	5.7875E−3	8.2658E−3	1.6805E−2
0.2	2.6097E−3	1.0767E−3	1.9063E−3	2.3403E−3	3.0586E−3	3.7661E−3	3.9981E−3	3.9267E−3	5.1666E−3	1.1581E−3
0.3	2.8468E−3	8.7155E−4	1.5326E−3	1.7868E−3	2.3576E−3	2.8825E−3	2.7549E−3	1.9657E−3	1.9447E−3	6.4023E−3
0.4	3.0282E−3	7.5225E−4	1.2490E−3	1.3226E−3	1.7342E−3	2.0403E−3	1.4875E−3	9.2722E−5	1.3891E−3	1.2842E−3
0.5	3.1322E−3	7.2105E−4	1.0555E−3	9.4823E−4	1.1879E−3	1.2373E−3	1.9461E−4	2.2433E−3	4.8194E−3	3.7539E−3
0.6	3.1383E−3	7.7914E−4	9.5098E−4	6.6340E−4	7.1781E−4	4.7117E−4	1.1238E−3	4.4781E−3	8.3267E−3	8.6876E−3
0.7	3.0277E−3	9.2665E−4	9.3356E−4	4.6693E−4	3.2177E−4	2.6046E−4	2.4666E−3	6.7864E−3	1.1887E−2	1.3488E−2
0.8	2.7843E−3	1.1626E−3	1.0005E−3	3.5681E−4	2.6502E−6	9.5989E−4	3.8313E−3	9.1541E−3	1.5473E−2	1.8125E−2
0.9	2.3953E−3	1.4847E−3	1.1484E−3	3.3029E−4	2.5848E−4	1.6291E−3	5.2136E−3	1.1564E−2	1.9051E−2	2.2563E−2

TABLE 6.2

Comparison of Approximate Solutions Obtained by the Two-Dimensional Chebyshev Wavelet Method and Homotopy Analysis Method for the Fractional Order Nonlinear SK Equation Given in Equation 6.15 at Various Points of x and t, Taking $\hbar = -1$ and $\alpha = 0.5$

	$t=0.05$		$t=0.1$		$t=0.15$		$t=0.2$		$t=0.25$		$t=0.3$		$t=0.35$		$t=0.4$	
x	u_{CWM}	u_{HAM}	u_{CWM}	u_{HAM}	u_{CWM}	u_{HAM}	u_{CWM}	u_{HAM}	u_{CWM}	u_{HAM}	u_{CWM}	u_{HAM}	u_{CWM}	u_{HAM}	u_{CWM}	u_{HAM}
0.1	0.166651	0.17821	0.171782	0.178058	0.174807	0.177938	0.17661	0.177336	0.177852	0.177748	0.178998	0.177669	0.180348	0.177598	0.182066	0.177534
0.2	0.169437	0.179118	0.173314	0.179009	0.175723	0.178924	0.177294	0.178853	0.178503	0.178792	0.179687	0.17874	0.18107	0.178696	0.182778	0.178657
0.3	0.172229	0.179707	0.174756	0.179643	0.176499	0.179592	0.177818	0.179552	0.17899	0.179518	0.180218	0.179492	0.181640	0.179471	0.183341	0.179457
0.4	0.175030	0.179975	0.176107	0.179954	0.177132	0.179938	0.178179	0.179927	0.179311	0.17992	0.180586	0.179917	0.182052	0.179919	0.183749	0.179925
0.5	0.177842	0.179919	0.177367	0.179942	0.177622	0.179959	0.178374	0.179976	0.179461	0.179993	0.180786	0.180012	0.182301	0.180034	0.183996	0.180058
0.6	0.180668	0.17954	0.178538	0.179606	0.177968	0.179655	0.178402	0.179698	0.179439	0.179738	0.180816	0.179777	0.182382	0.179816	0.184077	0.179855
0.7	0.183505	0.178839	0.17962	0.178948	0.178172	0.179029	0.178264	0.179097	0.179244	0.179158	0.180673	0.179214	0.182292	0.179268	0.183986	0.17932
0.8	0.186351	0.177824	0.180614	0.177973	0.178235	0.178084	0.177961	0.178176	0.178875	0.178256	0.180356	0.178329	0.182029	0.178396	0.183721	0.17846
0.9	0.189198	0.17650	0.18152	0.176689	0.178161	0.176629	0.177496	0.176944	0.178335	0.177043	0.179865	0.17713	0.181591	0.17721	0.183278	0.177283

TABLE 6.3

L_2 and L_∞ Error Norms for a Nonlinear Time-Fractional SK Equation Using the Two-Dimensional Chebyshev Wavelet Method at Various Points of t, Taking $\alpha=1$

t	L_2	L_∞
0	2.82595E−3	3.1383E−3
0.1	1.04820E−3	1.4847E−3
0.2	1.42858E−3	2.3696E−3
0.3	1.53123E−3	2.9815E−3
0.4	1.96462E−3	3.8367E−3
0.5	2.45110E−3	4.6925E−3
0.6	3.37089E−3	5.2186E−3
0.7	6.15247E−3	1.1564E−2
0.8	1.02032E−2	1.9051E−2
0.9	1.26293E−2	2.2563E−2

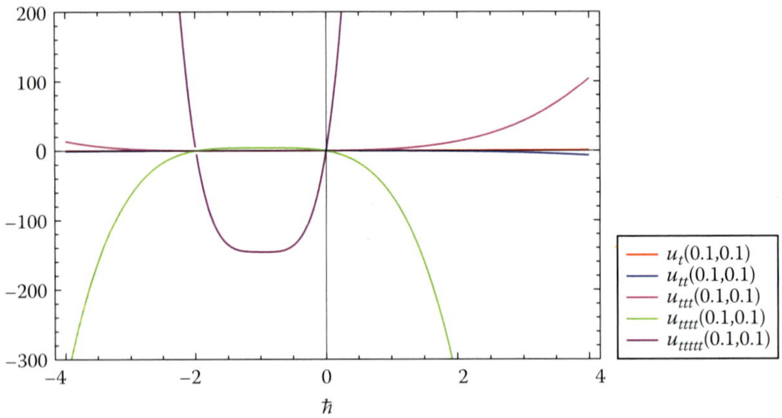

FIGURE 6.1
The ħ-curve for partial derivatives of $u(x,t)$ for the fifth-order HAM solution, taking $x=0.1$, $t=0.1$, and $\alpha=0.5$.

with initial condition [133]

$$u(x,0) = -k + k\sinh(x). \tag{6.34}$$

The Chebyshev wavelet solution of $u(x,t)$ is sought by assuming that $u(x,t)$ can be expanded in terms of a Chebyshev wavelet as

$$u(x,t) = \sum_{n=1}^{2^{k_1-1}}\sum_{i=0}^{M_1-1}\sum_{l=1}^{2^{k_2-1}}\sum_{j=0}^{M_2-1} c_{n,i,l,j}\psi_{n,i,l,j}(x,t), \tag{6.35}$$

where $n=1,\ldots,2^{k_1-1}$, $i=0,\ldots,M_1-1$, $l=1,\ldots,2^{k_2-1}$, $j=0,\ldots,M_2-1$.

Substituting Equation 6.35 in Equation 6.33, we will have

$$\begin{aligned}&\left(\sum_{n=1}^{2^{k_1-1}}\sum_{i=0}^{M_1-1}\sum_{l=1}^{2^{k_2-1}}\sum_{j=0}^{M_2-1} c_{n,i,l,j}\,{}_0^RD_t^{\alpha}\left(\psi_{n,i,l,j}(x,t)\right)\right)+2k\left[\sum_{n=1}^{2^{k_1-1}}\sum_{i=0}^{M_1-1}\sum_{l=1}^{2^{k_2-1}}\sum_{j=0}^{M_2-1} c_{n,i,l,j}\frac{\partial\psi_{n,i,l,j}(x,t)}{\partial x}\right]\\&-\frac{2}{3}\left(\sum_{n=1}^{2^{k_1-1}}\sum_{i=0}^{M_1-1}\sum_{l=1}^{2^{k_2-1}}\sum_{j=0}^{M_2-1} c_{n,i,l,j}\frac{\partial^3\psi_{n,i,l,j}(x,t)}{\partial x^2\partial t}\right)+3\left[\left(\sum_{n=1}^{2^{k_1-1}}\sum_{i=0}^{M_1-1}\sum_{l=1}^{2^{k_2-1}}\sum_{j=0}^{M_2-1} c_{n,i,l,j}\psi_{n,i,l,j}(x,t)\right)\right.\\&\left.\times\left(\sum_{n=1}^{2^{k_1-1}}\sum_{i=0}^{M_1-1}\sum_{l=1}^{2^{k_2-1}}\sum_{j=0}^{M_2-1} c_{n,i,l,j}\frac{\partial\psi_{n,i,l,j}(x,t)}{\partial x}\right)\right]+\left[\left(\sum_{n=1}^{2^{k_1-1}}\sum_{i=0}^{M_1-1}\sum_{l=1}^{2^{k_2-1}}\sum_{j=0}^{M_2-1} c_{n,i,l,j}\frac{\partial\psi_{n,i,l,j}(x,t)}{\partial x}\right)\right.\\&\left.\times\left(\sum_{n=1}^{2^{k_1-1}}\sum_{i=0}^{M_1-1}\sum_{l=1}^{2^{k_2-1}}\sum_{j=0}^{M_2-1} c_{n,i,l,j}\frac{\partial^2\psi_{n,i,l,j}(x,t)}{\partial x^2}\right)\right]+\frac{1}{2}\left[\left(\sum_{n=1}^{2^{k_1-1}}\sum_{i=0}^{M_1-1}\sum_{l=1}^{2^{k_2-1}}\sum_{j=0}^{M_2-1} c_{n,i,l,j}\psi_{n,i,l,j}(x,t)\right)\right.\\&\left.\times\left(\sum_{n=1}^{2^{k_1-1}}\sum_{i=0}^{M_1-1}\sum_{l=1}^{2^{k_2-1}}\sum_{j=0}^{M_2-1} c_{n,i,l,j}\frac{\partial^3\psi_{n,i,l,j}(x,t)}{\partial x^3}\right)\right]=0.\end{aligned} \tag{6.36}$$

Now, substituting the collocation points $x_l=(l-0.5)/(2^{k_1-1}M_1)$ and $t_r=(r-0.5)/(2^{k_2-1}M_2)$ for $l=1,2,\ldots,2^{k_1-1}M_1$ and $r=1,2,\ldots,2^{k_2-1}M_2$ in Equations 6.34 and 6.36, we have $(2^{k_1-1}M_1)(2^{k_2-1}M_2)$ equations in $(2^{k_1-1}M_1)(2^{k_2-1}M_2)$ unknowns in $c_{n,i,l,j}$. By solving this system of equations using mathematical software, the Chebyshev wavelet coefficients $c_{n,i,l,j}$ can be obtained.

6.7 Numerical Results and Discussion

The comparison of the absolute errors for a Riesz time-fractional CH equation (6.33) has been exhibited in Table 6.4, which is constructed using the results obtained by the Chebyshev wavelet method at different values of x and t taking $\alpha=0.5$. Similarly, Table 6.5 shows the comparison of the absolute errors for time-fractional CH equation (6.33) at various points of x and t taking $\alpha=0.75$. Again in order to examine the accuracy and reliability of the Chebyshev wavelets for solving fractional order CH equation, we compare the approximate numerical solutions of the Chebyshev wavelet with the approximate solutions obtained by HAM and VIM. Agreement between present numerical solutions with analytical solutions like VIM and HAM appear very satisfactory through the illustration in Tables 6.4 and 6.5. To show the accuracy of the proposed method, L_2 and L_∞ error norms for a fractional order nonlinear CH equation have been presented in Table 6.6.

Figures 6.2 through 6.4 demonstrate the graphical comparison of the numerical solutions obtained by two-dimensional Chebyshev wavelet approximation with regard to VIM and HAM, taking $\alpha=0.75$ and $\alpha=0.5$, respectively.

TABLE 6.4

Comparison of Absolute Errors Obtained by the Two-Dimensional Chebyshev Wavelet Method for the Fractional Nonlinear CH Equation Given in Equation 6.33 at Various Points of x and t, Taking $\alpha = 0.5$

	Comparison with Different Analytical Methods									
	$t=0.1$		$t=0.2$		$t=0.3$		$t=0.4$		$t=0.5$	
x	$\lvert u_{CWM}-u_{VIM}\rvert$	$\lvert u_{CWM}-u_{HAM}\rvert$	$\lvert u_{CWM}-u_{VIM}\rvert$	$\lvert u_{CWM}-u_{HAM}\rvert$	$\lvert u_{CWM}-u_{VIM}\rvert$	$\lvert u_{CWM}-u_{HAM}\rvert$	$\lvert u_{CWM}-u_{VIM}\rvert$	$\lvert u_{CWM}-u_{HAM}\rvert$	$\lvert u_{CWM}-u_{VIM}\rvert$	$\lvert u_{CWM}-u_{HAM}\rvert$
0.1	2.1329E−3	2.1361E−3	7.8868E−4	7.8402E−4	5.4917E−3	5.4858E−3	8.4921E−3	8.4852E−3	9.3582E−3	9.3504E−3
0.2	4.0685E−4	4.0868E−4	1.5069E−3	1.5042E−3	4.8677E−3	4.8643E−3	7.0508E−3	7.0468E−3	7.7726E−3	7.7680E−3
0.3	1.3197E−3	1.3192E−3	2.2497E−3	2.2491E−3	4.2747E−3	4.2739E−3	5.6335E−3	5.6326E−3	6.1955E−3	6.1945E−3
0.4	3.0276E−3	3.0287E−3	2.9895E−3	2.9912E−3	3.6831E−3	3.6852E−3	4.2131E−3	4.2155E−3	4.6052E−3	4.6080E−3
0.5	4.6974E−3	4.7003E−3	3.6982E−3	3.7024E−3	3.0627E−3	3.0681E−3	2.7621E−3	2.7684E−3	2.9790E−3	2.9862E−3
0.6	6.3093E−3	6.3142E−3	4.3472E−3	4.3545E−3	2.3829E−3	2.3921E−3	1.2524E−3	1.2634E−3	1.2937E−3	1.3062E−3
0.7	7.8429E−3	7.8503E−3	4.9074E−3	4.9184E−3	1.6122E−3	1.6261E−3	3.4466E−4	3.2816E−4	4.7449E−4	4.5561E−4
0.8	9.2769E−3	9.2874E−3	5.3489E−3	5.3644E−3	7.1856E−4	7.3824E−4	2.0588E−3	2.0354E−3	2.3504E−3	2.3236E−3
0.9	1.0589E−2	1.0603E−2	5.6408E−3	5.6619E−3	3.3100E−4	3.0414E−4	3.9205E−3	3.8884E−3	4.3597E−3	4.3229E−3

TABLE 6.5

Comparison of Absolute Errors Obtained by the Two-Dimensional Chebyshev Wavelet Method for the Fractional Nonlinear CH Equation Given in Equation 6.33 at Various Points of x and t, Taking $\alpha = 0.75$

	Comparison with Different Analytical Methods											
	$t=0.1$		$t=0.2$		$t=0.3$		$t=0.4$		$t=0.5$		$t=0.6$	
x	$\lvert u_{CWM}-u_{VIM}\rvert$	$\lvert u_{CWM}-u_{HAM}\rvert$	$\lvert u_{CWM}-u_{VIM}\rvert$	$\lvert u_{CWM}-u_{HAM}\rvert$	$\lvert u_{CWM}-u_{VIM}\rvert$	$\lvert u_{CWM}-u_{HAM}\rvert$	$\lvert u_{CWM}-u_{VIM}\rvert$	$\lvert u_{CWM}-u_{HAM}\rvert$	$\lvert u_{CWM}-u_{VIM}\rvert$	$\lvert u_{CWM}-u_{HAM}\rvert$	$\lvert u_{CWM}-u_{VIM}\rvert$	$\lvert u_{CWM}-u_{HAM}\rvert$
0.1	6.3182E−3	6.3150E−3	6.4241E−3	6.4194E−3	7.3456E−3	7.3408E−3	9.1567E−3	9.1498E−3	1.0373E−2	1.0365E−2	9.1687E−3	9.1599E−3
0.2	6.2304E−3	6.2285E−3	6.4316E−3	6.4290E−3	6.9095E−3	6.9062E−3	7.9849E−3	7.9809E−3	8.7183E−3	8.7137E−3	7.8194E−3	7.8144E−3
0.3	6.1456E−3	6.1451E−3	6.4473E−3	6.4466E−3	6.4856E−3	6.4848E−3	6.8251E−3	6.8242E−3	7.0637E−3	7.0627E−3	6.4553E−3	6.4542E−3
0.4	6.0535E−3	6.0547E−3	6.4548E−3	6.4565E−3	6.0540E−3	6.0561E−3	5.6571E−3	5.6596E−3	5.3955E−3	5.3983E−3	5.0703E−3	5.0735E−3
0.5	5.9436E−3	5.9465E−3	6.4379E−3	6.4421E−3	5.5938E−3	5.5991E−3	4.4598E−3	4.4661E−3	3.6984E−3	3.7056E−3	3.6578E−3	3.6659E−3
0.6	5.8047E−3	5.8097E−3	6.3792E−3	6.3865E−3	5.0829E−3	5.0922E−3	3.2114E−3	3.2223E−3	1.9567E−3	1.9692E−3	2.2104E−3	2.2243E−3
0.7	5.6251E−3	5.6325E−3	6.2610E−3	6.2720E−3	4.4990E−3	4.5130E−3	1.8896E−3	1.9061E−3	1.5382E−4	1.7269E−4	7.1993E−4	7.4103E−4
0.8	5.3923E−3	5.4028E−3	6.0645E−3	6.0800E−3	3.8136E−3	3.8383E−3	4.7116E−4	4.9455E−4	1.7276E−3	1.7008E−3	8.2260E−4	7.9259E−4
0.9	5.0929E−3	5.1071E−3	5.7699E−3	5.7910E−3	3.0173E−3	3.0441E−3	1.0680E−3	1.0360E−3	3.7057E−3	3.6689E−3	2.4271E−3	2.3859E−3

TABLE 6.6

L_2 and L_∞ Error Norms for a Nonlinear Time-Fractional CH Equation Using the Two-Dimensional Chebyshev Wavelet Methods at Various Points x, Taking $\alpha = 0.5$ and 0.75

	Error Analysis with Regard to VIM				Error Analysis with Regard to HAM			
	$\alpha=0.5$		$\alpha=0.75$		$\alpha=0.5$		$\alpha=0.75$	
x	L_2	L_∞	L_2	L_∞	L_2	L_∞	L_2	L_∞
0.1	6.38694E−3	9.3582E−3	7.11978E−3	1.0373E−2	6.38112E−3	9.3504E−3	7.11421E−3	1.0365E−2
0.2	5.40607E−3	7.7726E−3	7.40310E−3	8.7183E−3	5.39997E−3	7.7680E−3	7.39944E−3	8.7137E−3
0.3	4.55570E−3	6.1955E−3	6.57708E−3	7.0637E−3	4.55487E−3	6.1945E−3	6.57624E−3	7.0627E−3
0.4	3.90422E−3	4.6052E−3	5.79917E−3	6.4548E−3	3.90650E−3	4.6080E−3	5.80137E−3	6.4565E−3
0.5	3.53718E−3	4.6974E−3	5.08283E−3	6.4379E−3	3.54257E−3	4.7003E−3	5.08802E−3	6.4421E−3
0.6	3.53417E−3	6.3093E−3	4.45818E−3	6.3792E−3	3.54135E−3	6.3142E−3	4.46615E−3	6.3865E−3
0.7	3.90066E−3	7.8429E−3	3.98320E−3	6.2610E−3	3.90730E−3	7.8503E−3	3.99254E−3	6.2720E−3
0.8	4.56822E−3	9.2769E−3	3.74878E−3	6.0645E−3	4.57187E−3	9.2874E−3	3.75621E−3	6.0800E−3
0.9	5.46221E−3	1.0589E−2	3.85351E−3	5.7699E−3	5.46048E−3	1.0603E−2	3.85381E−3	5.7910E−3

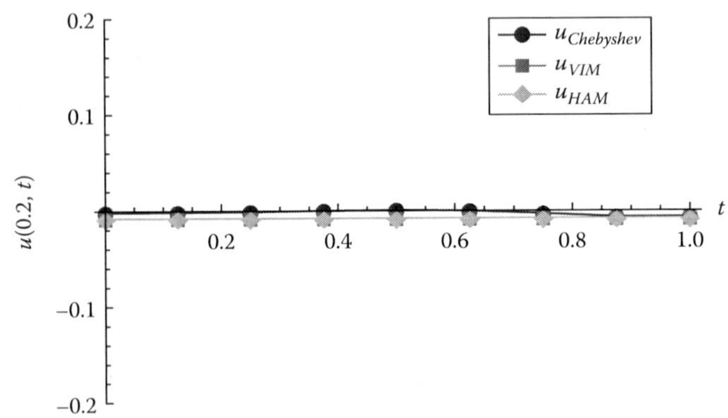

FIGURE 6.2
Comparison of numerical solutions of $u(0.2, t)$ obtained by the two-dimensional Chebyshev wavelet method with regard to VIM and HAM for $\alpha = 0.75$.

6.8 Implementation of the Two-Dimensional Chebyshev Wavelet Method for an Approximate Solution of a Riesz Space-Fractional SGE

In this section, two test examples have been considered with a purpose to exhibit the effectiveness and accuracy of the proposed Chebyshev wavelet method for numerical solution of the Riesz fractional SGE given in Equation 6.3. The numerical solutions acquired are compared with the exact solutions for classical case and also solutions obtained by modified HAM with Fourier transform (MHAM-FT) in a fractional order case [141].

Consider the nonlinear Riesz space-fractional SGE [141,142]

$$u_{tt} - {}^{R}D_x^{\alpha} u + \sin u = 0, \quad 1 < \alpha \leq 2 \tag{6.37}$$

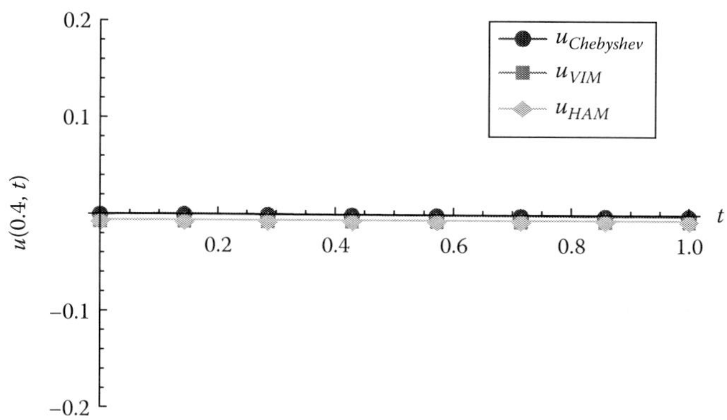

FIGURE 6.3
Comparison of numerical solutions of $u(0.4, t)$ obtained by the two-dimensional Chebyshev wavelet method with regard to VIM and HAM for $\alpha = 0.75$.

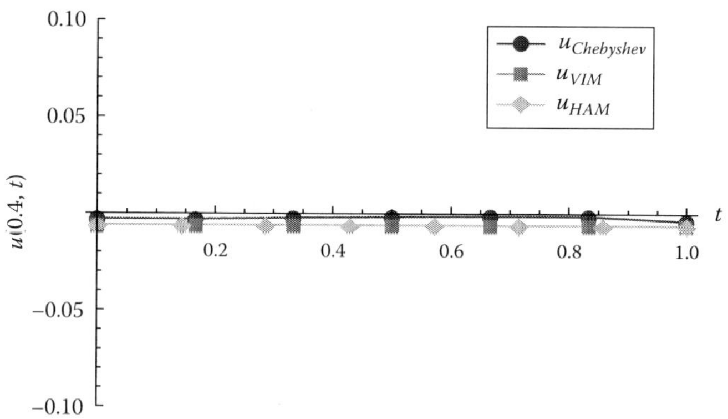

FIGURE 6.4
Comparison of numerical solutions of $u(0.4, t)$ obtained by the two-dimensional Chebyshev wavelet method with regard to VIM and HAM for $\alpha = 0.5$.

with initial conditions

$$u(x,0) = f(x) \quad \text{and} \quad u_t(x,0) = g(x). \tag{6.38}$$

The Chebyshev wavelet solution of $u(x, t)$ is sought through assuming that $u(x, t)$ can be expanded in terms of a Chebyshev wavelet as

$$u(x,t) = \sum_{n=1}^{2^{k1-1}} \sum_{i=0}^{M_1-1} \sum_{l=1}^{2^{k2-1}} \sum_{j=0}^{M_2-1} c_{n,i,l,j} \psi_{n,i,l,j}(x,t), \tag{6.39}$$

where $n = 1, \ldots, 2^{k_1-1}$, $i = 0, \ldots, M_1 - 1$, $l = 1, \ldots, 2^{k_2-1}$, $j = 0, \ldots, M_2 - 1$.

Now, applying Equation 6.39 in Equation 6.37, we have

$$\frac{\partial^2}{\partial t^2}\left(\sum_{n=1}^{2^{k_1-1}}\sum_{i=0}^{M_1-1}\sum_{l=1}^{2^{k_2-1}}\sum_{j=0}^{M_2-1} c_{n,i,l,j}\psi_{n,i,l,j}(x,t)\right) - {}^{R}D_x^{\alpha}\left(\sum_{n=1}^{2^{k_1-1}}\sum_{i=0}^{M_1-1}\sum_{l=1}^{2^{k_2-1}}\sum_{j=0}^{M_2-1} c_{n,i,l,j}\psi_{n,i,l,j}(x,t)\right)$$

$$+\sin\left(\sum_{n=1}^{2^{k_1-1}}\sum_{i=0}^{M_1-1}\sum_{l=1}^{2^{k_2-1}}\sum_{j=0}^{M_2-1} c_{n,i,l,j}\psi_{n,i,l,j}(x,t)\right) = 0. \tag{6.40}$$

Now, substituting the collocation points $x_l = (l-0.5)/(2^{k_1-1}M_1)$ and $t_r = (r-0.5)/(2^{k_2-1}M_2)$ for $l = 1,2,\ldots,2^{k_1-1}M_1$ and $r = 1,2,\ldots,2^{k_2-1}M_2$ in Equation 6.40, we have $(2^{k_1-1}M_1)(2^{k_2-1}M_2)$ equations in $(2^{k_1-1}M_1)(2^{k_2-1}M_2)$ unknowns in $c_{n,i,l,j}$. By solving this system of equations using mathematical software, the Chebyshev wavelet coefficients $c_{n,i,l,j}$ can be obtained.

Example 6.1

Consider nonlinear Riesz fractional SGE (6.37) subject to the following initial conditions [145,146]:

$$u(x,0) = 0, \quad u_t(x,0) = 4\,\mathrm{sech}(x). \tag{6.41}$$

The numerical solutions of Example 6.1 are presented in Tables 6.7 through 6.9. The results are compared with the exact solutions and also with solutions obtained by MHAM-FT [141]. It has been noticed from Tables 6.7 through 6.9 that the solutions obtained by the present method are in good agreement with the exact solutions and those with MHAM-FT [141].

Example 6.2

In this case, we shall find an approximate solution of the nonlinear Riesz fractional SGE (6.37) with given initial conditions [145,146]

$$u(x,0) = \pi + \varepsilon\cos(\mu x), \quad u_t(x,0) = 0 \tag{6.42}$$

where ε and μ are parameters. The numerical solutions of Example 6.2 are presented for $\mu = \sqrt{2}/2$, $\varepsilon = 0.001$ in Tables 6.10 through 6.12. The results are compared with the exact solutions as well as solutions obtained by MHAM-FT, which is discussed in the following section. It has been observed from Tables 6.10 through 6.12 that the solutions attained by the present method are in pretty good agreement with the exact solutions and those obtained by MHAM-FT [141].

6.9 Numerical Results and Discussion

In the present numerical investigation, in order to examine the accuracy and reliability of the proposed Chebyshev wavelet technique for solving fractional order SGE, the absolute errors have been compared with the exact solutions and those obtained by MHAM-FT for different values of x and t, taking $\hbar = -1$. In the case of $\alpha = 2$, using the results of Chebyshev wavelet method, the numerical solutions of the classical SGE (6.37) given in Examples 6.1 and 6.2 have been exhibited in Tables 6.7 and 6.10. Similarly, the absolute errors for the fractional SGE (6.37) given in Examples 6.1 and 6.2 have been demonstrated in Tables 6.9 through 6.12,

TABLE 6.7

Absolute Errors Obtained by the Two-Dimensional Chebyshev Wavelet Method with Regard to Exact Solutions for the Classical SGE (6.37) Given in Example 6.1 at Various Points of x and t, Taking $\alpha = 2$

	$\lvert u_{CWM} - u_{Exact} \rvert$									
x	$t=0.01$	$t=0.02$	$t=0.03$	$t=0.04$	$t=0.05$	$t=0.06$	$t=0.07$	$t=0.08$	$t=0.09$	$t=0.10$
0.01	6.7054E−6	2.5570E−5	5.5014E−5	9.3603E−5	1.4004E−4	1.9318E−4	2.5198E−4	3.1551E−4	3.8298E−4	4.5367E−4
0.02	6.6806E−6	2.5524E−5	5.4947E−5	9.3514E−5	1.3992E−4	1.9303E−4	2.5177E−4	3.1524E−4	3.8263E−4	4.5321E−4
0.03	6.6617E−6	2.5485E−5	5.4885E−5	9.3423E−5	1.3980E−4	1.9285E−4	2.5154E−4	3.1494E−4	3.8224E−4	4.5271E−4
0.04	6.6468E−6	2.5449E−5	5.4820E−5	9.3322E−5	1.3965E−4	1.9265E−4	2.5127E−4	3.1459E−4	3.8179E−4	4.5214E−4
0.05	6.6341E−6	2.5412E−5	5.4749E−5	9.3205E−5	1.3948E−4	1.9241E−4	2.5095E−4	3.1417E−4	3.8126E−4	4.5149E−4
0.06	6.6222E−6	2.5372E−5	5.4666E−5	9.3066E−5	1.3927E−4	1.9212E−4	2.5056E−4	3.1367E−4	3.8063E−4	4.5072E−4
0.07	6.6101E−6	2.5327E−5	5.4569E−5	9.2900E−5	1.3902E−4	1.9177E−4	2.5010E−4	3.1308E−4	3.7990E−4	4.4984E−4
0.08	6.5969E−6	2.5275E−5	5.4455E−5	9.2703E−5	1.3872E−4	1.9135E−4	2.4955E−4	3.1240E−4	3.7906E−4	4.4882E−4
0.09	6.5819E−6	2.5214E−5	5.4322E−5	9.2474E−5	1.3838E−4	1.9087E−4	2.4892E−4	3.1160E−4	3.7808E−4	4.4765E−4
0.10	6.5642E−6	2.5144E−5	5.4168E−5	9.2210E−5	1.3798E−4	1.9032E−4	2.4820E−4	3.1069E−4	3.7698E−4	4.4633E−4

TABLE 6.8

Absolute Errors Obtained by the Two-Dimensional Chebyshev Wavelet Method with Regard to MHAM Solutions for the Fractional SGE (6.37) Given in Example 6.1 at Various Points of x and t, Taking $\alpha = 1.75$

	$\lvert u_{CWM} - u_{MHAM} \rvert$									
x	$t=0.01$	$t=0.02$	$t=0.03$	$t=0.04$	$t=0.05$	$t=0.06$	$t=0.07$	$t=0.08$	$t=0.09$	$t=0.10$
0.01	8.9761E−4	3.5234E−3	7.7847E−3	1.3593E−2	2.0864E−2	2.9519E−2	3.9482E−2	5.0680E−2	6.3047E−2	7.6517E−2
0.02	8.7243E−4	3.4244E−3	7.5651E−3	1.3207E−2	2.0270E−2	2.8673E−2	3.8343E−2	4.9210E−2	6.1206E−2	7.4270E−2
0.03	8.4432E−4	3.3142E−3	7.3209E−3	1.2780E−2	1.9611E−2	2.7738E−2	3.7088E−2	4.7592E−2	5.9187E−2	7.1809E−2
0.04	8.1361E−4	3.1937E−3	7.0545E−3	1.2314E−2	1.8895E−2	2.6722E−2	3.5727E−2	4.5843E−2	5.7006E−2	6.9159E−2
0.05	7.8057E−4	3.0642E−3	6.7684E−3	1.1814E−2	1.8127E−2	2.5636E−2	3.4274E−2	4.3977E−2	5.4684E−2	6.6339E−2
0.06	7.4549E−4	2.9268E−3	6.4651E−3	1.1285E−2	1.7315E−2	2.4489E−2	3.2740E−2	4.2009E−2	5.2238E−2	6.3373E−2
0.07	7.0863E−4	2.7825E−3	6.1467E−3	1.0730E−2	1.6465E−2	2.3287E−2	3.1136E−2	3.9953E−2	4.9685E−2	6.0281E−2
0.08	6.7026E−4	2.6322E−3	5.8156E−3	1.0153E−2	1.5581E−2	2.2040E−2	2.9472E−2	3.7822E−2	4.7042E−2	5.7081E−2
0.09	6.3062E−4	2.4771E−3	5.4737E−3	9.5579E−2	1.4670E−2	2.0755E−2	2.7758E−2	3.5630E−2	4.4323E−2	5.3793E−2
0.10	5.8994E−4	2.3179E−3	5.1231E−3	8.9478E−2	1.3737E−2	1.9439E−2	2.6004E−2	3.3388E−2	4.1545E−2	5.0435E−2

TABLE 6.9

Absolute Errors Obtained by the Two-Dimensional Chebyshev Wavelet Method with Regard to MHAM Solutions for the Fractional SGE (6.37) Given in Example 6.1 at Various Points of x and t, Taking $\alpha = 1.5$

	$\lvert u_{CWM} - u_{MHAM} \rvert$									
x	$t=0.01$	$t=0.02$	$t=0.03$	$t=0.04$	$t=0.05$	$t=0.06$	$t=0.07$	$t=0.08$	$t=0.09$	$t=0.10$
0.01	4.1180E−4	1.6331E−3	3.6485E−3	6.4435E−3	1.0005E−2	1.4321E−2	1.9381E−2	2.5175E−2	3.1695E−2	3.8932E−2
0.02	4.5144E−4	1.7862E−3	3.9798E−3	7.0093E−3	1.0853E−2	1.5491E−2	2.0905E−2	2.7077E−2	3.3992E−2	4.1636E−2
0.03	4.8403E−4	1.9118E−3	4.2512E−3	7.4718E−3	1.1544E−2	1.6443E−2	2.2142E−2	2.8617E−2	3.5847E−2	4.3810E−2
0.04	5.1000E−4	2.0117E−3	4.4665E−3	7.8376E−3	1.2090E−2	1.7191E−2	2.3110E−2	2.9818E−2	3.7286E−2	4.5490E−2
0.05	5.2980E−4	2.0877E−3	4.6295E−3	8.1133E−3	1.2499E−2	1.7749E−2	2.3828E−2	3.0702E−2	3.8339E−2	4.6708E−2
0.06	5.4386E−4	2.1413E−3	4.7438E−3	8.3051E−3	1.2781E−2	1.8130E−2	2.4313E−2	3.1292E−2	3.9031E−2	4.7498E−2
0.07	5.5260E−4	2.1742E−3	4.8128E−3	8.4189E−3	1.2945E−2	1.8347E−2	2.4582E−2	3.1609E−2	3.9390E−2	4.7890E−2
0.08	5.5640E−4	2.1879E−3	4.8400E−3	8.4508E−3	1.3000E−2	1.8412E−2	2.4651E−2	3.1674E−2	3.9441E−2	4.7915E−2
0.09	5.5565E−4	2.1839E−3	4.8287E−3	8.4363E−3	1.2955E−2	1.8337E−2	2.4536E−2	3.1507E−2	3.9208E−2	4.7601E−2
0.10	5.5073E−4	2.1638E−3	4.7821E−3	8.3510E−3	1.2818E−2	1.8134E−2	2.4252E−2	3.1127E−2	3.8715E−2	4.6978E−2

TABLE 6.10

Comparison of Absolute Errors Obtained by the Two-Dimensional Chebyshev Wavelet Method with Regard to Exact Solutions and MHAM Solutions for the Classical SGE (6.37) Given in Example 6.2 at Various Points of x and t, Taking $\alpha = 2$

	$\lvert u_{CWM} - u_{MHAM} \rvert$									
x	$t=0.01$	$t=0.02$	$t=0.03$	$t=0.04$	$t=0.05$	$t=0.06$	$t=0.07$	$t=0.08$	$t=0.09$	$t=0.10$
0.01	3.359E−10	9.694E−12	5.667E−10	1.320E−9	2.255E−9	3.360E−9	4.619E−9	6.021E−9	7.553E−9	9.204E−9
0.02	2.329E−10	1.131E−10	6.708E−10	1.425E−9	2.363E−9	3.469E−9	4.732E−9	6.139E−9	7.676E−9	9.334E−9
0.03	1.456E−10	2.005E−10	7.588E−10	1.514E−9	2.453E−9	3.562E−9	4.827E−9	6.237E−9	7.779E−9	9.442E−9
0.04	7.237E−11	2.738E−10	8.323E−10	1.588E−9	2.528E−9	3.638E−9	4.905E−9	6.318E−9	7.863E−9	9.531E−9
0.05	1.161E−11	3.345E−10	8.930E−10	1.649E−9	2.589E−9	3.700E−9	4.969E−9	6.383E−9	7.931E−9	9.602E−9
0.06	3.809E−11	3.840E−10	9.423E−10	1.698E−9	2.638E−9	3.749E−9	5.019E−9	6.434E−9	7.984E−9	9.658E−9
0.07	7.807E−11	4.237E−10	9.814E−10	1.737E−9	2.677E−9	3.788E−9	5.057E−9	6.473E−9	8.025E−9	9.699E−9
0.08	1.095E−10	4.547E−10	1.012E−9	1.767E−9	2.706E−9	3.816E−9	5.085E−9	6.501E−9	8.053E−9	9.728E−9
0.09	1.336E−10	4.782E−10	1.034E−9	1.788E−9	2.726E−9	3.836E−9	5.104E−9	6.519E−9	8.071E−9	9.746E−9
0.10	1.514E−10	4.953E−10	1.051E−9	1.803E−9	2.740E−9	3.848E−9	5.114E−9	6.529E−9	8.079E−9	9.754E−9

TABLE 6.11

Absolute Errors Obtained by the Two-Dimensional Chebyshev Wavelet Method with Regard to MHAM Solutions for the Fractional SGE (6.37) Given in Example 6.2 at Various Points of x and t, Taking $\alpha = 1.75$

	$\lvert u_{CWM} - u_{MHAM} \rvert$							
x	$t = 0.01$	$t = 0.02$	$t = 0.03$	$t = 0.04$	$t = 0.05$	$t = 0.06$	$t = 0.07$	$t = 0.08$
0.01	7.80306E−4	3.12057E−3	7.02163E−3	1.24867E−2	1.95210E−2	2.81321E−2	3.83293E−2	5.01238E−2
0.02	9.25929E−4	3.67903E−3	8.22523E−3	1.45343E−2	2.25797E−2	3.23383E−2	4.37906E−2	5.69199E−2
0.03	1.05489E−3	4.17342E−3	9.29032E−3	1.63455E−2	2.52839E−2	3.60551E−2	4.86136E−2	6.29182E−2
0.04	1.16806E−3	4.60703E−3	1.02240E−2	1.79324E−2	2.76518E−2	3.93077E−2	5.28314E−2	6.81597E−2
0.05	1.26625E−3	4.98308E−3	1.10333E−2	1.93069E−2	2.97013E−2	4.21207E−2	5.64760E−2	7.26847E−2
0.06	1.35029E−3	5.30471E−3	1.17249E−2	2.04806E−2	3.14498E−2	4.45181E−2	5.95788E−2	7.65325E−2
0.07	1.42099E−3	5.57500E−3	1.23055E−2	2.14649E−2	3.29144E−2	4.65237E−2	6.21708E−2	7.97417E−2
0.08	1.47911E−3	5.79696E−3	1.27817E−2	2.22709E−2	3.41118E−2	4.81604E−2	6.42819E−2	8.23499E−2
0.09	1.52543E−3	5.97352E−3	1.31597E−2	2.29095E−2	3.50582E−2	4.94507E−2	6.59415E−2	8.43939E−2
0.10	1.56068E−3	6.10756E−3	1.34458E−2	2.33912E−2	3.57696E−2	5.04168E−2	6.71784E−2	8.59096E−2

TABLE 6.12

Absolute Errors Obtained by the Two-Dimensional Chebyshev Wavelet Method with Regard to MHAM Solutions for the Fractional SGE (6.37) Given in Example 6.2 at Various Points of x and t, Taking $\alpha = 1.5$

	$\lvert u_{CWM} - u_{MHAM} \rvert$									
x	$t = 0.01$	$t = 0.02$	$t = 0.03$	$t = 0.04$	$t = 0.05$	$t = 0.06$	$t = 0.07$	$t = 0.08$	$t = 0.09$	$t = 0.10$
0.01	1.0416E−3	3.9097E−3	8.2311E−3	1.3648E−2	1.9820E−2	2.6420E−2	3.3137E−2	3.9675E−2	4.5752E−2	5.1101E−2
0.02	8.0050E−4	2.9839E−3	6.2331E−3	1.0244E−2	1.4727E−2	1.9403E−2	2.4008E−2	2.8289E−2	3.2006E−2	3.4932E−2
0.03	5.7955E−4	2.1356E−3	4.4026E−3	7.1261E−2	1.0062E−2	1.2977E−2	1.5650E−2	1.7867E−2	1.9427E−2	2.0138E−2
0.04	3.7779E−4	1.3611E−3	2.7317E−3	4.2801E−3	5.8053E−3	7.1154E−3	8.0265E−3	8.3635E−3	7.9594E−3	6.6553E−3
0.05	1.9432E−4	6.5698E−4	1.2127E−3	1.6933E−3	1.9371E−3	1.7893E−3	1.1019E−3	2.6672E−4	2.4516E−3	5.5813E−3
0.06	2.8208E−5	1.9564E−5	1.6195E−4	6.4725E−4	1.5621E−3	3.0273E−3	5.1587E−3	8.0671E−3	1.1858E−2	1.6634E−2
0.07	1.2142E−4	5.5452E−4	1.3997E−3	2.7541E−3	4.7112E−3	7.3608E−3	1.0789E−2	1.5080E−2	2.0312E−2	2.6563E−2
0.08	2.5546E−4	1.0685E−3	2.5078E−3	4.6397E−3	7.5286E−3	1.1236E−2	1.5823E−2	2.1348E−2	2.7865E−2	3.5429E−2
0.09	3.7472E−4	1.5258E−3	3.4933E−3	6.3162E−3	1.0032E−2	1.4679E−2	2.0294E−2	2.6911E−2	3.4565E−2	4.3289E−2
0.10	4.8006E−4	1.9296E−3	4.3631E−3	7.7951E−3	1.2240E−2	1.7714E−2	2.4232E−2	3.1808E−2	4.0460E−2	5.0202E−2

which are devised using the results attained by the Chebyshev wavelet method and MHAM-FT at different values of x and t, taking $\alpha=1.75$ and 1.5, respectively. We observe that the proposed numerical simulation results are in good agreement with the exact solutions and those obtained by MHAM-FT [141] through illustrations in Tables 6.7 through 6.12.

6.10 Convergence Analysis of a Chebyshev Wavelet

Theorem 6.1 Convergence Theorem

If a continuous function, $u(x,t)\in L^2(\mathfrak{R}\times\mathfrak{R})$, defined on $[0,1)\times[0,1)$ is bounded, namely, $|u(x,t)|\le K$, then the Chebyshev wavelets expansion of $u(x,t)$ converges uniformly to it.

Proof

Let $u(x,t)$ be a function defined on $[0,1)\times[0,1)$ and $|u(x,t)|\le K$, K be a positive constant. The Chebyshev wavelet coefficients of continuous functions $u(x,t)$ are defined as

$$c_{n_1,m_1,n_2,m_2}=\int_0^1\int_0^1 u(x,t)\psi_{n_1,m_1}(x)\psi_{n_2,m_2}(t)\,dx\,dt$$

$$=\int_0^1\int_{I_1} u(x,t)2^{k_1/2}\bar{U}_{m_1}\left(2^{k_1}x-2n_1+1\right)\psi_{n_2,m_2}(t)\,dx\,dt,\quad \text{where } I_1=\left[\frac{n_1-1}{2^{k_1-1}},\frac{n_1}{2^{k_1-1}}\right).$$

$$=2^{k_1/2}\sqrt{\frac{2}{\pi}}\int_0^1\int_{I_1} u(x,t)U_{m_1}\left(2^{k_1}x-2n_1+1\right)\psi_{n_2,m_2}(t)\,dx\,dt.$$

Now, by change of variable $2^{k_1}x-2n_1+1=y$, we obtain

$$c_{n_1,m_1,n_2,m_2}=\frac{2^{k_1/2}}{2^{k_1}}\sqrt{\frac{2}{\pi}}\int_0^1\psi_{n_2,m_2}(t)\left(\int_{-1}^1 u\left(\frac{y+2n_1-1}{2^{k_1}},t\right)U_{m_1}(y)\,dy\right)dt.$$

Applying the mean value theorem of integral calculus, we have

$$c_{n_1,m_1,n_2,m_2}=\frac{1}{2^{k_1/2}}\sqrt{\frac{2}{\pi}}\int_0^1\psi_{n_2,m_2}(t)u\left(\frac{\xi+2n_1-1}{2^{k_1}},t\right)\left(\int_{-1}^1 U_{m_1}(y)\,dy\right)dt \quad \text{where } \xi\in(-1,1)$$

$$=\frac{1}{2^{k_1/2}}\sqrt{\frac{2}{\pi}}\int_0^1\psi_{n_2,m_2}(t)u\left(\frac{\xi+2n_1-1}{2^{k_1}},t\right)\left(\frac{T_{m_1+1}(y)}{m_1+1}\right)_{-1}^{1}dt$$

$$=\frac{1}{2^{k_1/2}}\sqrt{\frac{2}{\pi}}\left(\frac{1-(-1)^{m_1+1}}{m_1+1}\right)\int_0^1\psi_{n_2,m_2}(t)u\left(\frac{\xi+2n_1-1}{2^{k_1}},t\right)dt$$

$$=\frac{1}{2^{k_1/2}}\sqrt{\frac{2}{\pi}}\left(\frac{1-(-1)^{m_1+1}}{m_1+1}\right)\int_{I_2}u\left(\frac{\xi+2n_1-1}{2^{k_1}},t\right)2^{k_2/2}\bar{U}_{m_2}\left(2^{k_2}t-2n_2+1\right)dt,$$

where

$$I_2 = \left[\frac{n_2-1}{2^{k_2-1}}, \frac{n_2}{2^{k_2-1}}\right)$$

$$c_{n_1,m_1,n_2,m_2} = \frac{1}{2^{k_1/2}}\frac{2}{\pi}2^{k_2/2}\left(\frac{1-(-1)^{m_1+1}}{m_1+1}\right)\int_{I_2} u\left(\frac{\xi+2n_1-1}{2^{k_1}}, t\right)U_{m_2}\left(2^{k_2}t-2n_2+1\right)dt.$$

Again, by change of variable $2^{k_2}t-2n_2+1=w$, we get

$$\begin{aligned} c_{n_1,m_1,n_2,m_2} &= \frac{1}{2^{k_1/2}}\frac{2}{\pi}2^{k_2/2}\frac{1}{2^{k_2}}\left(\frac{1-(-1)^{m_1+1}}{m_1+1}\right)\int_{-1}^{1} u\left(\frac{\xi+2n_1-1}{2^{k_1}}, \frac{w+2n_2-1}{2^{k_2}}\right)U_{m_2}(w)dw \\ &= \frac{1}{2^{(k_1+k_2)/2}}\frac{2}{\pi}\left(\frac{1-(-1)^{m_1+1}}{m_1+1}\right)\int_{-1}^{1} u\left(\frac{\xi+2n_1-1}{2^{k_1}}, \frac{w+2n_2-1}{2^{k_2}}\right)U_{m_2}(w)dw. \end{aligned}$$

Applying the mean value theorem of integral calculus, we have

$$\begin{aligned} c_{n_1,m_1,n_2,m_2} &= \frac{1}{2^{(k_1+k_2)/2}}\frac{2}{\pi}\left(\frac{1-(-1)^{m_1+1}}{m_1+1}\right)u\left(\frac{\xi+2n_1-1}{2^{k_1}}, \frac{\eta+2n_2-1}{2^{k_2}}\right)\int_{-1}^{1} U_{m_2}(w)dw, \quad \text{where } \eta\in(-1,1) \\ &= \frac{1}{2^{(k_1+k_2)/2}}\frac{2}{\pi}\left(\frac{1-(-1)^{m_1+1}}{m_1+1}\right)u\left(\frac{\xi+2n_1-1}{2^{k_1}}, \frac{\eta+2n_2-1}{2^{k_2}}\right)\left(\frac{T_{m_2+1}(w)}{m_2+1}\right)_{-1}^{1} \\ &= \frac{1}{2^{(k_1+k_2)/2}}\frac{2}{\pi}\left(\frac{1-(-1)^{m_1+1}}{m_1+1}\right)\left(\frac{1-(-1)^{m_2+1}}{m_2+1}\right)u\left(\frac{\xi+2n_1-1}{2^{k_1}}, \frac{\eta+2n_2-1}{2^{k_2}}\right). \end{aligned}$$

Therefore,

$$\begin{aligned} \left|c_{n_1,m_1,n_2,m_2}\right| &= \frac{1}{2^{(k_1+k_2)/2}}\frac{2}{\pi}\left(\frac{1-(-1)^{m_1+1}}{m_1+1}\right)\left(\frac{1-(-1)^{m_2+1}}{m_2+1}\right)\left|u\left(\frac{\xi+2n_1-1}{2^{k_1}}, \frac{\eta+2n_2-1}{2^{k_2}}\right)\right| \\ &\le \frac{1}{2^{(k_1+k_2)/2}}\frac{2}{\pi}\left(\frac{1-(-1)^{m_1+1}}{m_1+1}\right)\left(\frac{1-(-1)^{m_2+1}}{m_2+1}\right)K. \end{aligned} \tag{6.43}$$

Therefore, $\sum_{n_1=0}^{\infty}\sum_{m_1=0}^{\infty}\sum_{n_2=0}^{\infty}\sum_{m_2=0}^{\infty} c_{n_1,m_1,n_2,m_2}$ is absolutely convergent.

Hence, the Chebyshev series expansion of $u(x,t)$ converges uniformly. □

Theorem 6.2 Error Estimate

If a continuous function, $u(x,t)\in L^2(\mathfrak{R}\times\mathfrak{R})$, defined on $[0,1)\times[0,1)$ is bounded, namely, $|u(x,t)|\le K$, then

$$\left\|\varepsilon_{u,k_1,M_1,k_2,M_2}\right\|_{L^2_\omega[[0,1)\times[0,1)]}$$

$$\leq\left[\sum_{n_1=2^{k_1-1}+1}^{\infty}\sum_{m_1=M_1}^{\infty}\sum_{n_2=2^{k_2-1}+1}^{\infty}\sum_{m_2=M_2}^{\infty}\left\{\frac{1}{2^{(k_1+k_2)/2}}\frac{2}{\pi}\left(\frac{1-(-1)^{m_1+1}}{m_1+1}\right)\left(\frac{1-(-1)^{m_2+1}}{m_2+1}\right)K\right\}^2\right]^{1/2}.$$

Proof

$$\left\|\varepsilon_{u,k_1,M_1,k_2,M_2}\right\|^2_{L^2_\omega[[0,1)\times[0,1)]}$$

$$=\int_0^1\int_0^1\left|u(x,t)-\sum_{n_1=1}^{2^{k_1-1}}\sum_{m_1=0}^{M_1-1}\sum_{n_2=1}^{2^{k_2-1}}\sum_{m_2=0}^{M_2-1}c_{n_1,m_1,n_2,m_2}\psi_{n_1,m_1}(x)\psi_{n_2,m_2}(t)\right|^2\omega_{n_1}(x)\omega_{n_2}(t)\,dx\,dt$$

$$=\int_0^1\int_0^1\left|\sum_{n_1=2^{k_1-1}+1}^{\infty}\sum_{m_1=M_1}^{\infty}\sum_{n_2=2^{k_2-1}+1}^{\infty}\sum_{m_2=M_2}^{\infty}c_{n_1,m_1,n_2,m_2}\psi_{n_1,m_1}(x)\psi_{n_2,m_2}(t)\right|^2\omega_{n_1}(x)\omega_{n_2}(t)\,dx\,dt$$

$$=\sum_{n_1=2^{k_1-1}+1}^{\infty}\sum_{m_1=M_1}^{\infty}\sum_{n_2=2^{k_2-1}+1}^{\infty}\sum_{m_2=M_2}^{\infty}\left|c_{n_1,m_1,n_2,m_2}\right|^2\int_0^1\int_0^1\left|\psi_{n_1,m_1}(x)\psi_{n_2,m_2}(t)\right|^2\omega_{n_1}(x)\omega_{n_2}(t)\,dx\,dt$$

$$=\sum_{n_1=2^{k_1-1}+1}^{\infty}\sum_{m_1=M_1}^{\infty}\sum_{n_2=2^{k_2-1}+1}^{\infty}\sum_{m_2=M_2}^{\infty}\left|c_{n_1,m_1,n_2,m_2}\right|^2. \tag{6.44}$$

Substituting Equation 6.43 of Theorem 6.1 in Equation 6.44, we obtain

$$\left\|\varepsilon_{u,k_1,M_1,k_2,M_2}\right\|^2_{L^2_\omega[[0,1)\times[0,1)]}\leq\sum_{n_1=2^{k_1-1}+1}^{\infty}\sum_{m_1=M_1}^{\infty}\sum_{n_2=2^{k_2-1}+1}^{\infty}\sum_{m_2=M_2}^{\infty}\left(\frac{1}{2^{(k_1+k_2)/2}}\frac{2}{\pi}\left(\frac{1-(-1)^{m_1+1}}{m_1+1}\right)\left(\frac{1-(-1)^{m_2+1}}{m_2+1}\right)K\right)^2.$$

This implies

$$\left\|\varepsilon_{u,k_1,M_1,k_2,M_2}\right\|_{L^2_\omega[[0,1)\times[0,1)]}$$

$$\leq\left[\sum_{n_1=2^{k_1-1}+1}^{\infty}\sum_{m_1=M_1}^{\infty}\sum_{n_2=2^{k_2-1}+1}^{\infty}\sum_{m_2=M_2}^{\infty}\left(\frac{1}{2^{(k_1+k_2)/2}}\frac{2}{\pi}\left(\frac{1-(-1)^{m_1+1}}{m_1+1}\right)\left(\frac{1-(-1)^{m_2+1}}{m_2+1}\right)K\right)^2\right]^{1/2}. \quad \square$$

6.11 Conclusion

The time-fractional fifth-order SK equation has been solved by using the two-dimensional Chebyshev wavelet method. The obtained results are then compared with exact solutions as well as with the HAM in a fractional order case. The obtained results demonstrate the accuracy, efficiency, and reliability of the proposed algorithm based on the two-dimensional Chebyshev wavelet method and its applicability to a nonlinear time-fractional SK equation. Agreement between present numerical results obtained by the Chebyshev wavelet method with HAM and exact solutions appear very satisfactory through illustrative results in Tables 6.1 through 6.3.

In the case of a nonlinear CH equation with a Riesz time-fractional derivative, the results obtained by the two-dimensional Chebyshev wavelet method are compared with the solutions obtained by HAM and VIM. The results demonstrated in Tables 6.4 through 6.6 show the accuracy, efficiency, and plausibility of the proposed algorithm based on the two-dimensional Chebyshev wavelet method and its applicability to a nonlinear CH equation with a Riesz time-fractional derivative. One can observe a pretty good agreement between the present numerical results obtained by the Chebyshev wavelet method with the HAM and VIM solutions through illustrations in Tables 6.4 through 6.6.

Next, the Chebyshev wavelet method has been successfully employed to demonstrate the approximate numerical solutions of the Riesz fractional SGE. The proposed wavelet method has been implemented for the first time to solve the fractional SGE numerically. Two test examples are given in order to show the validity and accuracy of this procedure. Also, the acquired results are compared with the exact solutions as well as with MHAM-FT [141], which reveals the efficiency and plausibility of the proposed Chebyshev wavelet method. The results exhibited in Tables 6.7 through 6.12 illustrate a pretty good agreement between the present numerical method with MHAM-FT and exact solution.

In this work, we consider the fractional order SK equation, fractional CH equation, and fractional SGEs. The motivation of the present work is to illustrate that the two-dimensional Chebyshev wavelet method is a powerful tool for solving the fractional SK equation, Riesz fractional CH equation, and Riesz fractional SGE. Finally, it is worthwhile to mention that the proposed method is a promising and powerful method for solving fractional differential equations in mathematical physics. Also, the present scheme is very simple, effective, and appropriate for obtaining numerical solutions of fractional differential equations. Analyzing the numerical results, it can be concluded that the two-dimensional Chebyshev wavelet method provides accurate numerical solutions for fractional differential equations.

7

Application of the Hermite Wavelet Method for Numerical Simulation of Fractional Differential Equations

7.1 Introduction

In this chapter, a new wavelet method based on the Hermite wavelet expansion together with operational matrices of fractional integration and derivative of wavelet functions is proposed to solve a time-fractional modified Fornberg–Whitham equation. The technique is also implemented for finding the numerical solution to a coupled system of nonlinear time-fractional Jaulent–Miodek equations. Consequently, the approximate solutions of time-fractional modified Fornberg–Whitham equations and fractional Jaulent–Miodek equations acquired by using the Hermite wavelet technique were compared with those derived by using the optimal homotopy asymptotic method (OHAM) and exact solutions.

Again, as the exact solution of fractional Fornberg–Whitham equation is unknown, we employ the first integral method (FIM) to determine exact solutions. The solitary wave solution of the fractional modified Fornberg–Whitham equation has been attained by using FIM. Analytical techniques such as FIM and OHAM are applied in order to determine the exact solutions of a fractional order modified Fornberg–Whitham equation.

The Fornberg–Whitham equation is given by [154]

$$\frac{\partial u}{\partial t}-\frac{\partial^3 u}{\partial x^2 \partial t}+\frac{\partial u}{\partial x}+u\frac{\partial u}{\partial x}=3\frac{\partial u}{\partial x}\frac{\partial^2 u}{\partial x^2}+u\frac{\partial^3 u}{\partial x^3}, \tag{7.1}$$

which was first proposed by Whitham in 1967 for studying the qualitative behavior of wave breaking [155]. In 1978, Fornberg and Whitham [156] obtained a peaked solution consisting of an arbitrary constant. Modifying the nonlinear term $u(\partial u/\partial x)$ in (7.1) to $u^2(\partial u/\partial x)$, He et al. proposed in [154] the modified Fornberg–Whitham equation as follows:

$$\frac{\partial u}{\partial t}-\frac{\partial^3 u}{\partial x^2 \partial t}+\frac{\partial u}{\partial x}+u^2\frac{\partial u}{\partial x}=3\frac{\partial u}{\partial x}\frac{\partial^2 u}{\partial x^2}+u\frac{\partial^3 u}{\partial x^3}, \quad t>0,\ x>0. \tag{7.2}$$

Consider the following time-fractional modified Fornberg–Whitham equation:

$$\frac{\partial^\alpha u}{\partial t^\alpha}-\frac{\partial^3 u}{\partial x^2 \partial t}+\frac{\partial u}{\partial x}+u^2\frac{\partial u}{\partial x}=3\frac{\partial u}{\partial x}\frac{\partial^2 u}{\partial x^2}+u\frac{\partial^3 u}{\partial x^3}. \tag{7.3}$$

Here, $0<\alpha\leq 1$ is the parameter representing the order of the fractional time derivative. The fractional derivative is considered in the Caputo sense. A great deal of research work has been invested in recent years for the study of classical order modified Fornberg–Whitham equations. Various methods such as the bifurcation theory and the method of phase portraits analysis [154], reduced differential transform method [157], and variational iteration method [158] have been developed independently for the solution of a modified Fornberg–Whitham equation. But according to the best possible information of the authors, the detailed study of the nonlinear fractional order modified Fornberg–Whitham equation is only the beginning.

In contemporary years, significant research has been done to study the classical Jaulent–Miodek equations. Various methods such as the unified algebraic method [124], Adomian decomposition method [159], tanh-sech method [160], homotopy perturbation method [161], exp-function method [162], and homotopy analysis method [163] had been implemented for solving coupled Jaulent–Miodek equations. But in keeping with the available information, the comprehensive analysis of the nonlinear fractional order coupled Jaulent–Miodek equation is only an initiation.

Consider the following time-fractional coupled Jaulent–Miodek equations:

$$\frac{\partial^\alpha u}{\partial t^\alpha}+\frac{\partial^3 u}{\partial x^3}+\frac{3}{2}v\frac{\partial^3 v}{\partial x^3}+\frac{9}{2}\frac{\partial v}{\partial x}\frac{\partial^2 v}{\partial x^2}-6u\frac{\partial u}{\partial x}-6uv\frac{\partial v}{\partial x}-\frac{3}{2}\frac{\partial u}{\partial x}v^2=0, \tag{7.4}$$

$$\frac{\partial^\alpha v}{\partial t^\alpha}+\frac{\partial^3 v}{\partial x^3}-6\frac{\partial u}{\partial x}v-6u\frac{\partial v}{\partial x}-\frac{15}{2}\frac{\partial v}{\partial x}v^2=0, \tag{7.5}$$

which is associated with energy-dependent Schrödinger potential [164–166]. Here, $0<\alpha\leq 1$ is the parameter representing the order of the fractional derivative, deemed in the Caputo sense. The present chapter lays emphasis on the implementation of the two-dimensional Hermite wavelet method to solve the problem of fractional differential equations. With a view to exhibit the capabilities of the proposed wavelet method, we employ the method to deal with fractional modified Fornberg–Whitham equations and fractional order coupled Jaulent–Miodek equations. The approximate solutions attained via the Hermite wavelet technique were compared with exact solutions and those derived by using OHAM in the case of fractional order.

7.2 Algorithm of Hermite Wavelet Method

The proposed numerical algorithm implemented in our numerical experiment is simple, very easy to implement, and it does not depend upon the mesh of discretized time and space. Furthermore, it is also efficient in computation.

Input: Consider the following general differential equation of the form:

$$\mathscr{A}u(x,t)=0, \tag{7.6}$$

which can be written as

$$\mathcal{L}u(x,t)+\mathcal{N}u(x,t)=g(x,t), \tag{7.7}$$

where
$\mathcal{L}$ is the linear operator
$\mathcal{N}$ is the nonlinear operator
$g(x,t)$ is the known function

Output: The approximate solution of $u(x,t)$.

Initial Step: Enter the values of M (order of Hermite polynomial) and k (level of resolution).

Step I: Construct the Hermite wavelet using the following formula:

$$\psi_{n_1,m_1,n_2,m_2}(x,t) = \begin{cases} AH_{m_1}\left(2^{k1}x-\hat{n}_1\right) H_{m_2}\left(2^{k_2}t-\hat{n}_2\right), & \frac{\hat{n}_1-1}{2^{k_1}} \leq x < \frac{\hat{n}_1+1}{2^{k_1}},\ \frac{\hat{n}_2-1}{2^{k_2}} \leq t < \frac{\hat{n}_2+1}{2^{k_2}} \\ 0, & \text{elsewhere} \end{cases} \tag{7.8}$$

where

$$A=\sqrt{\frac{1}{n_1!2^{n_1}\sqrt{\pi}}}\sqrt{\frac{1}{n_2!2^{n_2}\sqrt{\pi}}}2^{(k_1+k_2)/2}$$

m_1 and m_2 are order of Hermite polynomials
n_1 and n_2 are translation parameters
k_1 and k_2 are positive integers specifying level of resolution

Step II: Set

$$u(x,t)=\sum_{n_1=1}^{2^{k1-1}}\sum_{m_1=0}^{M_1-1}\sum_{n_2=1}^{2^{k2-1}}\sum_{m_2=0}^{M_2-1}a_{n_1,m_1,n_2,m_2}\psi_{n_1,m_1,n_2,m_2}(x,t), \tag{7.9}$$

and nonlinear term

$$\mathcal{N}u(x,t)=\sum_{n_1=1}^{2^{k1-1}}\sum_{m_1=0}^{M_1-1}\sum_{n_2=1}^{2^{k2-1}}\sum_{m_2=0}^{M_2-1}b_{n_1,m_1,n_2,m_2}\psi_{n_1,m_1,n_2,m_2}(x,t). \tag{7.10}$$

Step III: Compute

$$\mathcal{L}^{-1}\left[\mathcal{L}u(x,t)+\mathcal{N}u(x,t)\right]=\mathcal{L}^{-1}\left[g(x,t)\right]. \tag{7.11}$$

Step IV: For $i=1(1)M_1 2^{k_1-1}$ do
compute $x_i=(2i-1)/(2M_1 2^{k_1-1})$ (collocation points for spatial variable)
end
For $j=1(1)M_2 2^{k_2-1}$ do
calculate $t_j=(2j-1)/(2M_2 2^{k_2-1})$ (collocation points for temporal variable)
end

Step V: Substituting the collocation points x_i and t_j obtained in Step IV in Equation 7.11, obtain the system of algebraic equations in a_{n_1,m_1,n_2,m_2} and b_{n_1,m_1,n_2,m_2}.

Step VI: By solving the system of equations obtained in Step V using Newton's method, the Hermite wavelet coefficients a_{n_1,m_1,n_2,m_2} and b_{n_1,m_1,n_2,m_2} can be obtained.

Step VII: Substituting the value of a_{n_1,m_1,n_2,m_2} in Equation 7.9, obtain the approximate solution for $u(x,t)$.

Step VIII: Stop.

7.3 Application of Analytical and Numerical Methods for Solving a Time-Fractional Modified Fornberg–Whitham Equation

7.3.1 Two-Dimensional Hermite Wavelet Method for Solving a Nonlinear Time-Fractional Modified Fornberg–Whitham Equation

To exhibit the effectiveness and accuracy of the proposed numerical scheme, we consider the time-fractional modified Fornberg–Whitham equation. The numerical solutions thus obtained are compared with the exact solutions in case of classical order and with the solutions obtained by OHAM in case of fractional order, respectively.

Consider the nonlinear time-fractional modified Fornberg–Whitham equation

$$\frac{\partial^\alpha u}{\partial t^\alpha}-\frac{\partial^3 u}{\partial x^2\partial t}+\frac{\partial u}{\partial x}+u^2\frac{\partial u}{\partial x}=3\frac{\partial u}{\partial x}\frac{\partial^2 u}{\partial x^2}+u\frac{\partial^3 u}{\partial x^3},\quad t>0,\ x>0, \tag{7.12}$$

with initial condition

$$u(x,0)=\frac{3}{4}\left(\sqrt{15}-5\right)\operatorname{sech}^2\left(\frac{1}{20}\sqrt{10\left(5-\sqrt{15}\right)}\right)x. \tag{7.13}$$

The exact solution of Equation 7.12 is given by [154]

$$u(x,t)=\frac{3}{4}\left(\sqrt{15}-5\right)\operatorname{sech}^2\left[\frac{1}{20}\sqrt{10\left(5-\sqrt{15}\right)}\left(x-\left(5-\sqrt{15}\right)t\right)\right]. \tag{7.14}$$

The Hermite wavelet solution of $u(x,t)$ is sought by assuming that $u(x,t)$ can be expanded in terms of a Hermite wavelet as

$$u(x,t)=\sum_{n=1}^{2^{k_1-1}}\sum_{i=0}^{M_1-1}\sum_{l=1}^{2^{k_2-1}}\sum_{j=0}^{M_2-1}d_{n,i,l,j}\psi_{n,i,l,j}(x,t), \tag{7.15}$$

where $n=1,\ldots,2^{k_1-1}$, $i=0,\ldots,M_1-1$, $l=1,\ldots,2^{k_2-1}$, $j=0,\ldots,M_2-1$.

The nonlinear terms presented in Equation 7.12 can be approximated using a Hermite wavelet function as

$$u^2\frac{\partial u}{\partial x}=\sum_{n=1}^{2^{k_1-1}}\sum_{i=0}^{M_1-1}\sum_{l=1}^{2^{k_2-1}}\sum_{j=0}^{M_2-1}a_{n,i,l,j}\psi_{n,i,l,j}(x,t), \tag{7.16}$$

$$\frac{\partial u}{\partial x}\frac{\partial^2 u}{\partial x^2}=\sum_{n=1}^{2^{k_1-1}}\sum_{i=0}^{M_1-1}\sum_{l=1}^{2^{k_2-1}}\sum_{j=0}^{M_2-1}b_{n,i,l,j}\psi_{n,i,l,j}(x,t), \tag{7.17}$$

and

$$u\frac{\partial^3 u}{\partial x^3} = \sum_{n=1}^{2^{k1-1}}\sum_{i=0}^{M_1-1}\sum_{l=1}^{2^{k2-1}}\sum_{j=0}^{M_2-1} c_{n,i,l,j}\psi_{n,i,l,j}(x,t). \tag{7.18}$$

This implies

$$\left(\sum_{n=1}^{2^{k1-1}}\sum_{i=0}^{M_1-1}\sum_{l=1}^{2^{k2-1}}\sum_{j=0}^{M_2-1} d_{n,i,l,j}\psi_{n,i,l,j}(x,t)\right)^2 \times \left[\sum_{n=1}^{2^{k1-1}}\sum_{i=0}^{M_1-1}\sum_{l=1}^{2^{k2-1}}\sum_{j=0}^{M_2-1} d_{n,i,l,j}\frac{\partial \psi_{n,i,l,j}(x,t)}{\partial x}\right]$$

$$= \sum_{n=1}^{2^{k1-1}}\sum_{i=0}^{M_1-1}\sum_{l=1}^{2^{k2-1}}\sum_{j=0}^{M_2-1} a_{n,i,l,j}\psi_{n,i,l,j}(x,t). \tag{7.19}$$

$$\left(\sum_{n=1}^{2^{k1-1}}\sum_{i=0}^{M_1-1}\sum_{l=1}^{2^{k2-1}}\sum_{j=0}^{M_2-1} d_{n,i,l,j}\frac{\partial \psi_{n,i,l,j}(x,t)}{\partial x}\right) \times \left[\sum_{n=1}^{2^{k1-1}}\sum_{i=0}^{M_1-1}\sum_{l=1}^{2^{k2-1}}\sum_{j=0}^{M_2-1} d_{n,i,l,j}\frac{\partial^2 \psi_{n,i,l,j}(x,t)}{\partial x^2}\right]$$

$$= \sum_{n=1}^{2^{k1-1}}\sum_{i=0}^{M_1-1}\sum_{l=1}^{2^{k2-1}}\sum_{j=0}^{M_2-1} b_{n,i,l,j}\psi_{n,i,l,j}(x,t), \tag{7.20}$$

and

$$\left[\sum_{n=1}^{2^{k1-1}}\sum_{i=0}^{M_1-1}\sum_{l=1}^{2^{k2-1}}\sum_{j=0}^{M_2-1} d_{n,i,l,j}\psi_{n,i,l,j}(x,t)\right] \times \left(\sum_{n=1}^{2^{k1-1}}\sum_{i=0}^{M_1-1}\sum_{l=1}^{2^{k2-1}}\sum_{j=0}^{M_2-1} d_{n,i,l,j}\frac{\partial^3 \psi_{n,i,l,j}(x,t)}{\partial x^3}\right)$$

$$= \sum_{n=1}^{2^{k1-1}}\sum_{i=0}^{M_1-1}\sum_{l=1}^{2^{k2-1}}\sum_{j=0}^{M_2-1} c_{n,i,l,j}\psi_{n,i,l,j}(x,t). \tag{7.21}$$

Again applying J_t^α on both sides of Equation 7.12, we have

$$u(x,t) - u(x,0) = J_t^\alpha\left[\frac{\partial^3 u}{\partial x^2 \partial t} - \frac{\partial u}{\partial x} - u^2\frac{\partial u}{\partial x} + 3\frac{\partial u}{\partial x}\frac{\partial^2 u}{\partial x^2} + u\frac{\partial^3 u}{\partial x^3}\right]. \tag{7.22}$$

Putting Equations 7.15 through 7.18 in Equation 7.22, we have

$$\sum_{n=1}^{2^{k1-1}}\sum_{i=0}^{M_1-1}\sum_{l=1}^{2^{k2-1}}\sum_{j=0}^{M_2-1} d_{n,i,l,j}\psi_{n,i,l,j}(x,t) - u(x,0) = J_t^\alpha\left[\frac{\partial^3}{\partial x^2 \partial t}\left(\sum_{n=1}^{2^{k1-1}}\sum_{i=0}^{M_1-1}\sum_{l=1}^{2^{k2-1}}\sum_{j=0}^{M_2-1} d_{n,i,l,j}\psi_{n,i,l,j}(x,t)\right)\right.$$

$$-\frac{\partial}{\partial x}\left(\sum_{n=1}^{2^{k1-1}}\sum_{i=0}^{M_1-1}\sum_{l=1}^{2^{k2-1}}\sum_{j=0}^{M_2-1} d_{n,i,l,j}\psi_{n,i,l,j}(x,t)\right) - \sum_{n=1}^{2^{k1-1}}\sum_{i=0}^{M_1-1}\sum_{l=1}^{2^{k2-1}}\sum_{j=0}^{M_2-1} a_{n,i,l,j}\psi_{n,i,l,j}(x,t)$$

$$\left. +3\left(\sum_{n=1}^{2^{k1-1}}\sum_{i=0}^{M_1-1}\sum_{l=1}^{2^{k2-1}}\sum_{j=0}^{M_2-1} b_{n,i.l,j}\psi_{n,i,l,j}(x,t)\right) + \sum_{n=1}^{2^{k1-1}}\sum_{i=0}^{M_1-1}\sum_{l=1}^{2^{k2-1}}\sum_{j=0}^{M_2-1} c_{n,i,l,j}\psi_{n,i,l,j}(x,t)\right]. \tag{7.23}$$

Now, substituting the collocation points $x_l=(l-0.5)/(2^{k_1-1}M_1)$ and $t_r=(r-0.5)/(2^{k_2-1}M_2)$ for $l=1,2,\ldots,2^{k_1-1}M_1$ and $r=1,2,\ldots,2^{k_2-1}M_2$ in Equations 7.19 through 7.21 and 7.23, we have $4(2^{k_1-1}M_1)(2^{k_2-1}M_2)$ equations in $4(2^{k_1-1}M_1)(2^{k_2-1}M_2)$ unknowns in $a_{n,i,l,j}$, $b_{n,i,l,j}$, $c_{n,i,l,j}$, and $d_{n,i,l,j}$. By solving this system of equations using Newton's method, the Hermite wavelet coefficients $a_{n,i,l,j}$, $b_{n,i,l,j}$, $c_{n,i,l,j}$, and $d_{n,i,l,j}$ can be obtained.

7.3.2 Comparison with OHAM for the Solution of Time-Fractional Modified Fornberg–Whitham Equation

Using OHAM [30–32], the homotopy for Equation 7.12 can be written as

$$(1-p)L(\varphi(x,t;p))=H(p)\left[\frac{\partial^{\alpha}\varphi(x,t;p)}{\partial t^{\alpha}}-\frac{\partial^3}{\partial x^2\partial t}\varphi(x,t;p)+\frac{\partial}{\partial x}\varphi(x,t;p)\right.$$
$$\left.+\varphi(x,t;p)^2\frac{\partial\varphi(x,t;p)}{\partial x}-3\frac{\partial\varphi(x,t;p)}{\partial x}\frac{\partial^2\varphi(x,t;p)}{\partial x^2}-\varphi(x,t;p)\frac{\partial^3\varphi(x,t;p)}{\partial x^3}\right],\tag{7.24}$$

where

$$\varphi(x,t,p)=u_0(x,t)+\sum_{i=1}^{\infty}u_i(x,t)p^i\tag{7.25}$$

$$H(p)=C_1p+C_2P^2+C_3p^3+\cdots.\tag{7.26}$$

Substituting Equations 7.25 and 7.26 in Equation 7.24 and equating the coefficients of different powers in p, we have the following system of partial differential equations:

$$\text{Coefficients of } p^0:\ \frac{\partial^{\alpha}u_0(x,t)}{\partial t^{\alpha}}=0\tag{7.27}$$

$$\text{Coefficients of } p^1:\ \frac{\partial^{\alpha}u_1(x,t)}{\partial t^{\alpha}}-\frac{\partial^{\alpha}u_0(x,t)}{\partial t^{\alpha}}=C_1\left[\frac{\partial^{\alpha}u_0(x,t)}{\partial t^{\alpha}}-\frac{\partial^3u_0(x,t)}{\partial x^2\partial t}+\frac{\partial u_0(x,t)}{\partial x}\right.$$
$$\left.+(u_0(x,t))^2\frac{\partial u_0(x,t)}{\partial x}-3\frac{\partial u_0(x,t)}{\partial x}\frac{\partial^2u_0(x,t)}{\partial x^2}-u_0(x,t)\frac{\partial^3u_0(x,t)}{\partial x^3}\right]\tag{7.28}$$

$$\text{Coefficients of } p^2:\ \frac{\partial^{\alpha}u_2(x,t)}{\partial t^{\alpha}}-\frac{\partial^{\alpha}u_1(x,t)}{\partial t^{\alpha}}=C_1\left[\frac{\partial^{\alpha}u_1(x,t)}{\partial t^{\alpha}}-\frac{\partial^3u_1(x,t)}{\partial x^2\partial t}+\frac{\partial u_1(x,t)}{\partial x}\right.$$
$$+\left(2u_0(x,t)u_1(x,t)\frac{\partial u_0(x,t)}{\partial x}+(u_0(x,t))^2\frac{\partial u_1(x,t)}{\partial x}\right)$$
$$-3\left(\frac{\partial u_0(x,t)}{\partial x}\frac{\partial^2u_1(x,t)}{\partial x^2}+\frac{\partial u_1(x,t)}{\partial x}\frac{\partial^2u_0(x,t)}{\partial x^2}\right)$$
$$\left.-\left(u_0(x,t)\frac{\partial^3u_1(x,t)}{\partial x^3}+u_1(x,t)\frac{\partial^3u_0(x,t)}{\partial x^3}\right)\right]+C_2\left[\frac{\partial^{\alpha}u_0(x,t)}{\partial t^{\alpha}}-\frac{\partial^3u_0(x,t)}{\partial x^2\partial t}\right.$$
$$\left.+\frac{\partial u_0(x,t)}{\partial x}+(u_0(x,t))^2\frac{\partial u_0(x,t)}{\partial x}-3\frac{\partial u_0(x,t)}{\partial x}\frac{\partial^2u_0(x,t)}{\partial x^2}-u_0(x,t)\frac{\partial^3u_0(x,t)}{\partial x^3}\right]\tag{7.29}$$

$$\text{Coefficients of } p^3: \frac{\partial^\alpha u_3(x,t)}{\partial t^\alpha} - \frac{\partial^\alpha u_2(x,t)}{\partial t^\alpha} = C_1\left[\frac{\partial^\alpha u_2(x,t)}{\partial t^\alpha} - \frac{\partial^3 u_2(x,t)}{\partial x^2 \partial t} + \frac{\partial u_2(x,t)}{\partial x}\right.$$

$$+\left((u_1(x,t))^2 \frac{\partial u_0(x,t)}{\partial x} + 2u_0(x,t)u_2(x,t)\frac{\partial u_0(x,t)}{\partial x} + 2u_0(x,t)u_1(x,t)\frac{\partial u_1(x,t)}{\partial x} + (u_0(x,t))^2 \frac{\partial u_2(x,t)}{\partial x}\right)$$

$$-3\left(\frac{\partial u_0(x,t)}{\partial x}\frac{\partial^2 u_2(x,t)}{\partial x^2} + \frac{\partial u_1(x,t)}{\partial x}\frac{\partial^2 u_1(x,t)}{\partial x^2} + \frac{\partial u_2(x,t)}{\partial x}\frac{\partial^2 u_0(x,t)}{\partial x^2}\right)$$

$$\left.-\left(u_0(x,t)\frac{\partial^3 u_2(x,t)}{\partial x^3} + u_1(x,t)\frac{\partial^3 u_1(x,t)}{\partial x^3} + u_2(x,t)\frac{\partial^3 u_0(x,t)}{\partial x^3}\right)\right] + C_2\left[\frac{\partial^\alpha u_1(x,t)}{\partial t^\alpha}\right.$$

$$-\frac{\partial^3 u_1(x,t)}{\partial x^2 \partial t} + \frac{\partial u_1(x,t)}{\partial x} + \left(2u_0(x,t)u_1(x,t)\frac{\partial u_0(x,t)}{\partial x} + (u_0(x,t))^2 \frac{\partial u_1(x,t)}{\partial x}\right)$$

$$\left.-3\left(\frac{\partial u_0(x,t)}{\partial x}\frac{\partial^2 u_1(x,t)}{\partial x^2} + \frac{\partial u_1(x,t)}{\partial x}\frac{\partial^2 u_0(x,t)}{\partial x^2}\right) - \left(u_0(x,t)\frac{\partial^3 u_1(x,t)}{\partial x^3} + u_1(x,t)\frac{\partial^3 u_0(x,t)}{\partial x^3}\right)\right]$$

$$+C_3\left[\frac{\partial^\alpha u_0(x,t)}{\partial t^\alpha} - \frac{\partial^3 u_0(x,t)}{\partial x^2 \partial t} + \frac{\partial u_0(x,t)}{\partial x} + (u_0(x,t))^2 \frac{\partial u_0(x,t)}{\partial x}\right.$$

$$\left.-3\frac{\partial u_0(x,t)}{\partial x}\frac{\partial^2 u_0(x,t)}{\partial x^2} - u_0(x,t)\frac{\partial^3 u_0(x,t)}{\partial x^3}\right] \quad (7.30)$$

and so on.

For solving a fractional order modified Fornberg–Whitham equation using OHAM, we consider the following initial condition for Equation 7.12:

$$u(x,0) = \frac{3}{4}\left(\sqrt{15}-5\right)\operatorname{sech}^2\left(\frac{1}{20}\sqrt{10\left(5-\sqrt{15}\right)}\right)x.$$

Using the initial condition $u_0 = u(x,0)$ and solving Equations 7.27 through 7.30, we obtain the expressions for u_0, u_1, u_2, u_3, and so on.

Finally, the third-order approximate solution for a time-fractional modified Fornberg–Whitham equation is given by

$$u = u_0(x,t) + u_1(x,t) + u_2(x,t) + u_3(x,t). \quad (7.31)$$

The optimal values of the convergence control parameters $C_1, C_2,$ and C_3 can be obtained using the weighted residual least square method given in Equations 1.34 and 1.35.

7.4 Numerical Results and Discussion

The comparison of the absolute errors for the time-fractional modified Fornberg–Whitham Equation 7.12 have been exhibited in Tables 7.1 and 7.2, which are generated using the results obtained by the two-dimensional Hermite wavelet method and OHAM at different

TABLE 7.1

Absolute Errors Obtained by the Two-Dimensional Hermite Wavelet Method for the Nonlinear Modified Fornberg–Whitham Equation Given in Equation 7.12 at Various Points of x and t, Taking $\alpha = 1$

	$\lvert u_{Exact} - u_{Hermite\,wavelet} \rvert$								
x	$t=0.1$	$t=0.2$	$t=0.3$	$t=0.4$	$t=0.5$	$t=0.6$	$t=0.7$	$t=0.8$	$t=0.9$
0.1	3.7564E−4	5.5722E−4	5.1447E−4	2.1718E−4	3.6388E−4	1.2543E−3	2.4734E−3	4.0313E−3	5.9263E−3
0.2	3.6817E−4	5.4379E−4	5.0090E−4	2.1372E−4	3.4281E−4	1.1905E−3	2.3457E−3	3.8163E−3	5.6007E−3
0.3	3.6433E−4	5.3561E−4	4.9206E−4	2.1218E−4	3.2489E−4	1.1373E−3	2.2382E−3	3.6336E−3	5.3201E−3
0.4	3.6412E−4	5.3275E−4	4.8812E−4	2.1292E−4	3.0961E−4	1.0938E−3	2.1497E−3	3.4808E−3	5.0827E−3
0.5	3.6757E−4	5.3531E−4	4.8929E−4	2.1630E−4	2.9639E−4	1.0593E−3	2.0792E−3	3.3572E−3	4.8871E−3
0.6	3.7478E−4	5.4349E−4	4.9589E−4	2.2274E−4	2.8468E−4	1.0331E−3	2.0259E−3	3.2615E−3	4.7320E−3
0.7	3.8584E−4	5.5751E−4	5.0827E−4	2.3274E−4	2.7383E−4	1.0144E−3	1.9888E−3	3.1928E−3	4.6163E−3
0.8	4.0091E−4	5.7767E−4	5.2688E−4	2.4690E−4	2.6312E−4	1.0023E−3	1.9670E−3	3.1500E−3	4.5391E−3
0.9	4.2025E−4	6.0441E−4	5.5231E−4	2.6594E−4	2.5171E−4	9.9595E−4	1.9595E−3	3.1322E−3	4.4992E−3

TABLE 7.2

Absolute Errors Obtained by OHAM for the Modified Fornberg–Whitham Equation Given in Equation 7.12 at Various Points of x and t, Taking $\alpha=1$

	$\lvert u_{Exact}-u_{OHAM}\rvert$								
x	$t=0.1$	$t=0.2$	$t=0.3$	$t=0.4$	$t=0.5$	$t=0.6$	$t=0.7$	$t=0.8$	$t=0.9$
0.1	6.4610E−6	1.2928E−5	5.7106E−5	1.2329E−4	2.0699E−4	3.0203E−4	4.0061E−4	4.9330E−4	5.6917E−4
0.2	2.5099E−5	2.2342E−5	8.7492E−6	6.6918E−5	1.4918E−4	2.5087E−4	3.6563E−4	4.8544E−4	6.0074E−4
0.3	4.3236E−5	5.6305E−5	3.7197E−5	1.4365E−5	9.6935E−5	2.0734E−4	3.4071E−4	4.9050E−4	6.4853E−4
0.4	6.0707E−5	8.8650E−5	8.0295E−5	3.3830E−5	5.0831E−5	1.7204E−4	3.2644E−4	5.0897E−4	7.1289E−4
0.5	7.7360E−5	1.1908E−4	1.2014E−4	7.7185E−5	1.1402E−5	1.4551E−4	3.2329E−4	5.4121E−4	7.9400E−4
0.6	9.3052E−5	1.4735E−4	1.5637E−4	1.1527E−4	2.0905E−5	1.2816E−4	3.3161E−4	5.8741E−4	8.9183E−4
0.7	1.0765E−4	1.7320E−4	1.8867E−4	1.4773E−4	4.5730E−5	1.2029E−4	3.5157E−4	6.4760E−4	1.0061E−3
0.8	1.2105E−4	1.9644E−4	2.1678E−4	1.7429E−4	6.2814E−5	1.2211E−4	3.8324E−4	7.2163E−4	1.1365E−3
0.9	1.3314E−4	2.1689E−4	2.4049E−4	1.9473E−4	7.1997E−5	1.3366E−4	4.2653E−4	8.0916E−4	1.2824E−3

values of x and t, taking $\alpha=1$. In the present analysis, to examine the accuracy and reliability of the Hermite wavelets for solving a fractional order modified Fornberg–Whitham equation, we compare the approximate solution of Hermite wavelets with the third-order approximate solution obtained by OHAM. Tables 7.3 and 7.4 show the absolute errors of the fractional order modified Fornberg–Whitham Equation 7.12 at various points of x and t, taking $\alpha=0.75$ and 0.5, respectively. Agreement between present numerical results for a time-fractional modified Fornberg–Whitham equation obtained by Hermite wavelets and OHAM appears very satisfactory through illustrations in Tables 7.5 and 7.6. Table 7.7 shows the L_2 and L_∞ error norms for a fractional modified Fornberg–Whitham equation using two-dimensional Hermite wavelet methods and OHAM at various points of t, taking $\alpha=0.75$ and 0.5.

Figures 7.1 and 7.2 demonstrate the graphical comparison of the numerical solutions obtained by two-dimensional Hermite wavelet approximation with regard to OHAM for $\alpha=0.75$ and 0.5, respectively.

7.5 Application of Analytical Methods to Determine the Exact Solutions of a Time-Fractional Modified Fornberg–Whitham Equation

7.5.1 Implementation of the FIM for Solving a Fractional Modified Fornberg–Whitham Equation

Let us consider the following time-fractional modified Fornberg–Whitham equation:

$$\frac{\partial^\alpha u}{\partial t^\alpha}-\frac{\partial^\alpha}{\partial t^\alpha}\left(\frac{\partial^2 u}{\partial x^2}\right)+\frac{\partial u}{\partial x}+u^2\frac{\partial u}{\partial x}=3\frac{\partial u}{\partial x}\frac{\partial^2 u}{\partial x^2}+u\frac{\partial^3 u}{\partial x^3}. \tag{7.32}$$

Here, $0<\alpha\leq 1$ is the parameter representing the order of the fractional time derivative. In this present section, a new exact solution of a fractional modified Fornberg–Whitham equation has been derived using the FIM. The FIM has been implemented for finding the exact solution of the fractional modified Fornberg–Whitham equation given in Equation 7.32 with an intention to demonstrate the efficacy and accuracy of the proposed method. The exact solutions devised are compared with the numerical solution obtained by OHAM.

Let us introduce the following fractional complex transform in Equation 7.32:

$$u(x,t)=\Psi(\xi),\quad \xi=kx-\frac{\lambda t^\alpha}{\Gamma(\alpha+1)}, \tag{7.33}$$

where k and λ are constants.

By applying the fractional complex transform (7.33), Equation 7.32 can be transformed to the following nonlinear ordinary differential equation:

$$-\lambda\Psi_\xi(\xi)+\lambda k^2\Psi_{\xi\xi\xi}(\xi)+k\Psi_\xi(\xi)+k(\Psi(\xi))^2\Psi_\xi(\xi)=3k^3\Psi_\xi(\xi)\Psi_{\xi\xi}(\xi)+k^3\Psi(\xi)\Psi_{\xi\xi\xi}(\xi). \tag{7.34}$$

TABLE 7.3

Absolute Errors Obtained by the Two-Dimensional Hermite Wavelet Method and Third-Order OHAM Solution for the Fractional Order Nonlinear Modified Fornberg–Whitham Equation Given in Equation 7.12 at Various Points of x and t, Taking $\alpha = 0.75$

	$\lvert u_{Hermite\ wavelet} - u_{OHAM} \rvert$								
x	$t=0.1$	$t=0.2$	$t=0.3$	$t=0.4$	$t=0.5$	$t=0.6$	$t=0.7$	$t=0.8$	$t=0.9$
0.1	4.7565E−4	3.5641E−3	8.0016E−3	1.3396E−2	1.9495E−2	2.6133E−2	3.3182E−2	4.0509E−2	4.7923E−2
0.2	8.8635E−4	1.2693E−3	4.8602E−3	9.4511E−3	1.4770E−2	2.0648E−2	2.6963E−2	3.3585E−2	4.0320E−2
0.3	2.2468E−3	1.0167E−3	1.7374E−3	5.5349E−3	1.0084E−2	1.5214E−2	2.0804E−2	2.6727E−2	3.2788E−2
0.4	3.6037E−3	3.2904E−3	1.3619E−3	1.6542E−3	5.4468E−3	9.8382E−3	1.4712E−2	1.9946E−2	2.5339E−2
0.5	4.9553E−3	5.5482E−3	4.4327E−3	2.1843E−3	8.6443E−4	4.5308E−3	8.7013E−3	1.3255E−2	1.7987E−2
0.6	6.2997E−3	7.7868E−3	7.4700E−3	5.9744E−3	3.6545E−3	6.9903E−4	2.7806E−3	6.6656E−3	1.0747E−2
0.7	7.6352E−3	1.0003E−2	1.0469E−2	9.7098E−3	8.1023E−3	5.8419E−3	3.0385E−3	1.9102E−4	3.6320E−3
0.8	8.9603E−3	1.2193E−2	1.3425E−2	1.3384E−2	1.2471E−2	1.0888E−2	8.7451E−3	6.1561E−3	3.3425E−3
0.9	1.0273E−2	1.4356E−2	1.6335E−2	1.6992E−2	1.6754E−2	1.5830E−2	1.4328E−2	1.2363E−2	1.0162E−2

TABLE 7.4

Absolute Errors Obtained by the Two-Dimensional Hermite Wavelet Method and Third-Order OHAM Solution for the Fractional Order Nonlinear Modified Fornberg–Whitham Equation Given in Equation 7.12 at Various Points of x and t, Taking $\alpha = 0.5$

	$\lvert u_{OHAM} - u_{Hermite\,wavelet} \rvert$								
x	$t=0.1$	$t=0.2$	$t=0.3$	$t=0.4$	$t=0.5$	$t=0.6$	$t=0.7$	$t=0.8$	$t=0.9$
0.1	1.2334E−2	4.0688E−3	1.4027E−2	1.6573E−2	1.5687E−2	1.5562E−2	1.5805E−2	6.6256E−3	3.5980E−2
0.2	1.5565E−2	3.8814E−4	1.0784E−2	1.3953E−2	1.3347E−2	1.3079E−2	1.3374E−2	6.0014E−3	3.0299E−2
0.3	1.8769E−2	3.2892E−3	7.5422E−3	1.1332E−2	1.0998E−2	1.0573E−2	1.0901E−2	5.3094E−3	2.4724E−2
0.4	2.1934E−2	6.9543E−3	4.3072E−3	8.7171E−3	8.6477E−3	8.0497E−3	8.3899E−3	4.5590E−3	1.9229E−2
0.5	2.5043E−2	1.0597E−2	1.0867E−3	6.1147E−3	6.3028E−3	5.5169E−3	5.8492E−3	3.7602E−3	1.3790E−2
0.6	2.8082E−2	1.4209E−2	2.1116E−3	3.5329E−3	3.9725E−3	2.9834E−3	3.2865E−3	2.9235E−3	8.3783E−3
0.7	3.1037E−2	1.7780E−2	5.2800E−3	9.8022E−4	1.6663E−3	4.5848E−4	7.1012E−4	2.0604E−3	2.9665E−3
0.8	3.3892E−2	2.1298E−2	8.4099E−3	1.5347E−3	6.0631E−4	2.0482E−3	1.8707E−3	1.1828E−3	2.4738E−3
0.9	3.6632E−2	2.4754E−2	1.1492E−2	4.0027E−3	2.8348E−3	4.5263E−3	4.4464E−3	3.0333E−4	7.9720E−3

TABLE 7.5

Comparison of Approximate Solutions Obtained by the Two-Dimensional Hermite Wavelet Method and OHAM for the Fractional Order Nonlinear Modified Fornberg–Whitham Equation Given in Equation 7.12 at Various Points of x and t, Taking $\alpha = 0.75$

	$t=0.1$		$t=0.2$		$t=0.3$		$t=0.4$		$t=0.5$		$t=0.6$		$t=0.7$		$t=0.8$		$t=0.9$	
x	u_{HW}	u_{OHAM}	u_{HW}	u_{OHAM}	u_{HW}	u_{OHAM}	u_{HW}	u_{OHAM}	u_{HW}	u_{OHAM}	u_{HW}	u_{OHAM}	u_{HW}	u_{OHAM}	u_{HW}	u_{OHAM}	u_{HW}	u_{OHAM}
0.1	−0.84414	−0.84495	−0.84068	−0.84367	−0.83584	−0.84165	−0.83002	−0.83905	−0.82346	−0.83594	−0.81635	−0.83238	−0.80879	−0.82840	−0.80095	−0.82404	−0.79298	−0.81933
0.2	−0.84449	−0.84509	−0.84175	−0.84444	−0.83759	−0.84298	−0.83241	−0.84090	−0.82649	−0.83828	−0.81999	−0.83518	−0.81304	−0.83164	−0.80576	−0.82770	−0.79836	−0.82338
0.3	−0.84435	−0.84476	−0.84235	−0.84473	−0.83885	−0.84385	−0.83431	−0.84229	−0.82901	−0.84016	−0.82312	−0.83753	−0.81677	−0.83444	−0.81006	−0.83093	−0.80321	−0.82703
0.4	−0.84375	−0.84395	−0.84247	−0.84455	−0.83963	−0.84423	−0.83571	−0.84320	−0.83103	−0.84157	−0.82574	−0.83942	−0.81997	−0.83679	−0.81383	−0.83372	−0.80752	−0.83024
0.5	−0.84268	−0.84266	−0.84211	−0.84389	−0.83991	−0.84415	−0.83661	−0.84365	−0.83253	−0.84253	−0.82783	−0.84085	−0.82264	−0.83869	−0.81705	−0.83606	−0.81129	−0.83302
0.6	−0.84115	−0.84091	−0.84128	−0.84277	−0.83971	−0.84358	−0.83702	−0.84362	−0.83352	−0.84301	−0.82941	−0.84182	−0.82478	−0.84013	−0.81975	−0.83796	−0.81452	−0.83535
0.7	−0.83915	−0.83868	−0.83998	−0.84117	−0.83903	−0.84255	−0.83693	−0.84312	−0.83401	−0.84302	−0.83046	−0.84233	−0.82639	−0.84110	−0.82189	−0.83939	−0.81719	−0.83724
0.8	−0.83669	−0.83600	−0.83821	−0.83910	−0.83787	−0.84105	−0.83634	−0.84215	−0.83399	−0.84256	−0.83099	−0.84236	−0.82746	−0.84162	−0.82349	−0.84038	−0.81932	−0.83867
0.9	−0.83379	−0.83286	−0.83598	−0.83657	−0.83624	−0.83908	−0.83527	−0.84072	−0.83346	−0.84163	−0.83101	−0.84193	−0.82800	−0.84167	−0.82455	−0.84090	−0.82088	−0.83966

TABLE 7.6

Comparison of Approximate Solutions Obtained by the Two-Dimensional Hermite Wavelet Method and OHAM for the Fractional Order Nonlinear Modified Fornberg–Whitham Equation Given in Equation 7.12 at Various Points of x and t, Taking $\alpha = 0.5$

	$t=0.1$		$t=0.2$		$t=0.3$		$t=0.4$		$t=0.5$		$t=0.6$		$t=0.7$		$t=0.8$		$t=0.9$	
x	u_{HW}	u_{OHAM}	u_{HW}	u_{OHAM}	u_{HW}	u_{OHAM}	u_{HW}	u_{OHAM}	u_{HW}	u_{OHAM}	u_{HW}	u_{OHAM}	u_{HW}	u_{OHAM}	u_{HW}	u_{OHAM}	u_{HW}	u_{OHAM}
0.1	−0.85594	−0.84360	−0.83668	−0.84073	−0.82344	−0.83747	−0.81743	−0.83401	−0.81471	−0.83040	−0.81114	−0.82671	−0.80713	−0.82294	−0.81248	−0.81911	−0.85122	−0.81524
0.2	−0.85929	−0.84372	−0.84118	−0.84157	−0.82807	−0.83885	−0.82188	−0.83584	−0.81928	−0.83263	−0.81620	−0.82928	−0.81246	−0.82583	−0.81629	−0.82230	−0.84899	−0.81869
0.3	−0.86214	−0.84337	−0.84523	−0.84195	−0.83223	−0.83977	−0.82589	−0.83722	−0.82342	−0.83442	−0.82086	−0.83143	−0.81740	−0.82831	−0.81976	−0.82507	−0.84646	−0.82174
0.4	−0.86449	−0.84256	−0.84881	−0.84185	−0.83593	−0.84024	−0.82944	−0.83815	−0.82711	−0.83576	−0.82509	−0.83314	−0.82196	−0.83035	−0.82286	−0.82742	−0.84361	−0.82438
0.5	−0.86632	−0.84127	−0.85189	−0.84130	−0.83916	−0.84025	−0.83252	−0.83863	−0.83035	−0.83665	−0.82889	−0.83441	−0.82611	−0.83196	−0.82559	−0.82935	−0.84039	−0.82660
0.6	−0.86762	−0.83954	−0.85449	−0.84028	−0.84191	−0.83979	−0.83513	−0.83866	−0.83313	−0.83710	−0.83225	−0.83523	−0.82985	−0.83314	−0.82793	−0.83085	−0.83678	−0.82841
0.7	−0.86838	−0.83735	−0.85659	−0.83882	−0.84417	−0.83889	−0.83725	−0.83823	−0.83543	−0.83710	−0.83516	−0.83562	−0.83317	−0.83388	−0.82986	−0.83192	−0.83276	−0.82979
0.8	−0.86860	−0.83471	−0.85819	−0.83689	−0.84594	−0.83753	−0.83889	−0.83736	−0.83726	−0.83665	−0.83761	−0.83556	−0.83605	−0.83418	−0.83138	−0.83256	−0.82827	−0.83075
0.9	−0.86826	−0.83163	−0.85929	−0.83453	−0.84722	−0.83573	−0.84004	−0.83604	−0.83859	−0.83576	−0.83959	−0.83506	−0.83849	−0.83405	−0.83246	−0.83277	−0.82330	−0.83127

TABLE 7.7

L_2 and L_∞ Error Norms for a Fractional Order Nonlinear Modified Fornberg–Whitham Equation Using Two-Dimensional Hermite Wavelet Methods at Various Points of t, Taking $\alpha = 0.75$ and 0.5

	$\alpha = 0.75$		$\alpha = 0.5$	
t	L_2	L_∞	L_2	L_∞
0.1	1.80944E−2	3.6632E−2	7.80858E−2	1.0273E−2
0.2	2.39252E−2	2.4754E−2	4.21019E−2	1.4356E−2
0.3	2.69190E−2	1.4027E−2	2.49960E−2	1.6335E−2
0.4	3.00791E−2	1.6573E−2	2.72568E−2	1.6992E−2
0.5	3.52921E−2	1.5687E−2	2.62040E−2	1.9495E−2
0.6	4.31487E−2	1.5562E−2	2.55744E−2	2.6133E−2
0.7	5.34017E−2	1.5805E−2	2.62044E−2	3.3182E−2
0.8	6.55134E−2	6.6256E−3	1.29004E−2	4.0509E−2
0.9	7.87788E−2	3.5980E−2	5.94344E−2	4.7923E−2

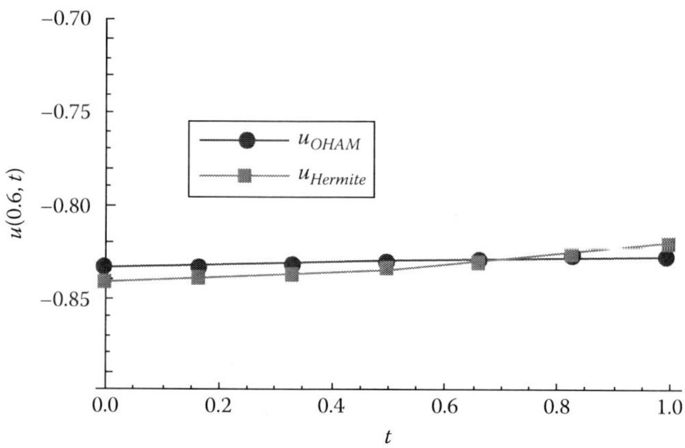

FIGURE 7.1
Comparison of the numerical solutions of $u(0.6, t)$ obtained by OHAM with regard to two-dimensional Hermite wavelet approximation for $\alpha = 0.75$.

Integrating Equation 7.34 once with respect to ξ, we obtain

$$\tilde{\xi}_0 - \lambda\Psi(\xi) + \lambda k^2\Psi_{\xi\xi}(\xi) + k\Psi(\xi) + \frac{k}{3}\left(\Psi(\xi)\right)^3 = \frac{3}{2}k^3\left(\Psi_\xi(\xi)\right)^2 + k^3\left(\Psi(\xi)\Psi_{\xi\xi}(\xi) - \frac{\left(\Psi_\xi(\xi)\right)^2}{2}\right), \tag{7.35}$$

where $\tilde{\xi}_0$ is an integration constant.

Using Equations 1.6 and 1.7, Equation 7.35 can be written as the following two-dimensional autonomous system:

$$\begin{aligned} \frac{dX(\xi)}{d\xi} &= Y(\xi), \\ \left(\lambda k^2 - k^3 X\right)\frac{dY(\xi)}{d\xi} &= \lambda X(\xi) - \frac{k}{3}\left(X(\xi)\right)^3 - kX(\xi) + k^3\left(Y(\xi)\right)^2 - \tilde{\xi}_0. \end{aligned} \tag{7.36}$$

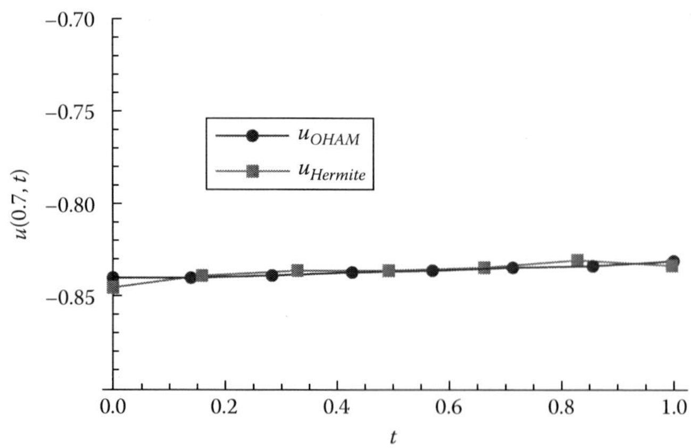

FIGURE 7.2
Comparison of the numerical solutions of $u(0.7,t)$ obtained by OHAM with regard to two-dimensional Hermite wavelet approximation for $\alpha=0.5$.

According to the FIM, we suppose that $X(\xi)$ and $Y(\xi)$ are the nontrivial solutions of Equation 7.36 and

$$Q(X,Y)=\sum_{i=0}^{m} a_i(X)Y^i$$

is an irreducible polynomial in the complex domain $C[X,Y]$ such that

$$Q[X(\xi),Y(\xi)]=\sum_{i=0}^{m} a_i(X(\xi))Y^i(\xi)=0, \tag{7.37}$$

where $a_i(X(\xi))$, $i=0,1,2,\ldots,m$ are polynomials in X and $a_m(X)\neq 0$. Equation 7.37 is called the first integral to Equation 7.36. Applying the division Theorem 1.1, there exists a polynomial $g(X)+h(X)Y$ in the complex domain $C[X,Y]$ such that

$$\frac{dQ}{d\xi}=\left(g(X)+h(X)Y\right)\sum_{i=0}^{m} a_i(X)Y^i. \tag{7.38}$$

Considering $m=1$ in Equation 7.38, we obtain

$$\frac{\partial Q}{\partial X}\frac{dX}{d\xi}+\frac{\partial Q}{\partial Y}\frac{dY}{d\xi}=\left(g(X)+h(X)Y\right)\left(a_0(X)+a_1(X)Y\right). \tag{7.39}$$

Equating the coefficients of Y^i, $i=0,1$ on both sides of Equation 7.39, we have

$$Y^0: a_1(X)\left(\lambda X(\xi)-\frac{k}{3}\left(X(\xi)\right)^3-kX(\xi)-\xi_0\right)=a_0(X)g(X)\left(\lambda k^2-k^3X\right), \tag{7.40}$$

$$Y^1: \dot{a}_0(X)(\lambda k^2 - k^3 X) = (a_0(X)h(X) + a_1(X)g(X))(\lambda k^2 - k^3 X), \tag{7.41}$$

$$Y^2: \dot{a}_1(X)(\lambda k^2 - k^3 X) = a_1(X)h(X)(\lambda k^2 - k^3 X) - a_1(X)k^3. \tag{7.42}$$

Since $a_i(X)$, $i=0,1$ are polynomials in X, from Equation 7.42 we infer that $a_1(X)$ is a constant and $h(X)=(k)/(\lambda - kX)$. For simplicity, we take $a_1(X)=1$. Then balancing the degrees of $a_0(X)$ and $g(X)$, Equation 7.41 indicates that $\deg(g(X)) \leq \deg(a_0(X))$; thus from Equation 7.41 we conclude that $\deg(g(X))=1$.

Now suppose that

$$g(X) = b_1 X + b_0, \quad a_0(X) = \frac{A_2}{2}X^2 + A_1 X + A_0 \quad (b_1 \neq 0, A_2 \neq 0), \tag{7.43}$$

where b_1, b_0, A_2, A_1, and A_0 are all constants to be determined later. Using Equation 7.41, we find that

$$b_0 = A_1 - \frac{kA_0}{\lambda}, \tag{7.44}$$

$$b_1 = A_2 - \frac{A_1 k}{\lambda} - \frac{k^2 A_0}{\lambda^2}. \tag{7.45}$$

Next, substituting $a_0(X)$ and $g(X)$ in Equation 7.40 and consequently equating the coefficients of X^i, $i=0,1,2,3,4$ to zero, we obtain the following system of nonlinear algebraic equations:

$$X^0: \quad b_0 A_0 \lambda k^2 + \xi_0 = 0 \tag{7.46}$$

$$X^1: \quad b_0 A_1 \lambda k^2 + b_1 A_0 \lambda k^2 - b_0 A_0 k^3 = \lambda - k \tag{7.47}$$

$$X^2: \quad \frac{b_0 A_2 \lambda k^2}{2} + b_1 A_1 \lambda k^2 - b_0 A_1 k^3 - A_0 b_1 k^3 = 0 \tag{7.48}$$

$$X^3: \quad \frac{b_1 A_2 \lambda k^2}{2} - \frac{b_0 A_2 k^3}{2} - b_1 A_1 k^3 = \frac{-k}{3} \tag{7.49}$$

$$X^4: \quad -\frac{b_1 A_2 k^3}{2} = 0 \tag{7.50}$$

Solving the preceding system of Equations 7.46 through 7.50 simultaneously, we get the following family of nontrivial solutions:

$$A_0 = \frac{-i\sqrt{\lambda}(\lambda^2 - 3\lambda k + 3k^2)}{3k\sqrt{k^4(k-\lambda)}}, \quad A_1 = \frac{-i\lambda^{3/2}}{3\sqrt{k^4(k-\lambda)}}, \quad A_2 = \frac{-2i\sqrt{\lambda}k}{3\sqrt{k^4(k-\lambda)}} \tag{7.51}$$

and

$$A_0 = \frac{i\sqrt{\lambda}\left(\lambda^2 - 3\lambda k + 3k^2\right)}{3k\sqrt{k^4(k-\lambda)}}, \quad A_1 = \frac{i\lambda^{3/2}}{3\sqrt{k^4(k-\lambda)}}, \quad A_2 = \frac{2i\sqrt{\lambda}k}{3\sqrt{k^4(k-\lambda)}}. \tag{7.52}$$

Case I: Substituting the values of A_0, A_1, and A_2 obtained in Equation 7.51 into Equation 7.37, we get

$$Y(\xi) = \frac{2i\sqrt{\lambda}k}{3\sqrt{k^4(k-\lambda)}}X^2 + \frac{i\lambda^{3/2}}{3\sqrt{k^4(k-\lambda)}}X + \frac{i\sqrt{\lambda}\left(\lambda^2 - 3\lambda k + 3k^2\right)}{3k\sqrt{k^4(k-\lambda)}},$$

or

$$\frac{dX(\xi)}{d\xi} = \frac{2i\sqrt{\lambda}k}{3\sqrt{k^4(k-\lambda)}}X^2 + \frac{i\lambda^{3/2}}{3\sqrt{k^4(k-\lambda)}}X + \frac{i\sqrt{\lambda}\left(\lambda^2 - 3\lambda k + 3k^2\right)}{3k\sqrt{k^4(k-\lambda)}}. \tag{7.53}$$

Solving Equation 7.53, we obtain the exact solution to Equation 7.32 as

$$\begin{aligned} u(x,t) &= \Psi(\xi) = X(\xi) \\ &= \frac{-1}{2k}\left[\lambda + i\sqrt{3}(\lambda - 2k)\tanh\left(\frac{(\lambda-2k)\left(\sqrt{\lambda}\sqrt{k^4(k-\lambda)}\xi + 3i\lambda k^5\eta_1 - 3ik^6\eta_1\right)}{2\sqrt{3}(\lambda-k)k^4}\right)\right] \end{aligned} \tag{7.54}$$

where η_1 is an arbitrary constant.

Case II: Substituting the values of A_0, A_1, and A_2 from Equation 7.52 into Equation 7.37, we get

$$Y(\xi) = -\frac{2i\sqrt{\lambda}k}{3\sqrt{k^4(k-\lambda)}}X^2 - \frac{i\lambda^{3/2}}{3\sqrt{k^4(k-\lambda)}}X - \frac{i\sqrt{\lambda}\left(\lambda^2 - 3\lambda k + 3k^2\right)}{3k\sqrt{k^4(k-\lambda)}}$$

or

$$\frac{dX(\xi)}{d\xi} = -\frac{2i\sqrt{\lambda}k}{3\sqrt{k^4(k-\lambda)}}X^2 - \frac{i\lambda^{3/2}}{3\sqrt{k^4(k-\lambda)}}X - \frac{i\sqrt{\lambda}\left(\lambda^2 - 3\lambda k + 3k^2\right)}{3k\sqrt{k^4(k-\lambda)}}. \tag{7.55}$$

Solving Equation 7.55, we obtain the exact solution to Equation 7.32 as

$$u(x,t)=\Psi(\xi)=X(\xi)$$
$$=\frac{i}{2k}\left[i\lambda+\sqrt{3}(\lambda-2k)\tanh\left(\frac{(\lambda-2k)\left(\sqrt{\lambda}\sqrt{k^4(k-\lambda)}\xi+3i\lambda k^5\eta_2-3ik^6\eta_2\right)}{2\sqrt{3}(\lambda-k)k^4}\right)\right] \tag{7.56}$$

where η_2 is an arbitrary constant.

The established solutions in Equations 7.54 and 7.56 have been checked by putting them into the Equation 7.34. Thus, the new exact solutions given in Equations 7.54 and 7.56 for time-fractional modified Fornberg–Whitham equation have been obtained for the first time in this present section.

7.5.2 Implementation of OHAM for Approximate Solution of Fractional Modified Fornberg–Whitham Equation

To exhibit the effectiveness and accuracy of the proposed scheme, we consider the fractional modified Fornberg–Whitham equation with an initial condition. The solutions thus obtained are compared with the exact solutions obtained by FIM.

Using OHAM [30–32], the homotopy for Equation 7.32 can be written as

$$(1-p)L(\varphi(x,t;p))=H(p)\left[\frac{\partial^\alpha\varphi(x,t;p)}{\partial t^\alpha}-\frac{\partial^\alpha}{\partial t^\alpha}\left(\frac{\partial^2\varphi(x,t;p)}{\partial x}\right)+\frac{\partial\varphi(x,t;p)}{\partial x}\right.$$
$$\left.+\varphi(x,t;p)^2\frac{\partial\varphi(x,t;p)}{\partial x}-3\frac{\partial\varphi(x,t;p)}{\partial x}\frac{\partial^2\varphi(x,t;p)}{\partial x^2}-\varphi(x,t;p)\frac{\partial^3\varphi(x,t;p)}{\partial x^3}\right], \tag{7.57}$$

where

$$\varphi(x,t;p)=u_0(x,t)+\sum_{i=1}^{\infty}u_i(x,t)p^i \tag{7.58}$$

$$H(p)=C_1p+C_2p^2+C_3p^3+\cdots. \tag{7.59}$$

Substituting Equations 7.58 and 7.59 in Equation 7.57 and equating the coefficients of different powers in p, we have the following system of partial differential equations:

$$\text{Coefficients of } p^0: \frac{\partial^\alpha u_0(x,t)}{\partial t^\alpha}=0 \tag{7.60}$$

$$\text{Coefficients of } p^1: \frac{\partial^\alpha u_1(x,t)}{\partial t^\alpha}-\frac{\partial^\alpha u_0(x,t)}{\partial t^\alpha}=C_1\left[\frac{\partial^\alpha u_0(x,t)}{\partial t^\alpha}-\frac{\partial^\alpha}{\partial t^\alpha}\left(\frac{\partial^2 u_0(x,t)}{\partial x^2}\right)+\frac{\partial u_0(x,t)}{\partial x}\right.$$
$$\left.+(u_0(x,t))^2\frac{\partial u_0(x,t)}{\partial x}-3\frac{\partial u_0(x,t)}{\partial x}\frac{\partial^2 u_0(x,t)}{\partial x^2}-u_0(x,t)\frac{\partial^3 u_0(x,t)}{\partial x^3}\right] \tag{7.61}$$

Coefficients of p^2: $$\frac{\partial^\alpha u_2(x,t)}{\partial t^\alpha}-\frac{\partial^\alpha u_1(x,t)}{\partial t^\alpha}=C_1\left[\frac{\partial^\alpha u_1(x,t)}{\partial t^\alpha}-\frac{\partial^\alpha}{\partial t^\alpha}\left(\frac{\partial^2 u_1(x,t)}{\partial x^2}\right)+\frac{\partial u_1(x,t)}{\partial x}\right.$$
$$+\left(2u_0(x,t)u_1(x,t)\frac{\partial u_0(x,t)}{\partial x}+\left(u_0(x,t)\right)^2\frac{\partial u_1(x,t)}{\partial x}\right)$$
$$\left.-3\left(\frac{\partial u_0(x,t)}{\partial x}\frac{\partial^2 u_1(x,t)}{\partial x^2}+\frac{\partial u_1(x,t)}{\partial x}\frac{\partial^2 u_0(x,t)}{\partial x^2}\right)-\left(u_0(x,t)\frac{\partial^3 u_1(x,t)}{\partial x^3}+u_1(x,t)\frac{\partial^3 u_0(x,t)}{\partial x^3}\right)\right]$$
$$+C_2\left[\frac{\partial^\alpha u_0(x,t)}{\partial t^\alpha}-\frac{\partial^\alpha}{\partial t^\alpha}\left(\frac{\partial^2 u_0(x,t)}{\partial x^2}\right)+\frac{\partial u_0(x,t)}{\partial x}+\left(u_0(x,t)\right)^2\frac{\partial u_0(x,t)}{\partial x}\right.$$
$$\left.-3\frac{\partial u_0(x,t)}{\partial x}\frac{\partial^2 u_0(x,t)}{\partial x^2}-u_0(x,t)\frac{\partial^3 u_0(x,t)}{\partial x^3}\right] \tag{7.62}$$

and so on.

For solving a fractional order modified Fornberg–Whitham equation using OHAM, we consider the following initial condition for Equation 7.32:

$$u(x,0)=\frac{-1}{2k}\left[\lambda+i\sqrt{3}(\lambda-2k)\tanh\left\{\frac{(\lambda-2k)\left(\sqrt{\lambda}\sqrt{k^4(k-\lambda)}(kx)+3i\lambda k^5\eta_1-3i\lambda^6\eta_1\right)}{2\sqrt{3}(\lambda-k)k^4}\right\}\right]. \tag{7.63}$$

Using the initial condition $u_0=u(x,0)$ and solving Equations 7.60 through Equation 7.62, we obtain the expressions for u_0, u_1, u_2, and so on.

Finally, the third-order approximate solution for a fractional modified Fornberg–Whitham equation is given by

$$u=u_0(x,t)+u_1(x,t)+u_2(x,t). \tag{7.64}$$

The optimal values of the convergence control parameters C_1 and C_2 can be obtained using the weighted residual least square method given in Equations 1.34 and 1.35.

7.6 Numerical Results and Discussion

The comparison of the absolute errors for a fractional modified Fornberg–Whitham equation (7.32) have been exhibited in Tables 7.8 through 7.10, which are generated by using the results obtained by FIM and OHAM at different values of α. In order to examine the

TABLE 7.8

Absolute Errors Obtained by OHAM for the Nonlinear Modified Fornberg–Whitham Equation Given in Equation 7.32 at Various Points of x and t, Taking $k=1$, $\lambda=2.5$, $\eta_1=1$, and $\alpha=1$

	$\lvert u_{Exact}-u_{OHAM}\rvert$									
x	$t=0.1$	$t=0.2$	$t=0.3$	$t=0.4$	$t=0.5$	$t=0.6$	$t=0.7$	$t=0.8$	$t=0.9$	$t=1.0$
0.1	9.3908E−4	8.9819E−4	2.8677E−5	1.5351E−3	3.6724E−3	6.2730E−3	9.2349E−3	1.2462E−2	1.5864E−2	1.9351E−2
0.2	1.1574E−3	1.2728E−3	5.0544E−4	1.0041E−3	3.1298E−3	5.7577E−3	8.7828E−3	1.2107E−2	1.5637E−2	1.9284E−2
0.3	1.3864E−3	1.6646E−3	1.0025E−3	4.5210E−4	2.5680E−3	5.2267E−3	8.3199E−3	1.1746E−2	1.5413E−2	1.9227E−2
0.4	16274E−3	2.0759E−3	1.5231E−3	1.2413E−4	1.9834E−3	4.6765E−3	7.8428E−3	1.1379E−2	1.5188E−2	1.9178E−2
0.5	1.8818E−3	2.5092E−3	2.0703E−3	7.2846E−4	1.3721E−3	4.1029E−3	7.3478E−3	1.1000E−2	1.4960E−2	1.9135E−2
0.6	2.1514E−3	2.9676E−3	2.6480E−3	1.3652E−3	7.2944E−4	3.5016E−3	6.8304E−3	1.0605E−2	1.4725E−2	1.9094E−2
0.7	2.4381E−3	3.4543E−3	3.2605E−3	2.0392E−3	5.0184E−5	2.8671E−3	6.2856E−3	1.0191E−2	1.4480E−2	1.9053E−2
0.8	2.7441E−3	3.9732E−3	3.9127E−3	2.7562E−3	6.7161E−4	2.1934E−3	5.7074E−3	9.7520E−3	1.4218E−2	1.9007E−2
0.9	3.0720E−3	4.5286E−3	4.6103E−3	3.5225E−3	1.4428E−3	1.4734E−3	5.0890E−3	9.2805E−3	1.3936E−2	1.8951E−2
1.0	3.4245E−3	5.1254E−3	5.3596E−3	4.3457E−3	2.2716E−3	6.9904E−4	4.4222E−3	8.7694E−3	1.3624E−2	1.8880E−2

TABLE 7.9

Absolute Errors Obtained by Third-Order OHAM for the Nonlinear Modified Fornberg–Whitham Equation Given in Equation 7.32 at Various Points of x and t, Taking $k=1$, $\lambda=2.5$, $\eta_1=1$, and $\alpha=0.75$

	$\lvert u_{Exact}-u_{OHAM}\rvert$									
x	$t=0.1$	$t=0.2$	$t=0.3$	$t=0.4$	$t=0.5$	$t=0.6$	$t=0.7$	$t=0.8$	$t=0.9$	$t=1.0$
0.1	1.5780E−3	8.3963E−4	7.3067E−4	2.7686E−3	5.0835E−3	7.5542E−3	1.0094E−2	1.2641E−2	1.5139E−2	1.7548E−2
0.2	1.9515E−3	1.3377E−3	1.9635E−4	2.2546E−3	4.6313E−3	7.1966E−3	9.8595E−3	1.2551E−2	1.5217E−2	1.7813E−2
0.3	2.3436E−3	1.8601E−3	3.6403E−4	1.1749E−3	4.1550E−3	6.8173E−3	9.6052E−3	1.2446E−2	1.5283E−2	1.8069E−2
0.4	2.7562E−3	2.4098E−3	9.5402E−4	1.1456E−3	3.6507E−3	6.4126E−3	9.3285E−3	1.2321E−2	1.5333E−2	1.8312E−2
0.5	3.1920E−3	2.9902E−3	1.5775E−3	5.4267E−4	3.1141E−3	5.9781E−3	9.0252E−3	1.2174E−2	1.5363E−2	1.8541E−2
0.6	3.6537E−3	3.6052E−3	2.2390E−3	9.8730E−5	2.5405E−3	5.5091E−3	8.6907E−3	1.2000E−2	1.5370E−2	1.8749E−2
0.7	4.1442E−3	4.2590E−3	2.9435E−3	7.8394E−4	1.9243E−3	5.0001E−3	8.3195E−3	1.1792E−2	1.5349E−2	1.8934E−2
0.8	4.6671E−3	4.9566E−3	3.6966E−3	1.5190E−3	1.2590E−3	4.4443E−3	7.9051E−3	1.1546E−2	1.5293E−2	1.9089E−2
0.9	5.2262E−3	5.7034E−3	4.5047E−3	2.3111E−3	5.3754E−4	3.8345E−3	7.4401E−3	1.1252E−2	1.5195E−2	1.9207E−2
1.0	5.8260E−3	6.5056E−3	5.3752E−3	3.1681E−3	2.4874E−3	3.1621E−3	6.9156E−3	1.0904E−2	1.5047E−2	1.9279E−2

TABLE 7.10

Absolute Errors Obtained by Third-Order OHAM for the Nonlinear Modified Fornberg–Whitham Equation Given in Equation 7.32 at Various Points of x and t, Taking $k=1$, $\lambda=2.5$, $\eta_1=1$, and $\alpha=0.5$

	$\lvert u_{Exact}-u_{OHAM}\rvert$									
x	$t=0.1$	$t=0.2$	$t=0.3$	$t=0.4$	$t=0.5$	$t=0.6$	$t=0.7$	$t=0.8$	$t=0.9$	$t=1.0$
0.1	2.2513E−3	7.4423E−4	1.0160E−3	2.7858E−3	4.4912E−3	6.1032E−3	7.6093E−3	9.0044E−3	1.0286E−2	1.1455E−2
0.2	2.7556E−3	1.2335E−3	6.1201E−4	2.5001E−3	4.3430E−3	6.1056E−3	7.7717E−3	9.3342E−3	1.0789E−2	1.2136E−2
0.3	3.2903E−3	1.7588E−3	1.6932E−4	2.1737E−3	4.1529E−3	6.0647E−3	7.8897E−3	9.6181E−3	1.1245E−2	1.2769E−2
0.4	3.8590E−3	2.3241E−3	3.1646E−4	1.8023E−3	3.9160E−3	5.9757E−3	7.9580E−3	9.8510E−3	1.1648E−2	1.3347E−2
0.5	4.4652E−3	2.9338E−3	8.5036E−4	1.3803E−3	3.6268E−3	5.8327E−3	7.9708E−3	1.0026E−2	1.1993E−2	1.3865E−2
0.6	5.1131E−3	3.5931E−3	1.4381E−3	9.0172E−4	3.2786E−3	5.6289E−3	7.9209E−3	1.0138E−2	1.2271E−2	1.4314E−2
0.7	5.8072E−3	4.3078E−3	2.0862E−3	3.5928E−4	2.8640E−3	5.3561E−3	7.7998E−3	1.0175E−2	1.2473E−2	1.4685E−2
0.8	6.5529E−3	5.0845E−3	2.8021E−3	2.5513E−4	2.3740E−3	5.0051E−3	7.5977E−3	1.0129E−2	1.2588E−2	1.4967E−2
0.9	7.3558E−3	5.9304E−3	3.5944E−3	9.5096E−4	1.7984E−3	4.5648E−3	7.3028E−3	9.9877E−3	1.2604E−2	1.5145E−2
1.0	8.2226E−3	6.8540E−3	4.4727E−3	1.7391E−3	1.1254E−3	4.0224E−3	6.9015E−3	9.7347E−3	1.2506E−2	1.5205E−2

TABLE 7.11

L_2 and L_∞ Error Norms for the Time-Fractional Nonlinear Modified Fornberg–Whitham Equation Given in Equation 7.32 at Various Points of t, Taking $\alpha=0.75$ and 0.5

	$\alpha=0.75$		$\alpha=0.5$	
t	L_2	L_∞	L_2	L_∞
0.1	1.19640E−2	5.8260E−3	1.68180E−2	8.2226E−3
0.2	1.22987E−2	6.5056E−3	1.46109E−2	6.8540E−3
0.3	8.97947E−3	5.3752E−3	7.03247E−3	4.4727E−3
0.4	5.83484E−3	3.1681E−3	5.36896E−3	2.7858E−3
0.5	1.02858E−2	5.0835E−3	1.06703E−2	4.4912E−3
0.6	1.82208E−2	7.5542E−3	1.74206E−2	6.1056E−3
0.7	2.77533E−2	1.0094E−2	2.42832E−2	7.9708E−3
0.8	3.78696E−2	1.2641E−2	3.10113E−2	1.0175E−2
0.9	4.82540E−2	1.5370E−2	3.75238E−2	1.2604E−2
1.0	5.87011E−2	1.9279E−2	4.37840E−2	1.5205E−2

accuracy and reliability of FIM for solving a fractional order nonlinear modified Fornberg–Whitham equation, we compare the exact solution obtained by using FIM with the third-order approximate OHAM solutions. Table 7.11 illustrates the L_2 and L_∞ error norms for the fractional modified Fornberg–Whitham equation given in Equation 7.32 at various points of t, taking $\alpha=0.5$ and 0.75. Figures 7.3 through 7.6 demonstrate the comparison of approximate solutions obtained by OHAM with FIM for a modified Fornberg–Whitham equation. It can be observed that the derived numerical simulation results are in good agreement with the exact solutions obtained by FIM through illustrations in tables and figures.

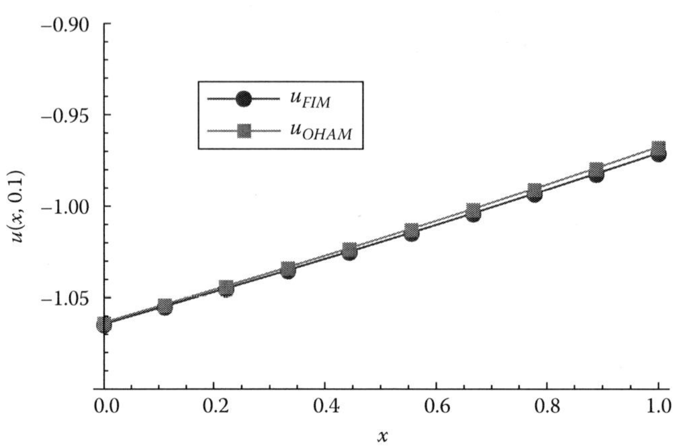

FIGURE 7.3
Comparison of the approximate solution obtained by OHAM with the exact solution obtained by FIM for a fractional modified Fornberg–Whitham equation at $t=0.1$, taking $\alpha=1$.

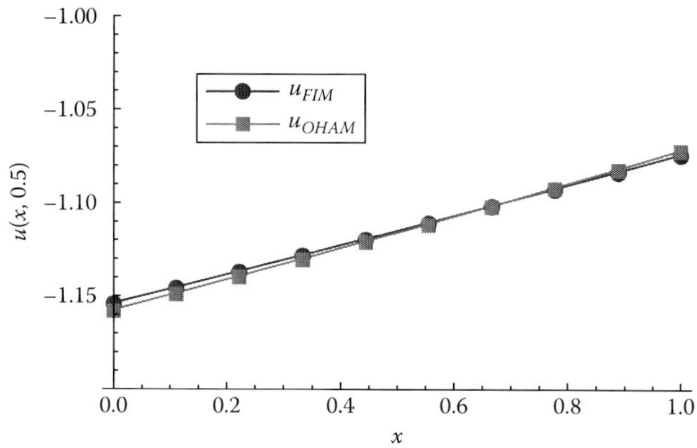

FIGURE 7.4
Comparison of the approximate solution obtained by OHAM with the exact solution obtained by FIM for a fractional modified Fornberg–Whitham equation at $t = 0.5$, taking $\alpha = 1$.

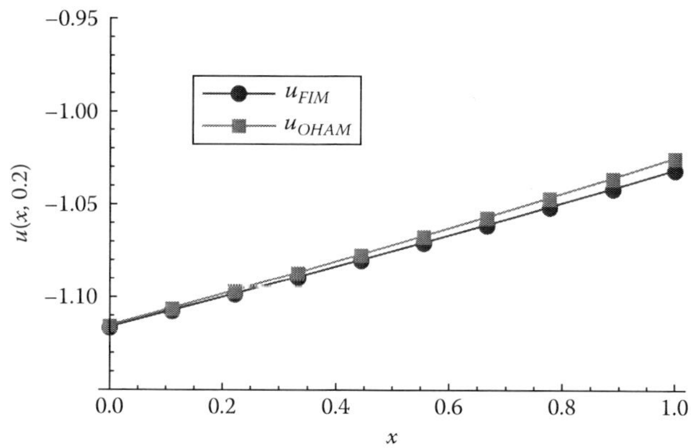

FIGURE 7.5
Comparison of the approximate solution obtained by OHAM with the exact solution obtained by FIM for a fractional modified Fornberg–Whitham equation at $t = 0.2$, taking $\alpha = 0.75$.

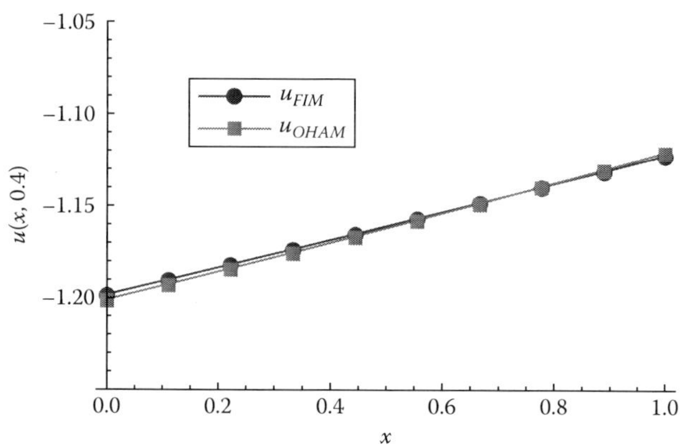

FIGURE 7.6
Comparison of the approximate solution obtained by OHAM with the exact solution obtained by FIM for a fractional modified Fornberg–Whitham equation $t = 0.4$, taking $\alpha = 0.5$.

7.7 Application of Analytical and Numerical Methods for Solving a Time-Fractional Coupled Jaulent–Miodek Equation

7.7.1 Two-Dimensional Hermite Wavelet Method for Solving Nonlinear Time-Fractional Coupled Jaulent–Miodek Equations

To demonstrate the accuracy and efficiency of the proposed numerical technique, we consider the time-fractional coupled Jaulent–Miodek equation. The numerical approximate solutions thus achieved are compared with the exact solutions in the case of classical order and with the solutions obtained by OHAM in the case of fractional order, respectively.

Consider the nonlinear time-fractional coupled Jaulent–Miodek equations

$$\frac{\partial^\alpha u}{\partial t^\alpha}+\frac{\partial^3 u}{\partial x^3}+\frac{3}{2}v\frac{\partial^3 v}{\partial x^3}+\frac{9}{2}\frac{\partial v}{\partial x}\frac{\partial^2 v}{\partial x^2}-6u\frac{\partial u}{\partial x}-6uv\frac{\partial v}{\partial x}-\frac{3}{2}\frac{\partial u}{\partial x}v^2=0, \tag{7.65}$$

$$\frac{\partial^\alpha v}{\partial t^\alpha}+\frac{\partial^3 v}{\partial x^3}-6\frac{\partial u}{\partial x}v-6u\frac{\partial v}{\partial x}-\frac{15}{2}\frac{\partial v}{\partial x}v^2=0, \tag{7.66}$$

with initial conditions [163]

$$u(x,0)=\frac{1}{8}\lambda^2\left[1-4\operatorname{sech}^2\left(\frac{\lambda x}{2}\right)\right], \tag{7.67}$$

$$v(x,0)=\lambda\operatorname{sech}\left(\frac{\lambda x}{2}\right), \tag{7.68}$$

where λ is an arbitrary constant.

For $\alpha=1$, the exact solutions of Equations 7.65 and 7.66 are given by [163]

$$u(x,t)=\frac{1}{8}\lambda^2\left[1-4\operatorname{sech}^2\left[\frac{1}{2}\lambda\left(x+\frac{1}{2}\lambda^2 t\right)\right]\right], \tag{7.69}$$

$$v(x,t)=\lambda\operatorname{sech}\left[\frac{1}{2}\lambda\left(x+\frac{1}{2}\lambda^2 t\right)\right]. \tag{7.70}$$

The Hermite wavelet solutions of $u(x,t)$ and $v(x,t)$ are sought by assuming that $u(x,t)$ and $v(x,t)$ can be expanded in terms of Hermite wavelets as

$$u(x,t)=\sum_{n=1}^{2^{k_1-1}}\sum_{i=0}^{M_1-1}\sum_{l=1}^{2^{k_2-1}}\sum_{j=0}^{M_2-1}a_{n,i,l,j}\psi_{n,i,l,j}(x,t), \tag{7.71}$$

$$v(x,t)=\sum_{n=1}^{2^{k_1-1}}\sum_{i=0}^{M_1-1}\sum_{l=1}^{2^{k_2-1}}\sum_{j=0}^{M_2-1}b_{n,i,l,j}\psi_{n,i,l,j}(x,t), \tag{7.72}$$

where $n=1,\ldots,2^{k_1-1}$, $i=0,\ldots,M_1-1$, $l=1,\ldots,2^{k_2-1}$, $j=0,\ldots,M_2-1$.

The nonlinear terms appearing in Equations 7.65 and 7.66 can be approximated using the Hermite wavelet function as

$$v\frac{\partial^3 v}{\partial x^3}=\sum_{n=1}^{2^{k_1-1}}\sum_{i=0}^{M_1-1}\sum_{l=1}^{2^{k_2-1}}\sum_{j=0}^{M_2-1}c_{n,i,l,j}\psi_{n,i,l,j}(x,t), \tag{7.73}$$

$$\frac{\partial v}{\partial x}\frac{\partial^2 v}{\partial x^2}=\sum_{n=1}^{2^{k_1-1}}\sum_{i=0}^{M_1-1}\sum_{l=1}^{2^{k_2-1}}\sum_{j=0}^{M_2-1}d_{n,i,l,j}\psi_{n,i,l,j}(x,t), \tag{7.74}$$

$$u\frac{\partial u}{\partial x}=\sum_{n=1}^{2^{k_1-1}}\sum_{i=0}^{M_1-1}\sum_{l=1}^{2^{k_2-1}}\sum_{j=0}^{M_2-1}e_{n,i,l,j}\psi_{n,i,l,j}(x,t), \tag{7.75}$$

$$uv\frac{\partial v}{\partial x}=\sum_{n=1}^{2^{k_1-1}}\sum_{i=0}^{M_1-1}\sum_{l=1}^{2^{k_2-1}}\sum_{j=0}^{M_2-1}f_{n,i,l,j}\psi_{n,i,l,j}(x,t), \tag{7.76}$$

$$\frac{\partial u}{\partial x}v^2=\sum_{n=1}^{2^{k_1-1}}\sum_{i=0}^{M_1-1}\sum_{l=1}^{2^{k_2-1}}\sum_{j=0}^{M_2-1}g_{n,i,l,j}\psi_{n,i,l,j}(x,t), \tag{7.77}$$

$$\frac{\partial u}{\partial x}v=\sum_{n=1}^{2^{k_1-1}}\sum_{i=0}^{M_1-1}\sum_{l=1}^{2^{k_2-1}}\sum_{j=0}^{M_2-1}h_{n,i,l,j}\psi_{n,i,l,j}(x,t), \tag{7.78}$$

$$u\frac{\partial v}{\partial x}=\sum_{n=1}^{2^{k_1-1}}\sum_{i=0}^{M_1-1}\sum_{l=1}^{2^{k_2-1}}\sum_{j=0}^{M_2-1}p_{n,i,l,j}\psi_{n,i,l,j}(x,t), \tag{7.79}$$

$$\frac{\partial v}{\partial x}v^2=\sum_{n=1}^{2^{k_1-1}}\sum_{i=0}^{M_1-1}\sum_{l=1}^{2^{k_2-1}}\sum_{j=0}^{M_2-1}q_{n,i,l,j}\psi_{n,i,l,j}(x,t). \tag{7.80}$$

This implies

$$\left(\sum_{n=1}^{2^{k_1-1}}\sum_{i=0}^{M_1-1}\sum_{l=1}^{2^{k_2-1}}\sum_{j=0}^{M_2-1}b_{n,i,l,j}\psi_{n,i,l,j}(x,t)\right)\times\left[\sum_{n=1}^{2^{k_1-1}}\sum_{i=0}^{M_1-1}\sum_{l=1}^{2^{k_2-1}}\sum_{j=0}^{M_2-1}b_{n,i,l,j}\frac{\partial^3\psi_{n,i,l,j}(x,t)}{\partial x^3}\right]$$
$$=\sum_{n=1}^{2^{k_1-1}}\sum_{i=0}^{M_1-1}\sum_{l=1}^{2^{k_2-1}}\sum_{j=0}^{M_2-1}c_{n,i,l,j}\psi_{n,i,l,j}(x,t), \tag{7.81}$$

$$\left(\sum_{n=1}^{2^{k_1-1}}\sum_{i=0}^{M_1-1}\sum_{l=1}^{2^{k_2-1}}\sum_{j=0}^{M_2-1}b_{n,i,l,j}\frac{\partial\psi_{n,i,l,j}(x,t)}{\partial x}\right)\times\left[\sum_{n=1}^{2^{k_1-1}}\sum_{i=0}^{M_1-1}\sum_{l=1}^{2^{k_2-1}}\sum_{j=0}^{M_2-1}b_{n,i,l,j}\frac{\partial^2\psi_{n,i,l,j}(x,t)}{\partial x^2}\right]$$
$$=\sum_{n=1}^{2^{k_1-1}}\sum_{i=0}^{M_1-1}\sum_{l=1}^{2^{k_2-1}}\sum_{j=0}^{M_2-1}d_{n,i,l,j}\psi_{n,i,l,j}(x,t), \tag{7.82}$$

$$\left(\sum_{n=1}^{2^{k_1-1}}\sum_{i=0}^{M_1-1}\sum_{l=1}^{2^{k_2-1}}\sum_{j=0}^{M_2-1}a_{n,i,l,j}\psi_{n,i,l,j}(x,t)\right)\times\left[\sum_{n=1}^{2^{k_1-1}}\sum_{i=0}^{M_1-1}\sum_{l=1}^{2^{k_2-1}}\sum_{j=0}^{M_2-1}a_{n,i,l,j}\frac{\partial\psi_{n,i,l,j}(x,t)}{\partial x}\right]$$

$$=\sum_{n=1}^{2^{k_1-1}}\sum_{i=0}^{M_1-1}\sum_{l=1}^{2^{k_2-1}}\sum_{j=0}^{M_2-1}e_{n,i,l,j}\psi_{n,i,l,j}(x,t), \tag{7.83}$$

$$\left(\sum_{n=1}^{2^{k_1-1}}\sum_{i=0}^{M_1-1}\sum_{l=1}^{2^{k_2-1}}\sum_{j=0}^{M_2-1}a_{n,i,l,j}\psi_{n,i,l,j}(x,t)\right)\times\left[\sum_{n=1}^{2^{k_1-1}}\sum_{i=0}^{M_1-1}\sum_{l=1}^{2^{k_2-1}}\sum_{j=0}^{M_2-1}b_{n,i,l,j}\psi_{n,i,l,j}(x,t)\right]$$

$$\times\left(\sum_{n=1}^{2^{k_1-1}}\sum_{i=0}^{M_1-1}\sum_{l=1}^{2^{k_2-1}}\sum_{j=0}^{M_2-1}b_{n,i,l,j}\frac{\partial\psi_{n,i,l,j}(x,t)}{\partial x}\right)=\sum_{n=1}^{2^{k_1-1}}\sum_{i=0}^{M_1-1}\sum_{l=1}^{2^{k_2-1}}\sum_{j=0}^{M_2-1}f_{n,i,l,j}\psi_{n,i,l,j}(x,t), \tag{7.84}$$

$$\left(\sum_{n=1}^{2^{k_1-1}}\sum_{i=0}^{M_1-1}\sum_{l=1}^{2^{k_2-1}}\sum_{j=0}^{M_2-1}a_{n,i,l,j}\frac{\partial\psi_{n,i,l,j}(x,t)}{\partial x}\right)\times\left[\sum_{n=1}^{2^{k_1-1}}\sum_{i=0}^{M_1-1}\sum_{l=1}^{2^{k_2-1}}\sum_{j=0}^{M_2-1}b_{n,i,l,j}\psi_{n,i,l,j}(x,t)\right]^2$$

$$=\sum_{n=1}^{2^{k_1-1}}\sum_{i=0}^{M_1-1}\sum_{l=1}^{2^{k_2-1}}\sum_{j=0}^{M_2-1}g_{n,i,l,j}\psi_{n,i,l,j}(x,t), \tag{7.85}$$

$$\left(\sum_{n=1}^{2^{k_1-1}}\sum_{i=0}^{M_1-1}\sum_{l=1}^{2^{k_2-1}}\sum_{j=0}^{M_2-1}a_{n,i,l,j}\frac{\partial\psi_{n,i,l,j}(x,t)}{\partial x}\right)\times\left[\sum_{n=1}^{2^{k_1-1}}\sum_{i=0}^{M_1-1}\sum_{l=1}^{2^{k_2-1}}\sum_{j=0}^{M_2-1}b_{n,i,l,j}\psi_{n,i,l,j}(x,t)\right]$$

$$=\sum_{n=1}^{2^{k_1-1}}\sum_{i=0}^{M_1-1}\sum_{l=1}^{2^{k_2-1}}\sum_{j=0}^{M_2-1}h_{n,i,l,j}\psi_{n,i,l,j}(x,t), \tag{7.86}$$

$$\left(\sum_{n=1}^{2^{k_1-1}}\sum_{i=0}^{M_1-1}\sum_{l=1}^{2^{k_2-1}}\sum_{j=0}^{M_2-1}a_{n,i,l,j}\psi_{n,i,l,j}(x,t)\right)\times\left(\sum_{n=1}^{2^{k_1-1}}\sum_{i=0}^{M_1-1}\sum_{l=1}^{2^{k_2-1}}\sum_{j=0}^{M_2-1}b_{n,i,l,j}\frac{\partial\psi_{n,i,l,j}(x,t)}{\partial x}\right)$$

$$=\sum_{n=1}^{2^{k_1-1}}\sum_{i=0}^{M_1-1}\sum_{l=1}^{2^{k_2-1}}\sum_{j=0}^{M_2-1}p_{n,i,l,j}\psi_{n,i,l,j}(x,t), \tag{7.87}$$

and

$$\left(\sum_{n=1}^{2^{k_1-1}}\sum_{i=0}^{M_1-1}\sum_{l=1}^{2^{k_2-1}}\sum_{j=0}^{M_2-1}b_{n,i,l,j}\frac{\partial\psi_{n,i,l,j}(x,t)}{\partial x}\right)\times\left[\sum_{n=1}^{2^{k_1-1}}\sum_{i=0}^{M_1-1}\sum_{l=1}^{2^{k_2-1}}\sum_{j=0}^{M_2-1}b_{n,i,l,j}\psi_{n,i,l,j}(x,t)\right]^2$$

$$=\sum_{n=1}^{2^{k_1-1}}\sum_{i=0}^{M_1-1}\sum_{l=1}^{2^{k_2-1}}\sum_{j=0}^{M_2-1}q_{n,i,l,j}\psi_{n,i,l,j}(x,t). \tag{7.88}$$

Again, employing J_t^α on both sides of Equations 7.65 and 7.66, we have

$$u(x,t)-u(x,0)=J_t^\alpha\left[-\frac{\partial^3 u}{\partial x^3}-\frac{3}{2}v\frac{\partial^3 v}{\partial x^3}-\frac{9}{2}\frac{\partial v}{\partial x}\frac{\partial^2 v}{\partial x^2}+6u\frac{\partial u}{\partial x}+6uv\frac{\partial v}{\partial x}+\frac{3}{2}\frac{\partial u}{\partial x}v^2\right], \tag{7.89}$$

$$v(x,t)-v(x,0)=J_t^\alpha\left[-\frac{\partial^3 v}{\partial x^3}+6\frac{\partial u}{\partial x}v+6u\frac{\partial v}{\partial x}+\frac{15}{2}\frac{\partial v}{\partial x}v^2\right]. \tag{7.90}$$

Putting Equations 7.73 through 7.80 in Equations 7.89 and 7.90, we have

$$\begin{aligned}\sum_{n=1}^{2^{k1-1}}\sum_{i=0}^{M_1-1}\sum_{l=1}^{2^{k2-1}}\sum_{j=0}^{M_2-1}a_{n,i,l,j}\psi_{n,i,l,j}(x,t)-u(x,0)&=J_t^\alpha\left[-\left(\sum_{n=1}^{2^{k1-1}}\sum_{i=0}^{M_1-1}\sum_{l=1}^{2^{k2-1}}\sum_{j=0}^{M_2-1}a_{n,i,l,j}\frac{\partial^3\psi_{n,i,l,j}(x,t)}{\partial x^3}\right)\right.\\&-\frac{3}{2}\left(\sum_{n=1}^{2^{k1-1}}\sum_{i=0}^{M_1-1}\sum_{l=1}^{2^{k2-1}}\sum_{j=0}^{M_2-1}c_{n,i,l,j}\psi_{n,i,l,j}(x,t)\right)-\frac{9}{2}\left(\sum_{n=1}^{2^{k1-1}}\sum_{i=0}^{M_1-1}\sum_{l=1}^{2^{k2-1}}\sum_{j=0}^{M_2-1}d_{n,i,l,j}\psi_{n,i,l,j}(x,t)\right)\\&+6\left(\sum_{n=1}^{2^{k1-1}}\sum_{i=0}^{M_1-1}\sum_{l=1}^{2^{k2-1}}\sum_{j=0}^{M_2-1}e_{n,i,l,j}\psi_{n,i,l,j}(x,t)\right)+6\left(\sum_{n=1}^{2^{k1-1}}\sum_{i=0}^{M_1-1}\sum_{l=1}^{2^{k2-1}}\sum_{j=0}^{M_2-1}f_{n,i,l,j}\psi_{n,i,l,j}(x,t)\right)\\&\left.+\frac{3}{2}\left(\sum_{n=1}^{2^{k1-1}}\sum_{i=0}^{M_1-1}\sum_{l=1}^{2^{k2-1}}\sum_{j=0}^{M_2-1}g_{n,i,l,j}\psi_{n,i,l,j}(x,t)\right)\right],\end{aligned} \tag{7.91}$$

$$\begin{aligned}\sum_{n=1}^{2^{k1-1}}\sum_{i=0}^{M_1-1}\sum_{l=1}^{2^{k2-1}}\sum_{j=0}^{M_2-1}b_{n,i,l,j}\psi_{n,i,l,j}(x,t)-v(x,0)&=J_t^\alpha\left[-\left(\sum_{n=1}^{2^{k1-1}}\sum_{i=0}^{M_1-1}\sum_{l=1}^{2^{k2-1}}\sum_{j=0}^{M_2-1}b_{n,i,l,j}\frac{\partial^3\psi_{n,i,l,j}(x,t)}{\partial x^3}\right)\right.\\&+6\left(\sum_{n=1}^{2^{k1-1}}\sum_{i=0}^{M_1-1}\sum_{l=1}^{2^{k2-1}}\sum_{j=0}^{M_2-1}h_{n,i,l,j}\psi_{n,i,l,j}(x,t)\right)+6\left(\sum_{n=1}^{2^{k1-1}}\sum_{i=0}^{M_1-1}\sum_{l=1}^{2^{k2-1}}\sum_{j=0}^{M_2-1}p_{n,i,l,j}\psi_{n,i,l,j}(x,t)\right)\\&\left.+\frac{15}{2}\left(\sum_{n=1}^{2^{k1-1}}\sum_{i=0}^{M_1-1}\sum_{l=1}^{2^{k2-1}}\sum_{j=0}^{M_2-1}q_{n,i,l,j}\psi_{n,i,l,j}(x,t)\right)\right].\end{aligned} \tag{7.92}$$

Now, substituting the collocation points $x_l=(l-0.5)/(2^{k_1-1}M_1)$ and $t_r=(r-0.5)/(2^{k_2-1}M_2)$ for $l=1,2,\ldots,2^{k_1-1}M_1$ and $r=1,2,\ldots,2^{k_2-1}M_2$ in Equations 7.81 through 7.88 and in (7.91) and (7.92), we have $10(2^{k_1-1}M_1)(2^{k_2-1}M_2)$ equations in $10(2^{k_1-1}M_1)(2^{k_2-1}M_2)$ unknowns. By solving these system of equations using Newton's method, the Hermite wavelet coefficients can be acquired.

7.7.2 Comparison with OHAM for the Solution of a Nonlinear Time-Fractional Coupled Jaulent–Miodek Equation

Implementing OHAM [30], the homotopy for Equations 7.65 and 7.66 can be written as

$$(1-p)L(\varphi(x,t;p)) = H(p)\left[\frac{\partial^\alpha \varphi(x,t;p)}{\partial t^\alpha} + \frac{\partial^3}{\partial x^3}\varphi(x,t;p) + \frac{3}{2}\psi(x,t;p)\frac{\partial^3 \psi(x,t;p)}{\partial x^3}\right.$$
$$+\frac{9}{2}\frac{\partial \psi(x,t;p)}{\partial x}\frac{\partial^2 \psi(x,t;p)}{\partial x^2} - 6\varphi(x,t;p)\frac{\partial \varphi(x,t;p)}{\partial x}$$
$$\left.-6\varphi(x,t;p)\psi(x,t;p)\frac{\partial \psi(x,t;p)}{\partial x} - \frac{3}{2}\frac{\partial \varphi(x,t;p)}{\partial x}(\psi(x,t;p))^2\right], \tag{7.93}$$

$$(1-p)L(\psi(x,t;p)) = \tilde{H}(p)\left[\frac{\partial^\alpha \psi(x,t;p)}{\partial t^\alpha} + \frac{\partial^3}{\partial x^3}\psi(x,t;p) - 6\frac{\partial \varphi(x,t;p)}{\partial x}\psi(x,t;p)\right.$$
$$\left.-6\varphi(x,t;p)\frac{\partial \psi(x,t;p)}{\partial x} - \frac{15}{2}\frac{\partial \psi(x,t;p)}{\partial x}(\psi(x,t;p))^2\right], \tag{7.94}$$

where

$$\varphi(x,t;p) = u_0(x,t) + \sum_{i=1}^{\infty} u_i(x,t)p^i \tag{7.95}$$

$$\psi(x,t;p) = v_0(x,t) + \sum_{i=1}^{\infty} v_i(x,t)p^i \tag{7.96}$$

$$H(p) = C_1 p + C_2 p^2 + C_3 p^3 + \cdots \tag{7.97}$$

$$\tilde{H}(p) = \tilde{C}_1 p + \tilde{C}_2 p^2 + \tilde{C}_3 p^3 + \cdots. \tag{7.98}$$

Substituting Equations 7.95 through 7.98 in Equations 7.93 and 7.94 and then comparing the coefficients of identical powers in p, we have the following system of equations:

$$\text{Coefficients of } p^0: \quad \frac{\partial^\alpha u_0(x,t)}{\partial t^\alpha} = 0, \tag{7.99}$$

$$\frac{\partial^\alpha v_0(x,t)}{\partial t^\alpha} = 0 \tag{7.100}$$

$$\text{Coefficients of } p^1: \quad \frac{\partial^\alpha u_1(x,t)}{\partial t^\alpha} - \frac{\partial^\alpha u_0(x,t)}{\partial t^\alpha} = C_1\left[\frac{\partial^\alpha u_0(x,t)}{\partial t^\alpha} + \frac{\partial^3 u_0(x,t)}{\partial x^3} + \frac{3}{2}v_0(x,t)\frac{\partial^3 v_0(x,t)}{\partial x^3}\right.$$
$$+\frac{9}{2}\frac{\partial v_0(x,t)}{\partial x}\frac{\partial^2 v_0(x,t)}{\partial x^2} - 6u_0(x,t)\frac{\partial u_0(x,t)}{\partial x}$$
$$\left.-6u_0(x,t)v_0(x,t)\frac{\partial v_0(x,t)}{\partial x} - \frac{3}{2}\frac{\partial u_0(x,t)}{\partial x}(v_0(x,t))^2\right], \tag{7.101}$$

$$\frac{\partial^\alpha v_1(x,t)}{\partial t^\alpha}-\frac{\partial^\alpha v_0(x,t)}{\partial t^\alpha}=\tilde{C}_1\left[\frac{\partial^\alpha v_0(x,t)}{\partial t^\alpha}+\frac{\partial^3}{\partial x^3}v_0(x,t)-6\frac{\partial u_0(x,t)}{\partial x}v_0(x,t)\right.$$
$$\left.-6u_0(x,t)\frac{\partial v_0(x,t)}{\partial x}-\frac{15}{2}\frac{\partial v_0(x,t)}{\partial x}(v_0(x,t))^2\right] \tag{7.102}$$

Coefficients of p^2: $\frac{\partial^\alpha u_2(x,t)}{\partial t^\alpha}-\frac{\partial^\alpha u_1(x,t)}{\partial t^\alpha}=C_1\left[\frac{\partial^\alpha u_1(x,t)}{\partial t^\alpha}+\frac{\partial^3 u_1(x,t)}{\partial x^3}+\frac{3}{2}\left(v_0(x,t)\frac{\partial^3 v_1(x,t)}{\partial x^3}\right.\right.$

$$\left.+v_1(x,t)\frac{\partial^3 v_0(x,t)}{\partial x^3}\right)+\frac{9}{2}\left(\frac{\partial v_0(x,t)}{\partial x}\frac{\partial^2 v_1(x,t)}{\partial x^2}+\frac{\partial v_1(x,t)}{\partial x}\frac{\partial^2 v_0(x,t)}{\partial x^2}\right)$$
$$-6\left(u_0(x,t)\frac{\partial u_1(x,t)}{\partial x}+u_1(x,t)\frac{\partial u_0(x,t)}{\partial x}\right)-6\left(u_1(x,t)v_0(x,t)\frac{\partial v_0(x,t)}{\partial x}\right.$$
$$\left.+u_0(x,t)v_1(x,t)\frac{\partial v_0(x,t)}{\partial x}+u_0(x,t)v_0(x,t)\frac{\partial v_1(x,t)}{\partial x}\right)$$
$$\left.-\frac{3}{2}\left(2v_0(x,t)v_1(x,t)\frac{\partial u_0(x,t)}{\partial x}+(v_0(x,t))^2\frac{\partial u_1(x,t)}{\partial x}\right)\right]$$
$$+C_2\left[\frac{\partial^\alpha u_0(x,t)}{\partial t^\alpha}\frac{\partial^3 u_0(x,t)}{\partial x^3}+\frac{3}{2}v_0(x,l)\frac{\partial^3 v_0(x,t)}{\partial x^3}+\frac{9}{2}\frac{\partial v_0(x,t)}{\partial x}\frac{\partial^2 v_0(x,t)}{\partial x^2}\right.$$
$$\left.-6u_0(x,t)\frac{\partial u_0(x,t)}{\partial x}-6u_0(x,t)v_0(x,t)\frac{\partial v_0(x,t)}{\partial x}-\frac{3}{2}\frac{\partial u_0(x,t)}{\partial x}(v_0(x,t))^2\right], \tag{7.103}$$

$$\frac{\partial^\alpha v_2(x,t)}{\partial t^\alpha}-\frac{\partial^\alpha v_1(x,t)}{\partial t^\alpha}=\tilde{C}_1\left[\frac{\partial^\alpha v_1(x,t)}{\partial t^\alpha}+\frac{\partial^3 v_1(x,t)}{\partial x^3}-6\left(\frac{\partial u_0(x,t)}{\partial x}v_1(x,t)+\frac{\partial u_1(x,t)}{\partial x}v_0(x,t)\right)\right.$$
$$-6\left(u_1(x,t)\frac{\partial v_0(x,t)}{\partial x}+u_0(x,t)\frac{\partial v_1(x,t)}{\partial x}\right)$$
$$-\frac{15}{2}\left(2v_0(x,t)v_1(x,t)\frac{\partial v_0(x,t)}{\partial x}+\left(v_0(x,t)\right)^2\frac{\partial v_1(x,t)}{\partial x}\right)$$
$$+\tilde{C}_2\left[\frac{\partial^\alpha v_0(x,t)}{\partial t^\alpha}+\frac{\partial^3}{\partial x^3}v_0(x,t)-6\frac{\partial u_0(x,t)}{\partial x}v_0(x,t)\right.$$
$$\left.-6u_0(x,t)\frac{\partial v_0(x,t)}{\partial x}-\frac{15}{2}\frac{\partial v_0(x,t)}{\partial x}\left(v_0(x,t)\right)^2\right] \tag{7.104}$$

and so on.

For solving fractional order coupled Jaulent–Miodek equations using OHAM, the following initial conditions for Equations 7.65 and 7.66 are considered:

$$u(x,0)=\frac{1}{8}\lambda^2\left[1-4\operatorname{sech}^2\left(\frac{\lambda x}{2}\right)\right],$$

$$v(x,0)=\lambda\operatorname{sech}\left(\frac{\lambda x}{2}\right).$$

Using the initial conditions $u_0=u(x,0)$ and $v_0=v(x,0)$ and solving Equations 7.99 through 7.104, we obtain the expressions for $u_0, v_0; u_1, v_1; u_2, v_2$; and so on.

Finally, the third-order approximate solutions for time-fractional nonlinear coupled Jaulent–Miodek equations are given by

$$u=u_0(x,t)+u_1(x,t)+u_2(x,t), \tag{7.105}$$

$$v=v_0(x,t)+v_1(x,t)+v_2(x,t). \tag{7.106}$$

The optimal values of the convergence control parameters C_1, C_2 and $\tilde{C}_1, \tilde{C}_2$ can be obtained using the weighted residual least square method given in Equations 1.34 and 1.35.

7.8 Numerical Results and Discussion

The comparison of the absolute errors for nonlinear time-fractional coupled Jaulent–Miodek equations (7.65) and (7.66) have been illustrated in Tables 7.12 and 7.13 that are generated through the results attained by the two-dimensional Hermite wavelet method and OHAM at various points of x and t, taking $\alpha=1$. In the present study, in order to inspect the accuracy and reliability of Hermite wavelets for solving the fractional order nonlinear coupled system of Jaulent–Miodek equations, we compare the numerical approximate solutions obtained by using Hermite wavelets with the third-order approximate OHAM solutions. Tables 7.14 through 7.17 illustrate the comparison of absolute errors of fractional order Jaulent–Miodek equations (7.65) and (7.66) at various points of x and t, taking $\alpha=0.75$ and 0.5, respectively. Agreements between present numerical results for a time-fractional Jaulent–Miodek equation obtained by Hermite wavelets and OHAM appear very satisfactory through illustrations in Tables 7.12 through 7.18, which also confirm the validity of the accurate solution by Hermite wavelets.

TABLE 7.12

Absolute Errors with Regard to $u(x,t)$ Obtained by the Hermite Wavelet Method for the Nonlinear System of Coupled Jaulent–Miodek Equations Given in Equations 7.65 and 7.66 at Various Points of x and t, Taking $\alpha=1$ and $\lambda=0.5$

	$\lvert u_{Exact}-u_{Hermite\ wavelet}\rvert$								
x	$t=0.1$	$t=0.2$	$t=0.3$	$t=0.4$	$t=0.5$	$t=0.6$	$t=0.7$	$t=0.8$	$t=0.9$
0.1	2.0553E−6	4.6970E−6	2.0543E−5	4.5657E−5	9.0317E−5	1.3421E−4	1.8769E−4	2.5054E−4	3.2257E−4
0.2	2.2375E−6	1.3360E−5	3.3614E−5	6.3236E−5	1.1255E−4	1.6141E−4	2.2021E−4	2.8892E−4	3.6751E−4
0.3	6.2475E−6	2.1113E−5	4.4884E−5	7.7842E−5	1.3034E−4	1.8239E−4	2.4438E−4	3.1645E−4	3.9868E−4
0.4	1.0252E−5	2.8524E−5	5.5227E−5	9.0679E−5	1.4527E−5	1.9909E−4	2.6257E−4	3.3593E−4	4.1938E−4
0.5	6.7072E−6	2.7007E−5	6.1236E−5	1.1011E−4	1.8039E−4	2.6055E−4	3.5746E−4	4.7156E−4	6.0328E−4
0.6	7.1378E−6	2.7087E−5	6.0652E−5	1.0857E−4	1.7763E−4	2.5639E−4	3.5173E−4	4.6421E−4	5.9434E−4
0.7	6.3626E−6	2.5083E−5	5.6901E−5	1.0257E−4	1.6888E−4	2.4445E−4	3.3619E−4	4.4468E−4	5.7054E−4
0.8	5.1634E−6	2.1991E−5	5.1218E−5	9.3599E−5	1.5592E−4	2.2684E−4	3.1326E−4	4.1581E−4	5.3510E−4
0.9	3.7368E−6	1.8227E−5	4.4262E−5	8.2587E−5	1.3998E−4	2.0510E−4	2.8485E−4	3.7985E−4	4.9074E−4

TABLE 7.13

Absolute Errors with Regard to $v(x,t)$ Obtained by the Hermite Wavelet Method for the Nonlinear System of Coupled Jaulent–Miodek Equations Given in Equations 7.65 and 7.66 at Various Points of x and t, Taking $\alpha=1$ and $\lambda=0.5$

	$\lvert v_{Exact}-v_{Hermite\,wavelet}\rvert$								
x	$t=0.1$	$t=0.2$	$t=0.3$	$t=0.4$	$t=0.5$	$t=0.6$	$t=0.7$	$t=0.8$	$t=0.9$
0.1	1.0266E−4	1.2531E−4	7.4403E−5	4.2725E−5	2.2305E−4	4.4542E−4	7.0631E−4	9.9217E−4	1.2894E−3
0.2	6.8598E−6	6.3948E−5	2.0545E−4	4.0998E−4	6.7411E−4	9.7665E−4	1.3136E−3	1.6714E−3	2.0365E−3
0.3	8.6641E−5	2.4761E−4	4.7556E−4	7.6247E−4	1.1045E−3	1.4804E−3	1.8858E−3	2.3069E−3	2.7302E−3
0.4	1.7814E−4	4.2632E−4	7.3694E−4	1.1016E−3	1.5161E−3	1.9591E−3	2.4257E−3	2.9022E−3	3.3748E−3
0.5	1.6264E−4	2.3867E−4	2.4153E−4	1.8420E−4	8.2495E−5	5.6018E−5	2.1508E−4	3.8165E−4	5.4268E−4
0.6	7.7242E−5	7.4979E−5	6.0988E−6	1.1639E−4	2.7664E−4	4.6748E−4	6.7260E−4	8.7936E−4	1.0751E−3
0.7	1.9921E−6	7.6519E−5	2.1056E−4	3.9113E−4	6.0232E−4	8.3738E−4	1.0799E−3	1.3178E−3	1.5387E−3
0.8	7.6073E−5	2.1710E−4	4.1005E−4	6.4194E−4	8.9692E−4	1.1685E−3	1.4405E−3	1.7010E−3	1.9382E−3
0.9	1.4522E−4	3.4725E−4	5.9317E−4	8.7004E−4	1.1620E−3	1.4631E−3	1.7569E−3	2.0321E−3	2.2773E−3

TABLE 7.14

Absolute Errors with Regard to $u(x,t)$ Obtained by the Two-Dimensional Hermite Wavelet Method and Third-Order OHAM Solution for the Fractional Order Nonlinear Coupled Jaulent–Miodek Equations Given in Equations 7.65 and 7.66 at Various Points of x and t, Taking $\alpha=0.75$ and $\lambda=0.5$

	$\lvert u_{OHAM}-u_{Hermite\ wavelet}\rvert$								
x	$t=0.1$	$t=0.2$	$t=0.3$	$t=0.4$	$t=0.5$	$t=0.6$	$t=0.7$	$t=0.8$	$t=0.9$
0.1	7.4968E−6	3.0040E−5	6.3702E−5	1.0598E−4	1.5534E−4	2.0969E−4	2.6883E−4	3.3038E−4	3.9195E−4
0.2	1.9727E−5	5.0514E−5	9.1979E−5	1.4208E−4	1.9931E−4	2.6235E−4	3.3101E−4	4.0334E−4	4.7734E−4
0.3	3.0934E−5	6.8532E−5	1.1601E−4	1.7181E−4	2.3442E−4	3.0327E−4	3.7815E−4	4.5748E−4	5.3958E−4
0.4	4.1653E−5	8.5048E−5	1.3716E−4	1.9693E−4	2.6291E−4	3.3516E−4	4.1348E−4	4.9660E−4	5.8308E−4
0.5	4.7032E−5	1.0898E−4	1.8956E−4	2.8591E−4	3.9484E−4	5.1702E−4	6.5070E−4	7.9455E−4	9.4703E−4
0.6	5.0687E−5	1.1373E−4	1.9513E−4	2.9216E−4	4.0149E−4	5.2437E−4	6.5901E−4	8.0432E−4	9.5894E−4
0.7	5.2128E−5	1.1459E−4	1.9494E−4	2.9053E−4	3.9790E−4	5.1884E−4	6.5148E−4	7.9493E−4	9.4801E−4
0.8	5.2327E−5	1.1287E−4	1.9061E−4	2.8301E−4	3.8642E−4	5.0316E−4	6.3127E−4	7.7000E−4	9.1831E−4
0.9	5.1682E−5	1.0929E−4	1.8321E−4	2.7101E−4	3.6885E−4	4.7954E−4	6.0099E−4	7.3260E−4	8.7341E−4

TABLE 7.15

Absolute Errors with Regard to $v(x,t)$ Obtained by the Two-Dimensional Hermite Wavelet Method and Third-Order OHAM Solution for the Fractional Order Nonlinear Coupled Jaulent–Miodek Equations Given in Equations 7.65 and 7.66 at Various Points of x and t, Taking $\alpha=0.75$ and $\lambda=0.5$

	$\lvert u_{OHAM}-u_{Hermite\,wavelet}\rvert$								
x	$t=0.1$	$t=0.2$	$t=0.3$	$t=0.4$	$t=0.5$	$t=0.6$	$t=0.7$	$t=0.8$	$t=0.9$
0.1	8.9842E−5	1.1334E−5	1.9121E−4	4.0028E−4	6.0268E−4	7.8294E−4	9.1713E−4	9.8878E−4	9.8134E−4
0.2	9.1969E−5	3.1053E−4	5.8586E−4	8.7629E−4	1.1504E−3	1.3940E−3	1.5852E−3	1.7089E−3	1.7502E−3
0.3	2.6796E−4	5.9763E−4	9.6132E−4	1.3253E−3	1.6626E−3	1.9604E−3	2.1985E−3	2.3636E−3	2.4420E−3
0.4	4.3878E−4	8.7377E−4	1.3192E−3	1.7495E−3	2.1422E−3	2.4855E−3	2.7614E−3	2.9580E−3	3.0627E−3
0.5	1.9169E−4	1.7339E−4	8.2319E−5	2.5428E−5	1.1164E−4	1.7211E−4	1.8761E−4	1.5108E−4	5.5148E−5
0.6	3.6408E−4	7.1595E−4	2.2882E−4	3.8801E−4	5.1629E−4	6.1082E−4	6.5499E−4	6.4337E−4	5.7004E−4
0.7	1.0684E−4	2.9537E−4	5.0987E−4	7.1171E−4	8.7304E−4	9.9249E−4	1.0560E−3	1.0596E−3	9.9889E−4
0.8	2.3933E−4	4.9968E−4	7.6306E−4	9.9920E−4	1.1852E−3	1.3211E−3	1.3953E−3	1.4052E−3	1.3476E−3
0.9	3.6156E−4	6.8549E−4	9.8986E−4	1.2525E−3	1.4554E−3	1.5998E−3	1.6766E−3	1.6845E−3	1.6216E−3

TABLE 7.16

Absolute Errors with Regard to $u(x,t)$ Obtained by the Two-Dimensional Hermite Wavelet Method and Third-Order OHAM Solution for the Fractional Order Nonlinear Coupled Jaulent–Miodek Equations Given in Equations 7.65 and 7.66 at Various Points of x and t, Taking $\alpha=0.5$ and $\lambda=0.5$

	$\lvert u_{OHAM}-u_{Hermite\,wavelet}\rvert$								
x	$t=0.1$	$t=0.2$	$t=0.3$	$t=0.4$	$t=0.5$	$t=0.6$	$t=0.7$	$t=0.8$	$t=0.9$
0.1	2.1552E−5	8.1878E−5	1.4911E−4	2.1501E−4	2.7542E−4	3.2160E−4	3.5273E−4	3.6046E−4	3.3639E−4
0.2	1.0292E−5	7.7946E−5	1.5643E−4	2.3633E−4	3.1247E−4	3.7782E−4	4.3104E−4	4.6468E−4	4.7123E−4
0.3	4.1238E−6	6.7652E−5	1.5385E−4	2.4384E−4	3.3140E−4	4.1103E−4	4.8085E−4	5.3418E−4	5.6425E−4
0.4	2.0429E−5	5.2912E−5	1.4387E−4	2.4062E−4	3.3588E−4	4.2555E−4	5.0718E−4	5.7474E−4	6.2205E−4
0.5	8.1331E−6	1.1511E−4	2.6479E−4	4.2458E−4	5.8491E−4	7.4584E−4	9.0415E−4	1.0570E−3	1.2014E−3
0.6	3.5413E−5	8.4721E−5	2.3575E−4	3.9943E−4	5.6478E−4	7.3306E−4	9.0046E−4	1.0645E−3	1.2226E−3
0.7	6.6534E−5	4.7447E−5	1.9658E−4	3.6070E−4	5.2736E−4	6.9890E−4	8.7092E−4	1.0413E−3	1.2076E−3
0.8	9.9929E−5	5.3824E−6	1.4987E−4	3.1142E−4	4.7612E−4	6.4732E−4	8.1999E−4	9.9225E−4	1.1618E−3
0.9	1.3460E−4	3.9949E−5	9.7606E−5	2.5404E−4	4.1400E−4	5.8175E−4	7.5163E−4	9.2194E−4	1.0905E−3

TABLE 7.17

Absolute Errors with Regard to $v(x,t)$ Obtained by the Two-Dimensional Hermite Wavelet Method and Third-Order OHAM Solution for the Fractional Order Nonlinear Coupled Jaulent–Miodek Equations Given in Equations 7.65 and 7.66 at Various Points of x and t, Taking $\alpha=0.5$ and $\lambda=0.5$

	$\lvert v_{OHAM}-v_{Hermite\,wavelet}\rvert$								
x	$t=0.1$	$t=0.2$	$t=0.3$	$t=0.4$	$t=0.5$	$t=0.6$	$t=0.7$	$t=0.8$	$t=0.9$
0.1	8.3521E−5	2.9312E−4	4.2261E−4	4.3707E−4	3.3456E−4	1.0091E−4	2.5336E−4	7.3818E−4	1.3635E−3
0.2	3.3416E−4	6.5094E−4	8.5761E−4	9.3554E−4	8.9197E−4	7.1374E−4	4.1479E−4	1.2595E−5	5.7631E−4
0.3	5.6957E−4	9.8211E−4	1.2548E−3	1.3850E−3	1.3888E−3	1.2538E−3	9.9728E−4	6.1325E−4	9.5719E−5
0.4	7.9133E−4	1.2891E−3	1.6175E−3	1.7897E−3	1.8301E−3	1.7270E−3	1.5006E−3	1.1468E−3	6.6099E−4
0.5	3.9482E−4	3.6236E−4	3.4634E−4	3.9120E−4	5.0371E−4	6.8476E−4	9.2508E−4	1.2283E−3	1.5984E−3
0.6	2.0871E−4	1.0875E−4	4.8599E−5	5.8024E−5	1.3694E−4	2.8552E−4	4.9242E−4	7.5994E−4	1.0908E−3
0.7	4.4730E−5	1.0899E−4	2.0229E−4	2.1800E−4	1.6237E−4	3.5946E−5	1.4821E−4	3.9122E−4	6.9475E−4
0.8	9.7652E−5	2.9380E−4	4.0997E−4	4.4120E−4	3.9923E−4	2.8531E−4	1.1391E−4	1.1510E−4	4.0251E−4
0.9	2.2031E−4	4.4782E−4	5.7730E−4	6.1511E−4	5.7787E−4	4.6746E−4	2.9953E−4	7.4704E−5	2.0705E−4

TABLE 7.18

L_2 and L_∞ Error Norms for Fractional Order Nonlinear Coupled Jaulent–Miodek Equation Using the Two-Dimensional Hermite Wavelet Method and OHAM at Various Points of t, Taking $\alpha = 0.5, 0.75$, and 1

	Error Analysis of $u(x, t)$ with Regard to a Hermite Wavelet						Error Analysis of $v(x, t)$ with Regard to a Hermite Wavelet					
	$\alpha = 1$		$\alpha = 0.75$		$\alpha = 0.5$		$\alpha = 1$		$\alpha = 0.75$		$\alpha = 0.5$	
t	L_2	L_∞	L_2	L_∞	L_2	L_∞	L_2	L_∞	L_2	L_∞	L_2	L_∞
0.1	1.8177E−5	2.8524E−5	1.2666E−4	5.2327E−5	1.8669E−4	4.3878E−4	3.3034E−4	1.7814E−4	1.2194E−3	4.3878E−4	1.1527E−3	7.9133E−4
0.2	6.4740E−5	1.0252E−5	2.7930E−4	1.1459E−4	2.1085E−4	8.7377E−4	7.0640E−4	4.2632E−4	1.6021E−3	8.7377E−4	1.8915E−3	1.2891E−3
0.3	1.4786E−4	6.1236E−5	4.7612E−4	1.9513E−4	5.3609E−4	1.3192E−3	1.1998E−3	7.3694E−4	2.2193E−3	1.3192E−3	2.4019E−3	1.6175E−3
0.4	2.6529E−4	1.1011E−4	7.1047E−4	2.9216E−4	9.2243E−4	1.7495E−3	1.8260E−3	1.1016E−3	2.9950E−3	1.7495E−3	2.6390E−3	1.7897E−3
0.5	4.1778E−4	1.8039E−4	9.7402E−4	4.0149E−4	1.3166E−3	2.1422E−3	2.5736E−3	1.5161E−3	3.6883E−3	2.1422E−3	3.6883E−3	1.8301E−3
0.6	6.3551E−4	2.6055E−4	1.2522E−3	5.2437E−4	1.7134E−3	2.4855E−3	3.4075E−3	1.9591E−3	4.2743E−3	2.4855E−3	2.4343E−3	1.7270E−3
0.7	8.6958E−4	3.5746E−4	1.5929E−3	6.5901E−4	2.1030E−3	2.7614E−3	4.3001E−3	2.4257E−3	4.7064E−3	2.7614E−3	2.1692E−3	1.5006E−3
0.8	1.1454E−3	4.7156E−4	1.9410E−3	8.0432E−4	2.4760E−3	2.9580E−3	5.1364E−3	2.9022E−3	4.9604E−3	2.9580E−3	2.1200E−3	1.2283E−3
0.9	1.4642E−3	6.0328E−4	2.3095E−3	9.5894E−4	2.8242E−3	3.0627E−3	6.1166E−3	3.3748E−3	5.0141E−3	3.0627E−3	2.6588E−3	1.5984E−3

7.9 Convergence of a Hermite Wavelet

Theorem 7.1 **Convergence Theorem**

If a continuous function, $u(x,t)\in L^2(\mathfrak{R}\times\mathfrak{R})$, defined on $[0,1)\times[0,1)$ is bounded, that is, $|u(x,t)|\le K$, then the Hermite wavelets expansion of $u(x,t)$ converges uniformly to it.

Proof

Let $u(x,t)$ be a function defined on $[0,1)\times[0,1)$ and $|u(x,t)|\le K$, where K is a positive constant.

The Hermite wavelet coefficients of continuous functions $u(x,t)$ are defined as

$$c_{ij}=\int_0^1\int_0^1 u(x,t)\psi_i(x)\psi_j(t)\,dx\,dt$$

$$=2^{k_1/2}\sqrt{\frac{1}{n_i!2^{n_1}\sqrt{\pi}}}\int_0^1\int_{I_1} u(x,t)H_{m_1}\left(2^{k_1}x-\hat{n}_1\right)\psi_j(t)\,dx\,dt \quad \text{where } I_1=\left[\frac{\hat{n}_1-1}{2^{k_1}},\frac{\hat{n}_1+1}{2^{k_1}}\right).$$

Now, by change of variable $2^{k_1}x-\hat{n}_1=y$, we get

$$c_{ij}=\frac{2^{k_1/2}}{2^{k_1}}\sqrt{\frac{1}{n_1!2^{n_1}\sqrt{\pi}}}\int_0^1\psi_j(t)\left(\int_{-1}^1 u\left(\frac{y+2n_1-1}{2^{k_1}},t\right)H_{m_1}(y)\,dy\right)dt.$$

Using the mean value theorem of integral calculus, we will have

$$c_{ij}=\frac{1}{2^{k_1/2}}\sqrt{\frac{1}{n_1!2^{n_1}\sqrt{\pi}}}\int_0^1\psi_j(t)u\left(\frac{\xi+2n_1-1}{2^{k_1}},t\right)\left(\int_{-1}^1 H_{m_1}(y)\,dy\right)dt,\quad \text{where}\,\xi\in(-1,1)$$

$$=\frac{1}{2^{k_1/2}}\sqrt{\frac{1}{n_1!2^{n_1}\sqrt{\pi}}}\int_0^1\psi_j(t)u\left(\frac{\xi+2n_1-1}{2^{k_1}},t\right)\left(\int_{-1}^1 \frac{H'_{m_1+1}(y)}{2(m_1+1)}\,dy\right)dt,$$

since $H'_{m_1+1}(y)=2(m_1+1)H_{m_1}(y)$

$$=\frac{1}{2^{k_1/2}}\sqrt{\frac{1}{n_1!2^{n_1}\sqrt{\pi}}}\int_0^1\psi_j(t)u\left(\frac{\xi+2n_1-1}{2^{k_1}},t\right)\left(\frac{H_{m_1+1}(y)}{2(m_1+1)}\right)_{-1}^{1}dt$$

$$=\frac{1}{2^{k_1/2}}\sqrt{\frac{1}{n_1!2^{n_1}\sqrt{\pi}}}\left(\frac{H_{m_1+1}(1)-H_{m_1+1}(-1)}{2(m_1+1)}\right)\int_0^1\psi_j(t)u\left(\frac{\xi+2n_1-1}{2^{k_1}},t\right)dt$$

$$=\frac{1}{2^{k_1/2}}\sqrt{\frac{1}{n_1!2^{n_1}\sqrt{\pi}}}\left(\frac{H_{m_1+1}(1)-H_{m_1+1}(-1)}{2(m_1+1)}\right)$$

$$\times\int_{I_2} u\left(\frac{\xi+2n_1-1}{2^{k_1}},t\right)2^{k_2/2}\sqrt{\frac{1}{n_2!2^{n_2}\sqrt{\pi}}}H_{m_2}\left(2^{k_2}t-\hat{n}_2\right)dt,$$

where $I_2 = \left[\frac{\hat{n}_2 - 1}{2^{k_2}}, \frac{\hat{n}_2 + 1}{2^{k_2}}\right)$.

$$c_{ij} = \frac{1}{2^{k_1/2}} 2^{k_2/2} \sqrt{\frac{1}{n_1! 2^{n_1} \sqrt{\pi}}} \sqrt{\frac{1}{n_2! 2^{n_2} \sqrt{\pi}}} \left(\frac{H_{m_1+1}(1) - H_{m_1+1}(-1)}{2(m_1+1)}\right)$$
$$\times \int_{I_2} u\left(\frac{\xi + 2n_1 - 1}{2^{k_1}}, t\right) H_{m_2}\left(2^{k_2} t - \hat{n}_2\right) dt.$$

Again by change of variable $2^{k_2} t - \hat{n}_2 = w$, we get

$$c_{ij} = \frac{1}{2^{k_1/2}} \frac{2^{k_2/2}}{2^{k_2}} \sqrt{\frac{1}{n_1! 2^{n_1} \sqrt{\pi}}} \sqrt{\frac{1}{n_2! 2^{n_2} \sqrt{\pi}}} \left(\frac{H_{m_1+1}(1) - H_{m_1+1}(-1)}{2(m_1+1)}\right)$$
$$\times \int_{-1}^{1} u\left(\frac{\xi + 2n_1 - 1}{2^{k_1}}, \frac{w + 2n_2 - 1}{2^{k_2}}\right) H_{m_2}(w) dw.$$

Using the mean value theorem of integral calculus, we have

$$c_{ij} = \frac{1}{2^{(k_1+k_2)/2}} \sqrt{\frac{1}{n_1! 2^{n_1} \sqrt{\pi}}} \sqrt{\frac{1}{n_2! 2^{n_2} \sqrt{\pi}}} \left(\frac{H_{m_1+1}(1) - H_{m_1+1}(-1)}{2(m_1+1)}\right)$$
$$\times u\left(\frac{\xi + 2n_1 - 1}{2^{k_1}}, \frac{\eta + 2n_2 - 1}{2^{k_2}}\right) \int_{-1}^{1} H_{m_2}(w) dw,$$

where $\eta \in (-1, 1)$

$$c_{ij} = \frac{1}{2^{(k_1+k_2)/2}} \sqrt{\frac{1}{n_1! 2^{n_1} \sqrt{\pi}}} \sqrt{\frac{1}{n_2! 2^{n_2} \sqrt{\pi}}} \left(\frac{H_{m_1+1}(1) - H_{m_1+1}(-1)}{2(m_1+1)}\right)$$
$$\times u\left(\frac{\xi + 2n_1 - 1}{2^{k_1}}, \frac{\eta + 2n_2 - 1}{2^{k_2}}\right) \int_{-1}^{1} \frac{H'_{m_2}(w)}{2(m_2+1)} dw \quad \text{since } H'_{m_2+1}(w) = 2(m_2+1) H_{m_2}(w)$$
$$= \frac{1}{2^{(k_1+k_2)/2}} \sqrt{\frac{1}{n_1! 2^{n_1} \sqrt{\pi}}} \sqrt{\frac{1}{n_2! 2^{n_2} \sqrt{\pi}}} \left(\frac{H_{m_1+1}(1) - H_{m_1+1}(-1)}{2(m_1+1)}\right)$$
$$\times u\left(\frac{\xi + 2n_1 - 1}{2^{k_1}}, \frac{\eta + 2n_2 - 1}{2^{k_2}}\right) \left(\frac{H_{m_2+1}(w)}{2(m_2+1)}\right)_{-1}^{1}$$
$$= \frac{1}{2^{(k_1+k_2)/2}} \sqrt{\frac{1}{n_1! 2^{n_1} \sqrt{\pi}}} \sqrt{\frac{1}{n_2! 2^{n_2} \sqrt{\pi}}} \left(\frac{H_{m_1+1}(1) - H_{m_1+1}(-1)}{2(m_1+1)}\right)$$
$$\times \left(\frac{H_{m_2+1}(1) - H_{m_2+1}(-1)}{2(m_2+1)}\right) \times u\left(\frac{\xi + 2n_1 - 1}{2^{k_1}}, \frac{\eta + 2n_2 - 1}{2^{k_2}}\right).$$

Therefore,

$$|c_{ij}| = \left| \frac{1}{2^{(k_1+k_2)/2}} \sqrt{\frac{1}{n_1!2^{n_1}\sqrt{\pi}}} \sqrt{\frac{1}{n_2!2^{n_2}\sqrt{\pi}}} \left(\frac{H_{m_1+1}(1) - H_{m_1+1}(-1)}{2(m_1+1)} \right) \right.$$

$$\left. \times \left(\frac{H_{m_2+1}(1) - H_{m_2+1}(-1)}{2(m_2+1)} \right) u\left(\frac{\xi + 2n_1 - 1}{2^{k_1}}, \frac{\eta + 2n_2 - 1}{2^{k_2}} \right) \right|$$

$$\leq \frac{1}{2^{(k_1+k_2)/2}} \sqrt{\frac{1}{n_1!2^{n_1}\sqrt{\pi}}} \sqrt{\frac{1}{n_2!2^{n_2}\sqrt{\pi}}} \left(\frac{H_{m_1+1}(1) - H_{m_1+1}(-1)}{2(m_1+1)} \right) \left(\frac{H_{m_2+1}(1) - H_{m_2+1}(-1)}{2(m_2+1)} \right) K,$$

since $u(x,t)$ is bounded.

Therefore $\sum_{i=0}^{\infty} \sum_{j=0}^{\infty} c_{ij}$ is absolutely convergent.

Hence from [44], the Hermite series expansion of $u(x,t)$ converges uniformly. □

7.10 Conclusion

In this chapter, the numerical solution of a time-fractional modified Fornberg–Whitham equation has been determined by using the two-dimensional Hermite wavelet method. The results thus obtained are then compared with exact solutions as well as with OHAM. The obtained results demonstrate the accuracy, efficiency, and reliability of the proposed algorithm based on the two-dimensional Hermite wavelet method and its applicability to a nonlinear time-fractional modified Fornberg–Whitham equation. Agreement between present numerical results obtained by the Hermite wavelet method with OHAM and exact solutions appear very satisfactory through illustrative results in Tables 7.1 through 7.7.

The solitary wave solution of the fractional modified Fornberg–Whitham equation has been obtained for the first time by using FIM. The fractional order modified Fornberg–Whitham equation has also been solved by using the OHAM. The acquired numerical approximate results are compared with the exact solutions obtained through FIM, which reveals the efficiency and plausibility of the proposed OHAM technique. The results exhibited in Tables 7.8 through 7.10 demonstrate a pretty good agreement between the present numerical methods with the exact solution. Finally, it is worthwhile to mention that the FIM is promising and powerful for solving nonlinear fractional differential equations in mathematical physics.

Next, the two-dimensional Hermite wavelet method has been implemented to a coupled fractional differential equation, namely, the fractional Jaulent–Miodek equation. The results thus obtained are then compared with exact solutions as well as with OHAM. The evaluated outcomes demonstrate the efficiency, accuracy, and reliability of the proposed algorithm based on the two-dimensional Hermite wavelet approach and its applicability to nonlinear time-fractional coupled Jaulent–Miodek equations. It can be observed that the agreement between proposed numerical results attained by the Hermite wavelet technique with OHAM and exact solutions appear very satisfactory by means of illustrative results in Tables 7.12 through 7.18. The present scheme is easy to implement, effective, and suitable for acquiring numerical solutions of nonlinear time-fractional coupled Jaulent–Miodek equations.

The application of the proposed numerical approach based on the two-dimensional Hermite wavelet method for the solutions of the time-fractional modified Fornberg–Whitham equation and fractional coupled Jaulent–Miodek equations quite satisfactorily justifies its simplicity, effectivity, and applicability.

8

Implementation of the Petrov–Galerkin Method for Solving Fractional Partial Differential Equations

8.1 Introduction

In the present chapter, the Petrov Galerkin method has been utilized for the numerical solution of the fractional Korteweg–de Vries–Burgers (KdVB) equation and fractional Sharma–Tasso–Olver (STO) equation. In the past few years, tremendous effort has been anticipated by the researchers on the study of nonlinear evolution equations that appeared in mathematical physics. The nonlinear fractional differential equations have been solved numerically through the Petrov–Galerkin approach by utilizing a linear hat function as the trial function and a quintic B-spline function as the test function.

In the present study, we consider one of the well-known equations, namely, the KdVB, which plays an essential role in both applied mathematics and physics. Especially Korteweg–de Vries (KdV) type equations have been paid more attention due to their various applications in plasma physics, solid-state physics, and quantum field theory.

The KdVB equation is a nonlinear partial differential equation of the form

$$u_t + \varepsilon u u_x - \nu u_{xx} + \mu u_{xxx} = 0, \tag{8.1}$$

which was first derived by Su and Gardner [167]. It arises in quite a lot of contexts as a model equation incorporating a few foremost physical phenomena, namely, dispersion, viscosity, and nonlinear advection. This equation arises within the description of long wave propagation in shallow water [168], propagation of waves in an elastic tube stuffed with a viscous fluid [169], and weakly nonlinear plasma waves with certain dissipative effects [170]. It additionally represents long wavelength approximations where the effect of the nonlinear advection uu_x is counterbalanced by means of the dispersion u_{xxx}.

The KdVB equation given in Equation 8.1 is a combination of Burgers' equation (when $\mu = 0$) and the KdV equation (when $\nu = 0$). In the year 1939, Burgers proposed an equation (known as Burgers' equation) for the study of turbulence and approximate theory of flow through a shock wave traveling in a viscous fluid [49,171]. The KdV equation was first suggested by Korteweg and de Vries [172]. This equation was used to study the change in shape of long waves moving in a rectangular channel [132,172]. When diffusion dominates dispersion, the numerical solutions of Equation 8.1 tend to behave like Burgers' equation solutions, and hence the steady-state solutions of the KdVB equation are monotonic shocks. However, when dispersion dominates, the KdV behavior is observed and the shocks are oscillatory.

A number of theoretical issues associated to the KdVB equation have received substantial attention. Many analytical and numerical methods have been proposed in the recent past for the study of the classical KdVB equation. Various methods such as the decomposition method [173], tanh method [174], hyperbolic tangent method and exponential rational function approach [175], septic B-spline method [176], radial basis functions [177], quartic B-spline Galerkin approach [178], and quintic B-spline finite elements [179] had been developed independently and had been used to acquire the exact as well as numerical solutions of the KdVB equation. However, as far as we know, no numerical works have been reported to solve the fractional KdVB equation. Methods such as the Adomian decomposition method [180] and homotopy perturbation method [181] were used to obtain the approximate solution of the fractional KdVB equation.

Let us consider the time-fractional KdVB equation [181–183] as follows:

$$D_t^{\alpha}u+\varepsilon uu_x-\nu u_{xx}+\mu u_{xxx}=0, \tag{8.2}$$

where

ε, v, and μ are constants

α denotes the order of fractional derivative whose range is $0<\alpha\leq 1$

In the present numerical scheme, the fractional derivative has been discretized by the Grünwald–Letnikov derivative and the fractional KdVB equation has been converted directly into finite difference equation. Then it has been adjusted in the form of the implicit finite difference scheme.

Next, the STO equation has been considered, which plays an essential role both in physics and applied mathematics. Remarkable effort has been anticipated by the researchers for studying nonlinear equations arising in mathematical physics. In recent years, STO equations [184,185] had been paid more attention due to their numerous implementations in mathematical physics.

Let us consider the time-fractional STO equation [186] as follows:

$$D_t^{\alpha}u+3au_x^2+3au^2u_x+3auu_{xx}+au_{xxx}=0, \tag{8.3}$$

where

$u(x,t)$ is the unknown function depending on the spatial variable x and the temporal variable t

a is a real parameter

Here, α denotes the order of fractional derivative whose range is $0<\alpha\leq 1$.

Equation 8.3 can also be written as [185]

$$D_t^{\alpha}u+\frac{3}{2}a\left(u^2\right)_{xx}+a\left(u^3\right)_x+au_{xxx}=0. \tag{8.4}$$

This equation contains both the linear dispersive term au_{xxx} and the double nonlinear terms $(u^2)_{xx}$ and $(u^3)_x$. The equation was first derived by Tasso [187] in 1976 as an example of odd members of the Burgers hierarchy by extending the linearization achieved through the Cole–Hopf ansatz to equations containing, as highest derivatives, odd space derivatives. In a subsequent report [188], the properties of the wave envelope solutions of this equation are investigated.

The STO equation appears in many scientific applications such as quantum field theory, plasma physics, dispersive wave phenomena, relativistic physics, nonlinear optics, and physical sciences [184]. It also arises as an evolution equation that possesses infinitely many symmetries [189]. Many analytical and numerical methods have been proposed in the recent past for the study of the classical STO equation. Various methods such as the Cole–Hopf transformations method [190], the Adomian decomposition method [186], the variational iteration method (VIM) [186], the homotopy perturbation method [186], Bäcklund transform method [185], exp-solution method [191], improved G'/G-expansion method [192], and tanh and extended tanh method [193] had been used to acquire exact as well as numerical solutions of the STO equation.

According to the best possible information of the authors, so far no numerical works have been reported to solve the fractional STO equation. The approximate analytical solution of fractional STO equation has been obtained by using the homotopy analysis method [194], the improved generalized tanh-coth method [195], the Adomian decomposition method [186], the VIM [186], and homotopy perturbation method [186]. In the present numerical scheme, the fractional derivative has been discretized by the Grünwald–Letnikov derivative and the fractional STO equation has been converted directly into a finite difference equation. Then it has been adjusted in the form of implicit finite difference scheme.

The present chapter lays emphasis on the application of the Petrov–Galerkin method for solving fractional differential equations such as the fractional KdVB equation and the fractional STO equation with a view to exhibit the capabilities of this method in handling the nonlinear equation. The main objective of this chapter is to establish the efficiency and accuracy of the Petrov–Galerkin method in solving fractional differential equations numerically by implementing a linear hat function as the trial function and a quintic B-spline function as the test function.

8.2 Implementation of the Petrov–Galerkin Method for the Numerical Solution of the Time-Fractional KdVB Equation

Let us consider the time-fractional KdVB equation [181–183] as

$$D_t^{\alpha}u+\varepsilon u u_x-\nu u_{xx}+\mu u_{xxx}=0, \tag{8.5}$$

with initial condition [196]

$$u(x,0)=\frac{-1}{50\varepsilon\nu\mu}\left[250\lambda\mu^2+6\nu^3\operatorname{sech}^2\left(\frac{\upsilon x}{10\mu}\right)\left(-1+\sinh\left(\frac{\upsilon x}{5\mu}\right)\right)\right]. \tag{8.6}$$

Equation 8.5 implies

$$D_t^{\alpha}u+\frac{\varepsilon}{2}\left(u^2\right)_x-\nu u_{xx}+\mu u_{xxx}=0,$$

where

ε, υ, and η are constants

α denotes the order of the fractional derivative whose range is $0<\alpha\leq 1$

The exact solution of Equation 8.5 is given by [196]

$$u(x,t)=\frac{-1}{50\varepsilon\upsilon\mu}\left[250\lambda\mu^2+6\nu^3\operatorname{sech}^2\left(\frac{1}{2}\left(\frac{\nu x}{5\mu}+\frac{\lambda t^\alpha}{\Gamma(\alpha+1)}\right)\right)\left(-1+\sinh\left(\frac{\upsilon x}{5\mu}+\frac{\lambda t^\alpha}{\Gamma(\alpha+1)}\right)\right)\right]. \quad (8.7)$$

The space interval $a_1 \le x \le b_1$ is discretized with $(N+1)$ uniform grid points $x_j = a_1 + jh$, $j=0,1,2,\ldots,N$ and the grid spacing is given by $h=(b_1-a_1)/N$. Let $U_j(t)$ denote the approximation to the exact solution $u(x_j,t)$. We sought the approximate solution of Equation 8.5 as

$$u(x,t)=\sum_{j=0}^{N}U_j(t)\varphi_j(x), \quad (8.8)$$

where $\varphi_j(x)$, $j=0,1,2,\ldots,N$ are trial functions.

In order to deal with the nonlinear term uu_x, the product approximation technique [191] is utilized as follows:

$$u^2(x,t)=\sum_{j=0}^{N}U_j^2(t)\varphi_j(x), \quad (8.9)$$

where $\varphi_j(x)$; $j=0,1,2,\ldots,N$ are the usual piecewise linear hat functions given by

$$\varphi_j(x)=\begin{cases}1+\dfrac{x-jh}{j}, & x\in[x_{j-1},x_j)\\ 1-\dfrac{x-jh}{j}, & x\in[x_j,x_{j+1})\\ 0, & \text{elsewhere.}\end{cases} \quad (8.10)$$

The unknown functions $U_j(t)$ are determined from the variational formulation

$$\left\langle D_t^\alpha u,\psi_j\right\rangle+\frac{\varepsilon}{2}\left\langle\left(u^2\right)_x,\psi_j\right\rangle-\nu\left\langle u_{xx},\psi_j\right\rangle+\mu\left\langle u_{xxx},\psi_j\right\rangle=0, \quad (8.11)$$

where $\psi_j(x)$, $j=0,1,2,\ldots,N$ are quintic B-spline functions taken as the test functions and are given by

$$\psi_j(x)=\frac{1}{h^5}\begin{cases}(x-x_{j-3})^5; & x\in[x_{j-3},x_{j-2})\\ (x-x_{j-3})^5-6(x-x_{j-2})^5; & x\in[x_{j-2},x_{j-1})\\ (x-x_{j-3})^5-6(x-x_{j-2})^5+15(x-x_{j-1})^5; & x\in[x_{j-1},x_j)\\ (x-x_{j-3})^5-6(x-x_{j-2})^5+15(x-x_{j-1})^5-20(x-x_j)^5; & x\in[x_j,x_{j+1})\\ (x-x_{j-3})^5-6(x-x_{j-2})^5+15(x-x_{j-1})^5-20(x-x_j)^5+15(x-x_{j+1})^5; & x\in[x_{j+1},x_{j+2})\\ (x-x_{j-3})^5-6(x-x_{j-2})^5+15(x-x_{j-1})^5-20(x-x_j)^5+15(x-x_{j+1})^5-6(x-x_{j+2})^5; & x\in[x_{j+2},x_{j+3})\\ 0; & \text{elsewhere.}\end{cases}$$

Integrating Equation 8.11 by parts, we obtain

$$\left\langle D_t^{\alpha} u, \psi_j \right\rangle + \frac{\varepsilon}{2}\left\langle \left(u^2\right)_x, \psi_j \right\rangle - \nu\left\langle u_x, \left(\psi_j\right)_x \right\rangle + \mu\left\langle u_x, \left(\psi_j\right)_{xx} \right\rangle = 0. \tag{8.12}$$

Each linear hat function covers two elements so that each subinterval $[x_j, x_{j+1}]$ is covered by two linear hat functions. On the other hand, each quantic B-spline covers six elements so that each subinterval $[x_j, x_{j+1}]$ is covered by six splines. In terms of local coordinate system given by

$$\xi = x - x_j, \quad 0 < \xi \leq h.$$

Both the linear hat functions φ_j and the quantic B-spline functions ψ_j over the element $[x_j, x_{j+1}]$ can be defined as follows:

$$\varphi_j = \frac{\xi}{h},$$
$$\varphi_{j+1} = 1 - \frac{\xi}{h},$$

and

$$\psi_{j-2} = 1 - 5\left(\frac{\xi}{h}\right) + 10\left(\frac{\xi}{h}\right)^2 - 10\left(\frac{\xi}{h}\right)^3 + 5\left(\frac{\xi}{h}\right)^4 - \left(\frac{\xi}{h}\right)^5,$$

$$\psi_{j-1} = 26 - 50\left(\frac{\xi}{h}\right) + 20\left(\frac{\xi}{h}\right)^2 + 20\left(\frac{\xi}{h}\right)^3 - 20\left(\frac{\xi}{h}\right)^4 + 5\left(\frac{\xi}{h}\right)^5,$$

$$\psi_j = 66 - 60\left(\frac{\xi}{h}\right)^2 + 30\left(\frac{\xi}{h}\right)^4 - 10\left(\frac{\xi}{h}\right)^5,$$

$$\psi_{j+1} = 26 + 50\left(\frac{\xi}{h}\right) + 20\left(\frac{\xi}{h}\right)^2 - 20\left(\frac{\xi}{h}\right)^3 - 20\left(\frac{\xi}{h}\right)^4 + 10\left(\frac{\xi}{h}\right)^5,$$

$$\psi_{j+2} = 1 + 5\left(\frac{\xi}{h}\right) + 10\left(\frac{\xi}{h}\right)^2 + 10\left(\frac{\xi}{h}\right)^3 + 5\left(\frac{\xi}{h}\right)^4 - 5\left(\frac{\xi}{h}\right)^5,$$

$$\psi_{j+3} = \left(\frac{\xi}{h}\right)^5.$$

From Equation 8.12, we have

$$\int_{x_0}^{x_N}\sum_{j=0}^{N} D_t^{\alpha} U_j(t)\varphi_j(x)\psi_j(x)\,dx + \frac{\varepsilon}{2}\int_{x_0}^{x_N}\sum_{j=0}^{N} U_j^2(t)\varphi_j'(x)\psi_j(x)\,dx$$
$$-\nu\int_{x_0}^{x_N}\sum_{j=0}^{N} U_j(t)\varphi_j'(x)\psi_j'(x)\,dx + \mu\int_{x_0}^{x_N}\sum_{j=0}^{N} U_j(t)\varphi_j'(x)\psi_j''(x)\,dx = 0. \tag{8.13}$$

Next, we set

$$I_1=\int_{x_0}^{x_N}\sum_{j=0}^{N}D_t^{\alpha}U_j(t)\varphi_j(x)\psi_j(x)dx,$$

$$I_2=\int_{x_0}^{x_N}\sum_{j=0}^{N}U_j^2(t)\varphi_j'(x)\psi_j(x)dx,$$

$$I_3=\int_{x_0}^{x_N}\sum_{j=0}^{N}U_j(t)\varphi_j'(x)\psi_j'(x)dx,$$

$$I_4=\int_{x_0}^{x_N}\sum_{j=0}^{N}U_j(t)\varphi_j'(x)\psi_j''(x)dx,$$

where $D_t^{\alpha}U_j(t)$ is defined by

$$D_t^{\alpha}U_j(t)=\lim_{\substack{\Delta t\to 0\\ \Delta tn=t-a}}(\Delta t)^{-\alpha}\sum_{r=0}^{n}\omega_r^{\alpha}U_j(t-\Delta tr).$$

Now

$$\begin{aligned}
I_1&=\int_{x_0}^{x_N}\sum_{j=0}^{N}D_t^{\alpha}U_j(t)\varphi_j(x)\psi_j(x)dx\\
&=\int_{x_{j-3}}^{x_{j-2}}(\Delta t)^{-\alpha}\sum_{r=0}^{n}\omega_r^{\alpha}U_{j-3}^{n-r}\varphi_{j-3}(x)\psi_{j-3}(x)dx+\int_{x_{j-3}}^{x_{j-2}}(\Delta t)^{-\alpha}\sum_{r=0}^{n}\omega_r^{\alpha}U_{j-2}^{n-r}\varphi_{j-2}(x)\psi_{j-3}(x)dx\\
&\quad+\int_{x_{j-2}}^{x_{j-1}}(\Delta t)^{-\alpha}\sum_{r=0}^{n}\omega_r^{\alpha}U_{j-2}^{n-r}\varphi_{j-2}(x)\psi_{j-2}(x)dx+\int_{x_{j-2}}^{x_{j-1}}(\Delta t)^{-\alpha}\sum_{r=0}^{n}\omega_r^{\alpha}U_{j-1}^{n-r}\varphi_{j-1}(x)\psi_{j-2}(x)dx\\
&\quad+\int_{x_{j-1}}^{x_{j}}(\Delta t)^{-\alpha}\sum_{r=0}^{n}\omega_r^{\alpha}U_{j-1}^{n-r}\varphi_{j-1}(x)\psi_{j-1}(x)dx+\int_{x_{j-1}}^{x_{j}}(\Delta t)^{-\alpha}\sum_{r=0}^{n}\omega_r^{\alpha}U_{j}^{n-r}\varphi_{j}(x)\psi_{j-1}(x)dx\\
&\quad+\int_{x_{j}}^{x_{j+1}}(\Delta t)^{-\alpha}\sum_{r=0}^{n}\omega_r^{\alpha}U_{j}^{n-r}\varphi_{j}(x)\psi_{j}(x)dx+\int_{x_{j}}^{x_{j+1}}(\Delta t)^{-\alpha}\sum_{r=0}^{n}\omega_r^{\alpha}U_{j+1}^{n-r}\varphi_{j+1}(x)\psi_{j}(x)dx\\
&\quad+\int_{x_{j+1}}^{x_{j+2}}(\Delta t)^{-\alpha}\sum_{r=0}^{n}\omega_r^{\alpha}U_{j+1}^{n-r}\varphi_{j+1}(x)\psi_{j+1}(x)dx+\int_{x_{j+1}}^{x_{j+2}}(\Delta t)^{-\alpha}\sum_{r=0}^{n}\omega_r^{\alpha}U_{j+2}^{n-r}\varphi_{j+2}(x)\psi_{j+1}(x)dx\\
&\quad+\int_{x_{j+2}}^{x_{j+3}}(\Delta t)^{-\alpha}\sum_{r=0}^{n}\omega_r^{\alpha}U_{j+2}^{n-r}\varphi_{j+2}(x)\psi_{j+2}(x)dx+\int_{x_{j+2}}^{x_{j+3}}(\Delta t)^{-\alpha}\sum_{r=0}^{n}\omega_r^{\alpha}U_{j+3}^{n-r}\varphi_{j+3}(x)\psi_{j+3}(x)dx\\
&=(\Delta t)^{-\alpha}\sum_{r=0}^{n}\omega_r^{\alpha}\left[\frac{h}{42}\left(U_{j-3}^{n-r}+120U_{j-2}^{n-r}+1191U_{j-1}^{n-r}+2416U_{j}^{n-r}+1191U_{j+1}^{n-r}+120U_{j+2}^{n-r}+U_{j+3}^{n-r}\right)\right].
\end{aligned}\tag{8.14}$$

Similarly,

$$I_2 = \frac{1}{6}\left[-\left(U_{j-3}^n\right)^2 - 56\left(U_{j-2}^n\right)^2 - 245\left(U_{j-1}^n\right)^2 + 245\left(U_{j+1}^n\right)^2 + 56\left(U_{j+2}^n\right)^2 + \left(U_{j+3}^n\right)^2\right], \quad (8.15)$$

$$I_3 = \frac{-1}{h}\left[U_{j-3}^n + 24U_{j-2}^n + 15U_{j-1}^n - 80U_j^n + 15U_{j+1}^n + 24U_{j+2}^n + U_{j+3}^n\right], \quad (8.16)$$

$$I_4 = \frac{1}{h^2}\left[-5U_{j-3}^n - 40U_{j-2}^n + 95U_{j-1}^n - 95U_{j+1}^n + 40U_{j+2}^n + 5U_{j+3}^n\right]. \quad (8.17)$$

Substituting Equations 8.14 through 8.17 in Equation 8.13, we have

$$\begin{aligned}
&(\Delta t)^{-\alpha}\sum_{r=0}^{n}\omega_r^\alpha\left[\frac{h}{42}\left(U_{j-3}^{n-r} + 120U_{j-2}^{n-r} + 1191U_{j-1}^{n-r} + 2416U_j^{n-r} + 1191U_{j+1}^{n-r} + 120U_{j+2}^{n-r} + U_{j+3}^{n-r}\right)\right] \\
&+\frac{\varepsilon}{12}\left[-\left(U_{j-3}^n\right)^2 - 56\left(U_{j-2}^n\right)^2 - 245\left(U_{j-1}^n\right)^2 + 245\left(U_{j+1}^n\right)^2 + 56\left(U_{j+2}^n\right)^2 + \left(U_{j+3}^n\right)^2\right] \\
&+\frac{\nu}{h}\left[U_{j-3}^n + 24U_{j-2}^n + 15U_{j-1}^n - 80U_j^n + 15U_{j+1}^n + 24U_{j+2}^n + U_{j+3}^n\right] \\
&+\frac{\mu}{h^2}\left[-5U_{j-3}^n - 40U_{j-2}^n + 95U_{j-1}^n - 95U_{j+1}^n + 40U_{j+2}^n + 5U_{j+3}^n\right] = 0, \quad (8.18)
\end{aligned}$$

where $j=1,2,\ldots,N$. Now to solve the preceding system, we assume U_j^n to be a discrete approximation to the exact solution $u(x_j, t_n)$. $U_j^n = 0$ for $j = -2, -1, 0, N+1, N+2$, and $N+3$. The system (8.18) is a three-time-level scheme, so we require two initial time levels, and for the computation, the exact values at time equals zero and time equals Δt are used for the required initial conditions. This nonlinear system (8.18) can be solved by Newton's method in order to compute the unknown approximate solutions U_j^n. Hence, the required solution of the time-fractional KdVB equation can be found.

8.3 Numerical Results and Discussion

The comparison of the absolute errors for a time-fractional KdVB equation (8.5) have been exhibited in Tables 8.1 and 8.2, which are generated using the results acquired by means of the Petrov–Galerkin method and the new method proposed in Reference [196] at different values of x and t, taking $\alpha = 0.75$ and 0.5, respectively. In this present analysis, in order to evaluate the accuracy and reliability of the Petrov–Galerkin for solving the fractional order KdVB equation, we compare L_2 and L_∞ error norms at various points of t, taking $\alpha=1$, 0.75, and 0.5 as illustrated in Tables 8.3 through 8.6. Agreement between the present numerical results for the time-fractional KdVB equation obtained by the Petrov–Galerkin method appears very satisfactory through illustrations in Tables 8.1 through 8.6. The computed outcomes exhibit that this proposed procedure can also be comfortably applied to such a variety of nonlinear equations and good accuracy can also be attained. With the aid of conducting a comparison between the absolute error for the obtained numerical results and the analytic solution of the KdVB equation, we will test the accuracy of the proposed procedure.

TABLE 8.1

Absolute Errors Obtained by the Petrov–Galerkin Method with Regard to the Solution Obtained by the New Method in Reference [196] for the Time-Fractional KdVB Equation Given in (8.5) at Various Points of x and t, Taking $\alpha=0.75$, $\lambda=1$, $\varepsilon=6$, $\nu=1$, and $\mu=2$

	$\lvert u_{PGM}-u_{Exact}\rvert$							
x	$t=0.2$	$t=0.3$	$t=0.4$	$t=0.5$	$t=0.6$	$t=0.7$	$t=0.8$	$t=0.9$
0.1	1.46833E−3	2.79680E−3	4.03999E−3	5.21834E−3	6.34158E−3	7.41499E−3	8.44173E−3	9.42393E−3
0.2	1.47268E−3	2.80387E−3	4.04863E−3	5.22763E−3	6.35075E−3	7.42339E−3	8.44882E−3	9.42925E−3
0.3	1.47692E−3	2.81075E−3	4.0570E−3	5.23656E−3	6.35949E−3	7.43129E−3	8.45536E−3	9.43397E−3
0.4	1.48107E−3	2.81745E−3	4.06509E−3	5.24514E−3	6.36780E−3	7.43871E−3	8.46143E−3	9.43807E−3
0.5	1.48512E−3	2.82396E−3	4.07291E−3	5.25337E−3	6.37568E−3	7.44562E−3	8.46676E−3	9.44157E−3
0.6	1.48908E−3	2.83028E−3	4.08045E−3	5.26124E−3	6.38314E−3	7.45204E−3	8.47164E−3	9.44446E−3
0.7	1.49293E−3	2.83640E−3	4.08772E−3	5.26875E−3	6.39017E−3	7.45797E−3	8.47595E−3	9.44674E−3
0.8	1.49669E−3	2.84234E−3	4.09472E−3	5.27591E−3	6.39676E−3	7.46341E−3	8.47972E−3	9.44842E−3
0.9	1.50034E−3	2.84809E−3	4.10143E−3	5.28272E−3	6.40293E−3	7.46834E−3	8.48293E−3	9.44949E−3
1.0	1.50390E−3	2.85364E−3	4.10787E−3	5.28916E−3	6.40867E−3	7.47279E−3	8.48559E−3	9.44996E−3

TABLE 8.2

Absolute Errors Obtained by the Petrov–Galerkin Method with Regard to the Solution Obtained by the New Method in Reference [196] for the Time-Fractional KdVB Equation Given in (8.5) at Various Points of x and t, Taking $\alpha=0.5$, $\lambda=10$, $\varepsilon=6$, $\nu=0.05$, and $\mu=0.1$

	$\lvert u_{PGM}-u_{Exact}\rvert$								
x	$t=0.1$	$t=0.2$	$t=0.3$	$t=0.4$	$t=0.5$	$t=0.6$	$t=0.7$	$t=0.8$	$t=0.9$
0.1	1.7701E−4	2.5821E−4	2.7527E−4	2.8030E−4	2.8209E−4	2.8281E−4	2.8313E−4	2.8328E−4	2.8335E−4
0.2	1.7476E−4	2.5594E−4	2.7284E−4	2.7781E−4	2.7958E−4	2.8030E−4	2.8062E−4	2.8076E−4	2.8084E−4
0.3	1.7328E−4	2.5369E−4	2.7042E−4	2.7535E−4	2.7710E−4	2.7781E−4	2.7812E−4	2.7827E−4	2.7834E−4
0.4	1.7181E−4	2.5146E−4	2.6802E−4	2.7290E−4	2.7464E−4	2.7534E−4	2.7565E−4	2.7579E−4	2.7587E−4
0.5	1.7035E−4	2.4924E−4	2.6565E−4	2.7047E−4	2.7219E−4	2.7289E−4	2.7319E−4	2.7334E−4	2.7341E−4
0.6	1.6890E−4	2.4704E−4	2.6329E−4	2.6806E−4	2.6977E−4	2.7046E−4	2.7076E−4	2.7090E−4	2.7097E−4
0.7	1.6746E−4	2.4486E−4	2.6094E−4	2.6568E−4	2.6736E−4	2.6804E−4	2.6834E−4	2.6848E−4	2.6856E−4
0.8	1.6603E−4	2.4269E−4	2.5862E−4	2.6331E−4	2.6497E−4	2.6565E−4	2.6595E−4	2.6609E−4	2.6616E−4
0.9	1.6460E−4	2.4055E−4	2.5632E−4	2.6096E−4	2.6261E−4	2.6327E−4	2.6357E−4	2.6371E−4	2.6378E−4
1.0	1.6319E−4	2.3842E−4	2.5403E−4	2.5862E−4	2.6026E−4	2.6092E−4	2.6121E−4	2.6135E−4	2.6142E−4

8.4 Implementation of the Petrov–Galerkin Method for the Numerical Solution of the Time-Fractional STO Equation

This section involves the numerical simulations by means of the proposed Petrov–Galerkin method for the time-fractional STO equation.

Let us consider the time-fractional STO equation [186] as

$$D_t^{\alpha}u+3au_x^2+3au^2u_x+3auu_{xx}+au_{xxx}=0, \tag{8.19}$$

TABLE 8.3

L_2 and L_∞ Error Norms for a Nonlinear KdVB Equation Using the Petrov–Galerkin Method at Various Points of t, Taking $\alpha=1$, $\varepsilon=6$, $\nu=0.0005$, and $\mu=0.1$

t	L_2	L_∞
0.1	1.29814E−10	8.90530E−11
0.2	3.24729E−10	2.22708E−10
0.3	4.55738E−10	3.12239E−10
0.4	6.51429E−10	4.46194E−10
0.5	7.83639E−10	5.36188E−10
0.6	9.80090E−10	6.70419E−10
0.7	1.11351E−9	7.60861E−10
0.8	1.31070E−9	8.95346E−10
0.9	1.44534E−9	9.86221E−10
1.0	1.61436E−9	9.16034E−10

TABLE 8.4

L_2 and L_∞ Error Norms for a Nonlinear Fractional KdVB Equation Using the Petrov–Galerkin Method at Various Points of t, Taking $\lambda=1$, $\varepsilon=6$, $\nu=1$, and $\mu=0.1$

	$\alpha=1$		$\alpha=0.75$		$\alpha=0.5$	
t	L_2	L_∞	L_2	L_∞	L_2	L_∞
0.1	9.40227E−5	6.92880E−5	2.70831E−3	8.69312E−4	3.76753E−3	1.20338E−3
0.2	2.33997E−4	1.72421E−4	7.40982E−3	2.37321E−3	9.22147E−3	2.93842E−3
0.3	3.21139E−4	2.36466E−4	1.16462E−2	3.72295E−3	1.34609E−2	4.28162E−3
0.4	4.56772E−4	3.36334E−4	1.55966E−2	4.97718E−3	1.70414E−2	5.41254E−3
0.5	5.37234E−4	3.95255E−4	1.93290E−2	6.15847E−3	2.01827E−2	6.40209E−3
0.6	6.68680E−4	4.92039E−4	2.28764E−2	7.27798E−3	2.29996E−2	7.28747E−3
0.7	7.42556E−4	5.45962E−4	2.62573E−2	8.34210E−3	2.55619E−2	8.09123E−3
0.8	8.70077E−4	6.39846E−4	2.94832E−2	9.35489E−3	2.79162E−2	8.83089E−3
0.9	9.37568E−4	6.88894E−4	3.25620E−2	1.03192E−2	3.00956E−2	9.51990E−3

TABLE 8.5

L_2 and L_∞ Error Norms for a Nonlinear Fractional KdVB Equation Using the Petrov–Galerkin Method at Various Points of t, Taking $\lambda=0.5$, $\varepsilon=6$, $\nu=0.05$, and $\mu=0.1$

	$\alpha=1$		$\alpha=0.75$		$\alpha=0.5$	
t	L_2	L_∞	L_2	L_∞	L_2	L_∞
0.1	1.47277E−6	1.08060E−6	6.58012E−5	2.11747E−5	9.01591E−5	2.89412E−5
0.2	3.67927E−6	2.69904E−6	1.78696E−4	5.74413E−5	2.20265E−4	7.06189E−5
0.3	5.13661E−6	3.76477E−6	2.79767E−4	8.98432E−5	3.21907E−4	1.03109E−4
0.4	7.33184E−6	5.37326E−6	3.74120E−4	1.20036E−4	4.08641E−4	1.30788E−4
0.5	8.77274E−6	6.42404E−6	4.63866E−4	1.48708E−4	4.85733E−4	1.55354E−4
0.6	1.09575E−5	8.02246E−6	5.50119E−4	1.76221E−4	5.55895E−4	1.77683E−4
0.7	1.23825E−5	9.05825E−6	6.33550E−4	2.02793E−4	6.20744E−4	1.98298E−4
0.8	1.45561E−5	1.06465E−6	7.14595E−4	2.28570E−4	6.81337E−4	2.17540E−4
0.9	1.59651E−5	1.16672E−5	7.93554E−4	2.53649E−4	7.38412E−4	2.35646E−4

TABLE 8.6

L_2 and L_∞ Error Norms for a Nonlinear Fractional KdVB Equation Using the Petrov–Galerkin Method at Various Points of t, Taking $\lambda=0.1$, $\varepsilon=6$, $\nu=5$, and $\mu=6$

	$\alpha=1$		$\alpha=0.75$		$\alpha=0.5$	
t	L_2	L_∞	L_2	L_∞	L_2	L_∞
0.1	3.47775E−3	2.43860E−3	2.17042E−3	7.07133E−4	2.90983E−3	9.47258E−3
0.2	8.56125E−3	6.00577E−3	5.83656E−3	1.90089E−3	7.04589E−3	2.29277E−3
0.3	1.08264E−2	7.56579E−3	9.06678E−3	2.95199E−3	1.02359E−2	3.32980E−3
0.4	1.53527E−2	1.07535E−2	1.20464E−2	3.92097E−3	1.29361E−2	4.20707E−3
0.5	1.65410E−2	1.15427E−2	1.48546E−2	4.83370E−3	1.53231E−2	4.98221E−3
0.6	2.07597E−2	1.45268E−2	1.75348E−2	5.70435E−3	1.74875E−2	5.68476E−3
0.7	2.09164E−2	1.46238E−2	2.01140E−2	6.54177E−3	1.94832E−2	6.33230E−3
0.8	2.52064E−2	1.75783E−2	2.26106E−2	7.35196E−3	2.13452E−2	6.93626E−3
0.9	2.41525E−2	1.69991E−2	2.50377E−2	8.13924E−3	2.30980E−2	7.50459E−3

with initial condition [186]

$$u(x,0)=\frac{2\lambda\left[w+\tanh(\lambda x)\right]}{1+w\tanh(\lambda x)}, \tag{8.20}$$

where λ and w are constants.

Equation 8.19 implies

$$D_t^\alpha u+\frac{3a}{2}\left(u^2\right)_{xx}+a\left(u^3\right)_x+au_{xxx}=0,$$

where α denotes the order of fractional derivative whose range is $0<\alpha\le 1$.

For $\alpha=1$, the exact solution of Equation 8.19 is given by [186]

$$u(x,t)=\frac{2\lambda\left[w+\tanh\left(\lambda\left(x-4a\lambda^2t\right)\right)\right]}{1+w\tanh\left(\lambda\left(x-4a\lambda^2t\right)\right)}. \tag{8.21}$$

The space interval $a_1\le x\le b_1$ is discretized with $(N+1)$ uniform grid points $x_j=a_1+jh$, $j=0,1,2,\ldots,N$ and the grid spacing is given by $h=(b_1-a_1)/N$. Let $U_j(t)$ denote the approximation to the exact solution $u(x_j,t)$. We sought the approximate solution of Equation 8.19 as

$$u(x,t)=\sum_{j=0}^{N}U_j(t)\varphi_j(x), \tag{8.22}$$

where $\varphi_j(x)$, $j=0,1,2,\ldots,N$ are trial functions.

In order to deal with the nonlinear terms u^2 and u^3, the product approximation technique [197] is utilized as follows:

$$u^2(x,t)=\sum_{j=0}^{N}U_j^2(t)\varphi_j(x),$$
$$u^3(x,t)=\sum_{j=0}^{N}U_j^3(t)\varphi_j(x), \tag{8.23}$$

where $\varphi_j(x)$, $j=0,1,2,\ldots,N$ are the usual piecewise linear hat functions given by

$$\varphi_j(x)=\begin{cases}1+\dfrac{x-jh}{j}, & x\in[x_{j-1},x_j)\\ 1-\dfrac{x-jh}{j}, & x\in[x_j,x_{j+1})\\ 0, & \text{elsewhere.}\end{cases} \tag{8.24}$$

The unknown functions $U_j(t)$ are determined from the variational formulation

$$\langle D_t^{\alpha}u,\psi_j\rangle+\frac{3a}{2}\left\langle (u^2)_{xx},\psi_j\right\rangle+a\left\langle (u^3)_x,\psi_j\right\rangle+a\langle u_{xxx},\psi_j\rangle=0, \tag{8.25}$$

where $\psi_j(x)$, $j=0,1,2,\ldots,N$ are quintic B-spline functions taken as the test functions and are given by

$$\psi_j(x)$$
$$=\frac{1}{h^5}\begin{cases}(x-x_{j-3})^5; & x\in[x_{j-3},x_{j-2})\\ (x-x_{j-3})^5-6(x-x_{j-2})^5; & x\in[x_{j-2},x_{j-1})\\ (x-x_{j-3})^5-6(x-x_{j-2})^5+15(x-x_{j-1})^5; & x\in[x_{j-1},x_j)\\ (x-x_{j-3})^5-6(x-x_{j-2})^5+15(x-x_{j-1})^5-20(x-x_j)^5; & x\in[x_j,x_{j+1})\\ (x-x_{j-3})^5-6(x-x_{j-2})^5+15(x-x_{j-1})^5-20(x-x_j)^5+15(x-x_{j+1})^5; & x\in[x_{j+1},x_{j+2})\\ (x-x_{j-3})^5-6(x-x_{j-2})^5+15(x-x_{j-1})^5-20(x-x_j)^5+15(x-x_{j+1})^5-6(x-x_{j+2})^5; & x\in[x_{j+2},x_{j+3})\\ 0 & \text{elsewhere.}\end{cases}$$

Integrating Equation 8.25 by parts, we obtain

$$\langle D_t^{\alpha}u,\psi_j\rangle-\frac{3a}{2}\left\langle (u^2)_x,(\psi_j)_x\right\rangle+a\left\langle (u^3)_x,\psi_j\right\rangle+a\left\langle u_x,(\psi_j)_{xx}\right\rangle=0. \tag{8.26}$$

Each linear hat function covers two elements so that each subinterval $[x_j,x_{j+1}]$ is covered by two linear hat functions. On the other hand, each quantic B-spline covers six elements so that each subinterval $[x_j,x_{j+1}]$ is covered by six splines. In terms of a local coordinate system given by

$$\xi=x-x_j,\quad 0<\xi\le h.$$

Both the linear hat functions φ_j and the quantic B-spline functions ψ_j over the element $[x_j, x_{j+1}]$ can be defined as follows:

$$\varphi_j = \frac{\xi}{h},$$
$$\varphi_{j+1} = 1 - \frac{\xi}{h},$$

and

$$\psi_{j-2} = 1 - 5\left(\frac{\xi}{h}\right) + 10\left(\frac{\xi}{h}\right)^2 - 10\left(\frac{\xi}{h}\right)^3 + 5\left(\frac{\xi}{h}\right)^4 - \left(\frac{\xi}{h}\right)^5,$$
$$\psi_{j-1} = 26 - 50\left(\frac{\xi}{h}\right) + 20\left(\frac{\xi}{h}\right)^2 + 20\left(\frac{\xi}{h}\right)^3 - 20\left(\frac{\xi}{h}\right)^4 + 5\left(\frac{\xi}{h}\right)^5,$$
$$\psi_j = 66 - 60\left(\frac{\xi}{h}\right)^2 + 30\left(\frac{\xi}{h}\right)^4 - 10\left(\frac{\xi}{h}\right)^5,$$
$$\psi_{j+1} = 26 + 50\left(\frac{\xi}{h}\right) + 20\left(\frac{\xi}{h}\right)^2 - 20\left(\frac{\xi}{h}\right)^3 - 20\left(\frac{\xi}{h}\right)^4 + 10\left(\frac{\xi}{h}\right)^5,$$
$$\psi_{j+2} = 1 + 5\left(\frac{\xi}{h}\right) + 10\left(\frac{\xi}{h}\right)^2 + 10\left(\frac{\xi}{h}\right)^3 + 5\left(\frac{\xi}{h}\right)^4 - 5\left(\frac{\xi}{h}\right)^5,$$
$$\psi_{j+3} = \left(\frac{\xi}{h}\right)^5.$$

From Equation 8.26, we have

$$\int_{x0}^{xN}\sum_{j=0}^{N} D_t^{\alpha}U_j(t)\varphi_j(x)\psi_j(x)dx - \frac{3a}{2}\int_{x0}^{xN}\sum_{j=0}^{N} U_j^2(t)\varphi_j'(x)\psi_j'(x)dx$$
$$+ a\int_{x0}^{xN}\sum_{j=0}^{N} U_j^3(t)\varphi_j'(x)\psi_j(x)dx + a\int_{x0}^{xN}\sum_{j=0}^{N} U_j(t)\varphi_j'(x)\psi_j''(x)dx = 0. \quad (8.27)$$

Next, we set

$$I_1 = \int_{x0}^{xN}\sum_{j=0}^{N} D_t^{\alpha}U_j(t)\varphi_j(x)\psi_j(x)dx,$$
$$I_2 = \int_{x0}^{xN}\sum_{j=0}^{N} U_j^2(t)\varphi_j'(x)\psi_j'(x)dx,$$
$$I_3 = \int_{x0}^{xN}\sum_{j=0}^{N} U_j^3(t)\varphi_j'(x)\psi_j(x)dx,$$
$$I_4 = \int_{x0}^{xN}\sum_{j=0}^{N} U_j(t)\varphi_j'(x)\psi_j''(x)dx,$$

where $D_t^\alpha U_j(t)$ is defined by

$$D_t^\alpha U_j(t) = \lim_{\substack{\Delta t \to 0 \\ \Delta tn = t-a}} (\Delta t)^{-\alpha} \sum_{r=0}^{n} \omega_r^\alpha U_j(t - \Delta tr).$$

Now

$$\begin{aligned}
I_1 &= \int_{x0}^{xN} \sum_{j=0}^{N} D_t^\alpha U_j(t)\varphi_j(x)\psi_j(x)dx \\
&= \int_{x_{j-3}}^{x_{j-2}} (\Delta t)^{-\alpha} \sum_{r=0}^{n} \omega_r^\alpha U_{j-3}^{n-r}\varphi_{j-3}(x)\psi_{j-3}(x)dx + \int_{x_{j-3}}^{x_{j-2}} (\Delta t)^{-\alpha} \sum_{r=0}^{n} \omega_r^\alpha U_{j-2}^{n-r}\varphi_{j-2}(x)\psi_{j-3}(x)dx \\
&\quad + \int_{x-2}^{x_{j-1}} (\Delta t)^{-\alpha} \sum_{r=0}^{n} \omega_r^\alpha U_{j-2}^{n-r}\varphi_{j-2}(x)\psi_{j-2}(x)dx + \int_{x-2}^{x_{j-1}} (\Delta t)^{-\alpha} \sum_{r=0}^{n} \omega_r^\alpha U_{j-1}^{n-r}\varphi_{j-1}(x)\psi_{j-2}(x)dx \\
&\quad + \int_{x_{j-1}}^{x_j} (\Delta t)^{-\alpha} \sum_{r=0}^{n} \omega_r^\alpha U_{j-1}^{n-r}\varphi_{j-1}(x)\psi_{j-1}(x)dx + \int_{x_{j-1}}^{x_j} (\Delta t)^{-\alpha} \sum_{r=0}^{n} \omega_r^\alpha U_{j}^{n-r}\varphi_{j}(x)\psi_{j-1}(x)dx \\
&\quad + \int_{x_j}^{x_{j+1}} (\Delta t)^{-\alpha} \sum_{r=0}^{n} \omega_r^\alpha U_{j}^{n-r}\varphi_{j}(x)\psi_{j}(x)dx + \int_{x_j}^{x_{j+1}} (\Delta t)^{-\alpha} \sum_{r=0}^{n} \omega_r^\alpha U_{j+1}^{n-r}\varphi_{j+1}(x)\psi_{j}(x)dx \\
&\quad + \int_{x_{j+1}}^{x_{j+2}} (\Delta t)^{-\alpha} \sum_{r=0}^{n} \omega_r^\alpha U_{j+1}^{n-r}\varphi_{j+1}(x)\psi_{j+1}(x)dx + \int_{x_{j+1}}^{x_{j+2}} (\Delta t)^{-\alpha} \sum_{r=0}^{n} \omega_r^\alpha U_{j+2}^{n-r}\varphi_{j+2}(x)\psi_{j+1}(x)dx \\
&\quad + \int_{x_{j+2}}^{x_{j+3}} (\Delta t)^{-\alpha} \sum_{r=0}^{n} \omega_r^\alpha U_{j+2}^{n-r}\varphi_{j+2}(x)\psi_{j+2}(x)dx + \int_{x_{j+2}}^{x_{j+3}} (\Delta t)^{-\alpha} \sum_{r=0}^{n} \omega_r^\alpha U_{j+3}^{n-r}\varphi_{j+3}(x)\psi_{j+3}(x)dx \\
&= (\Delta t)^{-\alpha} \sum_{r=0}^{n} \omega_r^\alpha \left[\frac{h}{42}\left(U_{j-3}^{n-r} + 120U_{j-2}^{n-r} + 1191U_{j-1}^{n-r} + 2416U_j^{n-r} + 1191U_{j+1}^{n-r} + 120U_{j+2}^{n-r} + U_{j+3}^{n-r} \right)\right].
\end{aligned} \tag{8.28}$$

Similarly,

$$I_2 = \frac{-1}{h}\left[\left(U_{j-3}^n\right)^2 + 24\left(U_{j-2}^n\right)^2 + 15\left(U_{j-1}^n\right)^2 - 80\left(U_j^n\right)^2 + 15\left(U_{j+1}^n\right)^2 + 24\left(U_{j+2}^n\right)^2 + \left(U_{j+3}^n\right)^2 \right], \tag{8.29}$$

$$I_3 = \frac{1}{6}\left[-\left(U_{j-3}^n\right)^3 - 56\left(U_{j-2}^n\right)^3 - 245\left(U_{j-1}^n\right)^3 + 245\left(U_{j+1}^n\right)^3 + 56\left(U_{j+2}^n\right)^3 + \left(U_{j+3}^n\right)^3 \right], \tag{8.30}$$

$$I_4 = \frac{1}{h^2}\left[-5U_{j-3}^n - 40U_{j-2}^n + 95U_{j-1}^n - 95U_{j+1}^n + 40U_{j+2}^n + 5U_{j+3}^n\right]. \tag{8.31}$$

Substituting Equations 8.28 through 8.31 in Equation 8.27, we have

$$\begin{aligned}
&(\Delta t)^{-\alpha}\sum_{r=0}^{n}\omega_r^{\alpha}\left[\frac{h}{42}\left(U_{j-3}^{n-r} + 120U_{j-2}^{n-r} + 1191U_{j-1}^{n-r} + 2416U_j^{n-r} + 1191U_{j+1}^{n-r} + 120U_{j+2}^{n-r} + U_{j+3}^{n-r}\right)\right]\\
&+\frac{3a}{2h}\left[\left(U_{j-3}^n\right)^2 + 24\left(U_{j-2}^n\right)^2 + 15\left(U_{j-1}^n\right)^2 - 80\left(U_j^n\right)^2 + 15\left(U_{j+1}^n\right)^2 + 24\left(U_{j+2}^n\right)^2 + \left(U_{j+3}^n\right)^2\right]\\
&+\frac{a}{6}\left[-\left(U_{j-3}^n\right)^3 - 56\left(U_{j-2}^n\right)^3 - 245\left(U_{j-1}^n\right)^3 + 245\left(U_{j+1}^n\right)^3 + 56\left(U_{j+2}^n\right)^3 + \left(U_{j+3}^n\right)^3\right]\\
&+\frac{a}{h^2}\left[-5U_{j-3}^n - 40U_{j-2}^n + 95U_{j-1}^n - 95U_{j+1}^n + 40U_{j+2}^n + 5U_{j+3}^n\right] = 0,
\end{aligned} \tag{8.32}$$

where $j = 1, 2, 3, 4, \ldots, N$. Now to solve the preceding system, we assume U_j^n to be a discrete approximation to the exact solution $u(x_j, t_n)$. $U_j^n = 0$ for $j = -2, -1, 0, N+1, N+2$, and $N + 3$. The system (8.32) is a three-time-level scheme, so we require two initial time levels, and for the computation, the exact values at time equals zero and time equals Δt are used for the required initial conditions. This nonlinear system (8.32) can be solved by Newton's method in order to compute the unknown approximate solutions U_j^n. Hence, the required solution of the time-fractional STO equation can be found.

8.5 Numerical Results and Discussion

In the present analysis, the absolute errors for an STO equation (8.19) have been exhibited in Table 8.7 in the case of integer order $\alpha = 1$. In order to evaluate the accuracy and reliability of the Petrov–Galerkin for solving the fractional order STO equation, the absolute errors obtained by the Petrov–Galerkin and VIM have been presented in Tables 8.8 and 8.9 for various points of x and t, taking $\alpha = 0.75$ and 0.5, respectively. The comparison of L_2 and L_∞ error norms at various points of t, taking $\alpha = 0.75$ and 0.5, has been illustrated in Table 8.10. Agreement between present numerical results for the time-fractional STO equation obtained by the Petrov–Galerkin method appears very satisfactory through illustrations in Tables 8.7 through 8.10. The computed outcomes exhibit that this proposed procedure can also be comfortably applied to such variety of nonlinear equations, and good accuracy can also be attained. The accuracy of the proposed procedure can be examined with the aid of conducting a comparison between the absolute errors obtained by proposed numerical results and the VIM solutions of the fractional STO equation. A comprehensible inference can be drawn from the numerical results that the proposed Petrov–Galerkin method imparts highly accurate numerical solutions for nonlinear fractional differential equations.

TABLE 8.7

Absolute Errors Obtained by the Petrov–Galerkin Method with Regard to the Exact Solution for the Time-Fractional STO Equation Given in (8.19) at Various Points of x and t, Taking $\alpha=1$, $\lambda=0.01$, $w=0.05$, and $a=1$

	$\lvert u_{PGM}-u_{Exact}\rvert$									
x	$t=0.01$	$t=0.02$	$t=0.03$	$t=0.04$	$t=0.05$	$t=0.06$	$t=0.07$	$t=0.08$	$t=0.09$	$t=0.1$
1	2.23959E−5	4.42244E−5	6.54935E−5	8.62037E−5	1.06349E−4	1.25918E−4	1.44892E−4	1.63248E−4	1.80957E−4	1.9799E−4
2	2.58227E−5	5.16002E−5	7.73625E−5	1.03126E−4	1.28891E−4	1.54645E−4	1.80359E−4	2.05993E−4	2.31490E−4	2.5678E−4
3	1.55609E−5	3.15260E−5	4.80049E−5	6.50892E−5	8.28516E−5	1.01344E−4	1.20597E−4	1.40617E−4	1.61390E−4	1.8287E−4
4	1.07875E−5	2.15582E−5	3.25306E−5	4.39120E−5	5.58949E−5	6.86546E−5	8.23457E−5	9.71000E−5	1.13024E−4	1.3019E−4
5	1.02889E−5	1.94437E−5	2.77437E−5	3.54742E−5	4.29222E−5	5.03723E−5	5.81029E−5	6.63819E−5	7.54633E−5	8.5583E−5
6	1.39288E−5	2.52770E−5	3.42716E−5	4.11663E−5	4.62383E−5	4.97847E−5	5.21189E−5	5.35662E−5	5.44601E−5	5.5137E−5
7	2.31580E−5	4.24684E−5	5.79771E−5	6.97697E−5	7.79699E−5	8.27384E−5	8.42708E−5	8.27945E−5	7.85668E−5	7.1871E−5
8	4.14883E−5	7.91986E−5	1.12952E−4	1.42603E−4	1.68044E−4	1.89204E−4	2.06047E−4	2.18575E−4	2.26826E−4	2.3087E−4
9	7.40691E−5	1.46667E−4	2.17516E−4	2.86355E−4	3.52940E−4	4.17049E−4	4.78481E−4	5.37061E−4	5.92634E−4	6.4507E−4

TABLE 8.8

Absolute Errors Obtained by the Petrov–Galerkin Method with Regard to the Solution Obtained by VIM for the Time-Fractional STO Equation Given in (8.19) at Various Points of x and t, Taking $a=1$, $\lambda=0.01$, $w=0.05$, and $\alpha=0.75$

	$\lvert u_{PGM}-u_{VIM}\rvert$								
x	$t=0.1$	$t=0.2$	$t=0.3$	$t=0.4$	$t=0.5$	$t=0.6$	$t=0.7$	$t=0.8$	$t=0.9$
1	1.0614E−8	2.1103E−8	3.0835E−8	4.0017E−8	4.8760E−8	5.7131E−8	6.5178E−8	7.2937E−8	8.0437E−8
2	1.0601E−8	2.1075E−8	3.0795E−8	3.9965E−8	4.8696E−8	5.7056E−8	6.5093E−8	7.2843E−8	8.0332E−8
3	1.0584E−8	2.1044E−8	3.0749E−8	3.9905E−8	4.8623E−8	5.6971E−8	6.4996E−8	7.2734E−8	8.0212E−8
4	1.0566E−8	2.1008E−8	3.0696E−8	3.9838E−8	4.8541E−8	5.6874E−8	6.4886E−8	7.2610E−8	8.0076E−8
5	1.0546E−8	2.0968E−8	3.0638E−8	3.9762E−8	4.8449E−8	5.6767E−8	6.4763E−8	7.2473E−8	7.9924E−8
6	1.0524E−8	2.0925E−8	3.0574E−8	3.9679E−8	4.8348E−8	5.6648E−8	6.4627E−8	7.2321E−8	7.9757E−8
7	1.0501E−8	2.0877E−8	3.0504E−8	3.9588E−8	4.8237E−8	5.6518E−8	6.4480E−8	7.2156E−8	7.9575E−8
8	1.0474E−8	2.0825E−8	3.0428E−8	3.9490E−8	4.8117E−8	5.6378E−8	6.4319E−8	7.1977E−8	7.9377E−8
9	1.0446E−8	2.0769E−8	3.0347E−8	3.9384E−8	4.7988E−8	5.6227E−8	6.4147E−8	7.1784E−8	7.9164E−8
10	1.0416E−8	2.0709E−8	3.0260E−8	3.9271E−8	4.7850E−8	5.6065E−8	6.3962E−8	7.1577E−8	7.8937E−8

TABLE 8.9

Absolute Errors Obtained by the Petrov–Galerkin Method with Regard to the Solution Obtained by VIM for the Time-Fractional STO Equation Given in (8.19) at Various Points of x and t, Taking $a=1$, $\lambda=0.01$, $w=0.05$, and $\alpha=0.5$

	$\lvert u_{PGM}-u_{VIM}\rvert$								
x	$t=0.1$	$t=0.2$	$t=0.3$	$t=0.4$	$t=0.5$	$t=0.6$	$t=0.7$	$t=0.8$	$t=0.9$
1	1.2512E−8	2.4987E−8	3.6440E−8	4.7065E−8	5.6977E−8	6.6251E−8	7.4944E−8	8.3098E−8	9.0749E−8
2	1.2495E−8	2.4954E−8	3.6392E−8	4.7004E−8	5.6903E−8	6.6165E−8	7.4847E−8	8.2990E−8	9.0632E−8
3	1.2477E−8	2.4917E−8	3.6338E−8	4.6934E−8	5.6818E−8	6.6066E−8	7.4735E−8	8.2866E−8	9.0496E−8
4	1.2456E−8	2.4875E−8	3.6276E−8	4.6854E−8	5.6722E−8	6.5954E−8	7.4608E−8	8.2726E−8	9.0343E−8
5	1.2432E−8	2.4828E−8	3.6208E−8	4.6766E−8	5.6614E−8	6.5829E−8	7.4467E−8	7.2569E−8	9.0171E−8
6	1.2406E−8	2.4776E−8	3.6132E−8	4.6668E−8	5.6496E−8	6.5692E−8	7.4311E−8	7.2396E−8	8.9983E−8
7	1.2378E−8	2.4719E−8	3.6049E−8	4.6561E−8	5.6367E−8	6.5541E−8	7.4141E−8	7.2208E−8	8.9777E−8
8	1.2347E−8	2.4658E−8	3.5960E−8	4.6445E−8	5.6227E−8	6.5379E−8	7.3957E−8	7.2004E−8	8.9554E−8
9	1.2314E−8	2.4592E−8	3.5863E−8	4.6321E−8	5.6076E−8	6.5204E−8	7.3758E−8	7.1784E−8	8.9314E−8
10	1.2278E−8	2.4521E−8	3.5760E−8	4.6188E−8	5.5915E−8	6.5016E−8	7.3546E−8	7.1549E−8	8.9057E−8

TABLE 8.10

L_2 and L_∞ Error Norms for a Time-Fractional STO Equation Using the Petrov–Galerkin Method at Various Points of t, Taking $\lambda = a = 1$ and $w = 0.5$

	$\alpha = 0.75$		$\alpha = 0.5$	
t	L_2	L_∞	L_2	L_∞
0.1	3.32917E−8	1.06143E−8	3.92436E−8	1.25119E−8
0.2	6.61903E−8	2.11033E−8	7.83723E−8	2.49873E−8
0.3	9.67138E−8	3.08351E−8	1.14294E−7	3.64403E−8
0.4	1.25515E−7	4.00176E−8	1.47621E−7	4.70657E−8
0.5	1.52935E−7	4.87599E−8	1.78709E−7	5.69776E−8
0.6	1.79190E−7	5.71309E−8	2.07798E−7	6.62518E−8
0.7	2.04431E−7	6.51783E−8	2.23007E−7	7.49444E−8
0.8	2.28769E−7	7.29378E−8	2.60638E−7	8.30987E−8
0.9	2.52290E−7	8.04370E−8	2.84636E−7	9.07498E−8

8.6 Conclusion

In the present chapter, we have presented the Petrov–Galerkin method for solving the time-fractional KdVB equation and time-fractional STO equation numerically by implementing a linear hat function as the trial function and a quintic B-spline function as the test function. This numerical method seems to be competent of producing numerical solutions of high accuracy for the fractional differential equations. A comparison between the numerical results is carried out to illustrate the pertinent feature of the proposed algorithm. The acquired numerical approximate solutions preserve good accuracy when compared with the exact solutions. The accuracy of the method is assessed in terms of L_2 and L_∞ error norms. A comprehensible inference can be drawn from the numerical results that the proposed Petrov–Galerkin method imparts highly accurate numerical solutions for nonlinear fractional differential equations.

References

1. L. Debnath, *Wavelet Transforms and Their Applications*. Birkhäuser, Boston, MA, 2002.
2. J. Morlet, G. Arens, I. Fourgeau, and D. Giard, Wave propagation and sampling theory, *Geophysics*, 47, 203–236, 1982.
3. A. Grossmann, and J. Morlet, Decomposition of Hardy functions into square integrable wavelets of constant shape, *SIAM Journal of Mathematical Analysis*, 15, 723–736, 1984.
4. J. Stromberg, A modified Franklin system and higher order systems of *Rn* as unconditional bases for Hardy spaces, in W. Beckner et al. (eds.), *Conference in Harmonic Analysis in Honor of A. Zygmund*, Chicago, IL, vol. 2, pp. 475–493. Wadsworth Mathematics Series, 1982.
5. Y. Meyer, Principe d'incertitude, bases hilbertiennes et algèbres d'operateurs, *Seminaire Bourbaki*, 28, 209–223, 1986.
6. S. G. Mallat, Multiresolution approximations and wavelet orthonormal bases of $L^2(\Re)$, *Transactions of the American Mathematical Society*, 315(1), 69–87, 1989.
7. I. Daubechies, *Ten Lectures on Wavelets*. Society for Industrial and Applied Mathematics, Philadelphia, PA, 1992.
8. C. K. Chui, *An Introduction to Wavelets*. Academic Press, London, U.K., 1992.
9. G. Beylkin, R. Coifman, and V. Rokhlin, Fast wavelet transforms and numerical algorithms, *Communications on Pure and Applied Mathematics*, 44(2), 141–183, 1991.
10. J. H. He, A new approach to nonlinear partial differential equations, *Communications in Nonlinear Science and Numerical Simulation*, 2(4), 230–235, 1997.
11. J. H. He, Variational iteration method—A kind of non-linear analytical technique: Some examples, *International Journal of Nonlinear Mechanics*, 34, 699–708, 1999.
12. J. H. He, A review on some new recently developed nonlinear analytical techniques, *International Journal of Nonlinear Sciences and Numerical Simulation*, 1, 51–70, 2000.
13. M. Tatari, and M. Dehghan, On the convergence of He's variational iteration method, *Journal of Computational and Applied Mathematics*, 207, 121–128, 2007.
14. Z. Feng, On explicit exact solutions to the compound Burgers–KdV equation, *Physics Letters A*, 293, 57–66, 2002.
15. Z. Feng, and X. Wang, The first integral method to the two-dimensional Burgers–Korteweg–de Vries equation, *Physics Letters A*, 308, 173–178, 2003.
16. Z. Feng, and G. Chenb, Solitary wave solutions of the compound Burgers–Korteweg–de Vries equation, *Physica A*, 352, 419–435, 2005.
17. K. R. Raslan, The first integral method for solving some important nonlinear partial differential equations, *Nonlinear Dynamics*, 53, 281–286, 2008.
18. Z. Feng, Traveling wave behavior for a generalized Fisher equation, *Chaos, Solitons and Fractals*, 38, 481–488, 2008.
19. Z. Feng, and R. Knobel, Traveling waves to a Burgers–Korteweg–de Vries-type equation with higher-order nonlinearities, *Journal of Mathematical Analysis and Applications*, 328, 1435–1450, 2007.
20. W. H. Su, X. J. Yang, H. Jafari, and D. Baleanu, Fractional complex transform method for wave equations on Cantor sets within local fractional differential operator, *Advances in Difference Equations*, 2013(97), 1–8, 2013.
21. X. J. Yang, D. Baleanu, and H. M. Srivastava, *Local Fractional Integral Transforms and Their Applications*. Academic Press/Elsevier, London, U.K., 2015.
22. T. R. Ding, and C. Z. Li, *Ordinary Differential Equations*. Peking University Press, Peking, China, 1996.
23. N. Bourbaki, *Commutative Algebra*. Addison-Wesley, Paris, France, 1972.

24. Z. Feng, and X. Wang, Explicit exact solitary wave solutions for the Kundu equation and the derivative Schrödinger equation, *Physica Scripta*, 64(1), 7–14, 2001.
25. J. H. He, Homotopy perturbation method: A new nonlinear analytical technique, *Applied Mathematics and Computation*, 135, 73–79, 2003.
26. J. H. He, Homotopy perturbation technique, *Computer Methods in Applied Mechanics and Engineering*, 178, 257–262, 1999.
27. A. Ghorbani, Beyond Adomian polynomials: He polynomials, *Chaos, Solitons and Fractals*, 39, 1486–1492, 2009.
28. V. Marinca, N. Herisanu, and I. Nemes, Optimal homotopy asymptotic method with application to thin film flow, *Central European Journal of Physics*, 6(3), 648–653, 2008.
29. V. Marinca, N. Herisanu, C. Bota, and B. Marinca, An optimal homotopy asymptotic method applied to the steady flow of fourth-grade fluid past a porous plate, *Applied Mathematics Letters*, 22(2), 245–251, 2009.
30. V. Marinca, and N. Herisanu, An optimal homotopy asymptotic method for solving nonlinear equations arising in heat transfer, *International Communications in Heat and Mass Transfer*, 35, 710–715, 2008.
31. R. A. Shah, S. Islam, A. M. Siddiqui, and A. Haroon, Optimal homotopy asymptotic method solution of unsteady second grade fluid in wire coating analysis, *Journal of the Korea Society for Industrial and Applied Mathematics*, 15(3), 201–222, 2011.
32. M. Ghoreishi, A. I. B. M. Ismail, A. K. Alomari, and A. S. Bataineh, The comparison between homotopy analysis method and optimal homotopy asymptotic method for nonlinear age-structured population models, *Communications in Nonlinear Science and Numerical Simulation*, 17, 1163–1177, 2012.
33. S. Liao, On the homotopy analysis method for nonlinear problems, *Applied Mathematics and Computation*, 147, 499–513, 2004.
34. S. Saha Ray, and A. Patra, Application of homotopy analysis method and Adomian decomposition method for the solution of neutron diffusion equation in the hemisphere and cylindrical reactors, *Journal of Nuclear Engineering & Technology*, 1(2–3), 1–12, 2011.
35. S. Liao, An optimal homotopy analysis approach for strongly nonlinear differential equations, *Communications in Nonlinear Science and Numerical Simulation*, 15, 2003–2016, 2010.
36. A. Haar, Zur theorie der orthogonalen Funktionsysteme, *Mathematical Analysis*, 69, 331–371, 1910.
37. Ü. Lepik, Numerical solution of evolution equations by the Haar Wavelet method, *Applied Mathematics and Computation*, 185, 695–704, 2007.
38. S. Saha Ray, On Haar wavelet operational matrix of general order and its application for the numerical solution of fractional Bagley Torvik equation, *Applied Mathematics and Computation*, 218, 5239–5248, 2012.
39. C. F. Chen, and C. H. Hsiao, Haar wavelet method for solving lumped and distributed parameter-systems, *IEE Proceedings—Control Theory and Applications*, 144(1), 87–94, 1997.
40. S. Saha Ray, and A. Patra, Haar wavelet operational methods for the numerical solutions of Fractional order nonlinear oscillatory Van der Pol system, *Applied Mathematics and Computation*, 220, 659–667, 2013.
41. I. Podlubny, *Fractional Differential Equations*. Academic Press, New York, 1999.
42. M. Rehman, and R. A. Khan, The Legendre wavelet method for solving fractional differential equations, *Communications in Nonlinear Science and Numerical Simulation*, 16, 4163–4173, 2011.
43. M. Razzaghi, and S. Yousefi, The Legendre wavelets operational matrix of integration, *International Journal of Systems Science*, 32, 495–502, 2001.
44. N. Liu, and E. B. Lin, Legendre wavelet method for numerical solutions of partial differential equations, *Numerical Methods for Partial Differential Equations*, 26(1), 81–94, 2010.
45. L. Zhu, and Q. Fan, Solving fractional nonlinear Fredholm integro-differential equations by the second kind Chebyshev wavelet, *Communications in Nonlinear Science and Numerical Simulation*, 17, 2333–2341, 2012.

46. Y. Wang, and Q. Fan, The second kind Chebyshev wavelet method for solving fractional differential equations, *Applied Mathematics and Computation*, 218, 8592–8601, 2012.
47. U. Saeed, and M. Rehman, Hermite wavelet method for fractional delay differential equations, *Journal of Difference Equations*, 2014, 1–8, Article ID 359093, 2014.
48. A. Ali, M. A. Iqbal, and S. T. Mohyud-Din, Hermite wavelets method for boundary value problems, *International Journal of Modern Applied Physics*, 3(1), 38–47, 2013.
49. A. M. Wazwaz, *Partial Differential Equations and Solitary Waves Theory*. Springer/HEP, Berlin, Germany, 2009.
50. A. Dogan, A Galerkin finite element approach to Burger's equation, *Applied Mathematics and Computation*, 157, 331–346, 2004.
51. A. H. Khater, R. S. Temsah, and M. M. Hassan, A Chebyshev spectral collocation method for solving Burgers' type equations, *Journal of Computational and Applied Mathematics*, 222, 333–350, 2008.
52. A. Korkmaz, A. M. Aksoy, and I. Dag, Quartic B-spline differential quadrature method, *International Journal Nonlinear Science*, 11(4), 403–411, 2011.
53. B. Saka, and I. Dag, Quartic B-spline collocation method to the numerical solutions of the Burgers' equation, *Chaos, Solitons and Fractals*, 32, 1125–1137, 2007.
54. T. Ozis, E. N. Aksan, and A. Ozdes, A finite element approach for solution of Burgers' equation, *Applied Mathematics and Computation*, 139, 417–428, 2003.
55. L. A. Hassanien, A. A. Salama, and H. A. Hosham, Fourth-order finite difference method for solving Burgers' equation, *Applied Mathematics and Computation*, 170, 781–800, 2005.
56. S. Kutulay, A. R. Bahadir, and A. Odes, Numerical solution of the one-dimensional Burgers' equation: Explicit and exact-explicit finite difference methods, *Journal of Computational and Applied Mathematics*, 103, 251–261, 1999.
57. S. Kutulay, A. Esen, and I. Dag, Numerical solutions of the Burgers' equation by the least-squares quadratic B-spline finite element method, *Journal of Computational and Applied Mathematics*, 167, 21–33, 2004.
58. D. Irk, Sextic B-spline collocation method for the Modified Burgers' equation, *Kybernetes*, 38(9), 1599–1620, 2009.
59. A. G. Bratsos, and L. A. Petrakis, An implicit numerical scheme for the modified Burgers' equation, *International Journal for Numerical Methods in Biomedical Engineering*, 27(2), 232–237, 2011.
60. T. Roshan, and K. S. Bhamra, Numerical solution of the modified Burgers' equation by Petrov-Galerkin method, *Applied Mathematics and Computation*, 218, 3673–3679, 2011.
61. M. A. Ramadan, and T. S. El-Danaf, Numerical treatment for the modified Burgers' equation, *Mathematics and Computers in Simulation*, 70, 90–98, 2005.
62. R. P. Zhang, X. J. Yu, and G. Z. Zhao, Modified Burgers' equation by the local discontinuous Galerkin method, *Chinese Physics B*, 22(3), 030210 (1–5), 2013.
63. Y. Duan, R. Liu, and Y. Jiang, Lattice Boltzmann model for the modified Burgers' equation, *Applied Mathematics and Computation*, 202, 489–497, 2008.
64. H. N. A. Ismail, K. Raslan, and A. A. Abd Rabboh, Adomian decomposition method for Burger's-Huxley and Burger's-Fisher equations, *Applied Mathematics and Computation*, 159, 291–301, 2004.
65. M. Javidi, A numerical solution of the generalized Burgers-Huxley equation by spectral collocation method, *Applied Mathematics and Computation*, 172(2), 338–344, 2006.
66. A. M. Wazwaz, Analytic study on Burgers, Fisher, Huxley equations and combined forms of these equations, *Applied Mathematics and Computation*, 195, 754–761, 2008.
67. S. H. Hashemi, H. R. M. Daniali, and D. D. Ganji, Numerical simulation of the generalized Huxley equation by He's homotopy perturbation method, *Applied Mathematics and Computation*, 192(1), 157–161, 2007.
68. X. W. Zhou, Exp-Function method for solving Huxley equation, *Mathematical Problems in Engineering*, 2008, 1–7, Article ID 538489, 2008.
69. I. Hashim, M. S. M. Noorani, and B. Batiha, A note on the Adomian decomposition method for the generalized Huxley equation, *Applied Mathematics and Computation*, 181(2), 1439–1445, 2006.

70. B. Batiha, M. S. M. Noorani, and I. Hashim, Numerical simulation of the generalized Huxley equation by He's variational iteration method, *Applied Mathematics and Computation*, 186(2), 1322–1325, 2007.
71. M. Sari, and G. Gürarslan, Numerical solutions of the generalized Burgers-Huxley equation by a Differential Quadrature method, *Mathematical Problems in Engineering*, 2009, 1–11, Article ID 370765, 2009.
72. X. Y. Wang, Z. S. Zhu, and Y. K. Lu, Solitary wave solutions of the generalized Burgers-Huxley equation, *Journal of Physics A: Mathematical and General*, 23, 271–274, 1990.
73. T. S. El-Danaf, Solitary wave solutions for the generalized Burgers-Huxley equation, *International Journal of Nonlinear Sciences and Numerical Simulation*, 8(3), 315–318, 2007.
74. Z. Yan, The modified KdV equation with variable coefficients: Exact uni/bi-variable travelling wave-like solutions, *Applied Mathematics and Computation*, 203, 106–112, 2008.
75. M. K. Kadalbajoo, and A. Awasthi, A numerical method based on Crank-Nicolson scheme for Burgers' equation, *Applied Mathematics and Computation*, 182, 1430–1442, 2006.
76. R. Jiwari, A Haar wavelet quasilinearization approach for numerical simulation of Burgers' equation, *Computer Physics Communications*, 183, 2413–2423, 2012.
77. W. L. Wood, An exact solution for Burger's equation, *Communications in Numerical Methods in Engineering*, 22(7), 797–798, 2006.
78. K. Rahman, N. Helil, and R. Yimin, Some new semi-implicit finite difference schemes for numerical solution of Burgers equation, *Proceedings of the International Conference on Computer Application and System Modeling*, 14(5622119), 451–455, 2010. DOI: 10.1109/ICCASM.2010.5622119.
79. R. C. Mittal, and R. K. Jain, Numerical solutions of nonlinear Burgers' equation with modified cubic B-splines collocation method, *Applied Mathematics and Computation*, 218, 7839–7855, 2012.
80. A. Bekir, On travelling wave solutions to combined KdV-mKdV equation and modified Burgers-KdV equation, *Communications in Nonlinear Science and Numerical Simulation*, 14, 1038–1042, 2009.
81. A. M. Wazwaz, A study on compacton-like solutions for the modified KdV and fifth order KdV-like equations, *Applied Mathematics and Computation*, 147, 439–447, 2004.
82. D. Kaya, An application for the higher order modified KdV equation by decomposition method, *Communications in Nonlinear Science and Numerical Simulation*, 10, 693–702, 2005.
83. J. Zhang, Y. Wu, and X. Li, Quasi-periodic solution of the (2+1)-dimensional Boussinesq-Burgers soliton equation, *Physica A: Statistical Mechanics and Its Applications*, 319(1), 213–232, 2003.
84. L. Zhang, L. F. Zhang, and C. Li, Some new exact solutions of Jacobian elliptic function about the generalized Boussinesq equation and Boussinesq-Burgers equation, *Chinese Physics B*, 17(2), 403–410, 2008.
85. A. S. A. Rady, and M. Khalfallah, On soliton solutions for Boussinesq-Burgers equations, *Communications in Nonlinear Science and Numerical Simulation*, 15, 886–894, 2010.
86. P. Wang, B. Tian, W. Liu, X. Lü, and Y. Jiang, Lax pair, Bäcklund transformation and multi-soliton solutions for the Boussinesq–Burgers equations from shallow water waves, *Applied Mathematics and Computation*, 218, 1726–1734, 2011.
87. A. Chen, and X. Li, Darboux transformation and soliton solutions for Boussinesq–Burgers equation, *Chaos, Solitons and Fractals*, 27(1), 43–49, 2006.
88. V. Merinca, and N. Herisanu, Determination of periodic solutions for the motion of a particle on a rotating parabola by means of the optimal homotopy asymptotic method, *Journal of Sound and Vibration*, 329, 1450–1459, 2010.
89. J. Biazar, and H. Aminikhan, Study of convergence of homotopy perturbation method for systems of partial differential equations, *Computers and Mathematics with Applications*, 58, 2221–2230, 2009.
90. S. G. Samko, A. A. Kilbas, and O. I. Marichev, *Fractional Integrals and Derivatives: Theory and Applications*. Taylor & Francis, London, U.K., 1993.
91. A. A. Kilbas, H. M. Srivastava, and J. J. Trujillo, *Theory and Applications of Fractional Differential Equations*, North-Holland Mathematical Studies 204. Elsevier (North-Holland) Science Publishers, Amsterdam, the Netherlands, 2006.

92. K. S. Miller, and B. Ross, *An Introduction to the Fractional Calculus and Fractional Differential Equations*. Wiley, New York, 1993.
93. M. Caputo, Linear models of dissipation whose Q is almost frequency independent. Part II, *Geophysical Journal of the Royal Astronomical Society*, 13, 529–539, 1967 [Reprinted in *Fractional Calculus and Applied Analysis* 11, 4–14, 2008].
94. H. J. Holmgren, Om differentialkalkylen med indices af hvad natur som helst, *Kungliga Svenska Vetenskaps-Akademins Handlingar*, 5(11), 1–83, 1865, Stockholm (in Swedish).
95. B. Riemann, Versuch einer allgemeinen Auffassung der Integration und Differentiation, *Gesammelte Werke und Wissenschaftlicher Nachlass*, Leipzig, Germany, 1876, pp. 331–344.
96. H. Risken, *The Fokker-Planck Equation: Methods of Solution and Applications*. Springer, Berlin, Germany, 1989.
97. J. Biazar, P. Gholamin, and K. Hosseini, Variational iteration method for solving Fokker-Planck equation, *Journal of the Franklin Institute*, 347, 1137–1147, 2010.
98. A. Yildirm, Application of the homotopy perturbation method for the Fokker-Planck equation, *International Journal for Numerical Methods in Biomedical Engineering*, 26, 1144–1154, 2010.
99. R. Metzler, and J. Klafter, The random walk's guide to anomalous diffusion: A fractional dynamics approach, *Physics Reports*, 339, 1–77, 2000.
100. T. D. Frank, *Nonlinear Fokker-Planck Equations: Fundamentals and Applications*. Springer, Berlin, Germany, 2005.
101. S. Saha Ray, Exact solutions for time-fractional diffusion-wave equations by decomposition method, *Physica Scripta*, 75, 53–61, 2007.
102. L. Yan, Numerical solutions of fractional Fokker-Planck equations using iterative Laplace transform method, *Abstract and Applied Analysis*, 2013, 1–7, Article ID 465160, 2013.
103. Z. Odibat, and S. Momani, Numerical solution of Fokker-Planck equation with space- and time-fractional derivatives, *Physics Letters A*, 369, 349–358, 2007.
104. S. K. Vanani, and A. Aminataei, A numerical algorithm for the space and time fractional Fokker-Planck equation, *International Journal of Numerical Methods for Heat & Fluid Flow*, 22(8), 1037–1052, 2012.
105. A. Yildirm, Analytical approach to Fokker-Planck equation with space- and time-fractional derivatives by means of the homotopy perturbation method, *Journal of King Saud University (Science)*, 22, 257–264, 2010.
106. J. Zhang, and G. Yan, A lattice Boltzmann model for the Burgers-Fisher equation, *Chaos*, 20, 023129(1–12), 2010.
107. Y. Liu, General solution of space fractional Fisher's nonlinear diffusion equation, *Journal of Fractional Calculus and Applications*, 1(2), 1–8, 2011.
108. A. M. Wazwaz, and A. Gorguis, An analytic study of Fisher's equation by using Adomian decomposition method, *Applied Mathematics and Computation*, 154, 609–620, 2004.
109. L. Wei, Y. He, A. Yildirim, and S. Kumar, Numerical algorithm based on an implicit fully discrete local discontinuous Galerkin method for the time-fractional KdV-Burgers-Kuramoto equation, *Journal of Applied Mathematics and Mechanics*, 93(1), 14–28, 2013.
110. T. Kawahara, Formation of saturated solitons in a nonlinear dispersive system with instability and dissipation, *Physical Review Letters*, 51, 381–383, 1983.
111. J. Topper, and T. Kawahara, Approximate equations for long nonlinear waves on a viscous fluid, *Journal of the Physical Society of Japan*, 44, 663–666, 1978.
112. B. Cohen, J. Krommes, W. Tang, and M. Rosenbluth, Non-linear saturation of the dissipative trapped-ion mode by mode coupling, *Nuclear Fusion*, 16, 971–992, 1976.
113. F. Huang, and S. Liu, Physical mechanism and model of turbulent cascades in a barotropic atmosphere, *Advances in Atmospheric Sciences*, 21, 34–40, 2004.
114. L. Song, and H. Zhang, Application of homotopy analysis method to fractional KdV–Burgers–Kuramoto equation, *Physics Letters A*, 367, 88–94, 2007.
115. M. Safari, D. D. Ganji, and M. Moslemi, Application of He's variational iteration method and Adomian's decomposition method to the fractional KdV–Burgers–Kuramoto equation, *Computers and Mathematics with Applications*, 58, 2091–2097, 2009.

116. Y. Pomeau, A. Ramani, and B. Grammaticos, Structural stability of the Korteweg-de Vries solitons under a singular perturbation, *Physica D: Nonlinear Phenomena*, 31(1), 127–134, 1988.
117. L. Wei, Y. He, and Y. Zhang, Numerical analysis of the fractional seventh-order KDV equation using an Implicit Fully Discrete Local Discontinuous Galerkin method, *International Journal of Numerical Analysis and Modeling*, 10(2), 430–444, 2013.
118. S. T. Mohyud-Din, M. A. Noor, and K. I. Noor, Travelling wave solutions of seventh-order generalized KdV equations by variational iteration method using Adomian's polynomial, *International Journal of Modern Physics B*, 23(15), 3265–3277, 2009.
119. S. T. Mohyud-Din, M. A. Noor, and K. I. Noor, Traveling wave solutions of seventh-order generalized KdV equations using He's polynomials, *International Journal of Nonlinear Sciences and Numerical Simulation*, 10(2), 227–233, 2009.
120. S. T. Mohyud-Din, M. A. Noor, and F. Jabeen, Modified variational iteration method for traveling wave solutions of seventh-order generalized KdV equations, *International Journal for Computational Methods in Engineering Science and Mechanics*, 12, 107–113, 2011.
121. N. Shang, and B. Zheng, Exact solutions for three fractional partial differential equations by the (G′/G) method, *International Journal of Applied Mathematics*, 43(3), 1–6, 2013.
122. D. J. Kaup, On the inverse scattering for cubic eigenvalue problems of the class equations, *Studies in Applied Mathematics*, 62, 189–195, 1980.
123. B. A. Kupershmidt, A super Korteweg-de-Vries equation: An integrable system, *Physics Letters A*, 102, 213–218, 1994.
124. E. Fan, Uniformly constructing a series of explicit exact solutions to nonlinear equations in mathematical physics, *Chaos, Solitons and Fractals*, 16, 819–839, 2003.
125. M. Inc, On numerical soliton solution of the Kaup–Kupershmidt equation and convergence analysis of the decomposition method, *Applied Mathematics and Computation*, 172(1), 72–85, 2006.
126. S. A. Yousefi, Numerical solution of a model describing biological species living together by using Legendre Multiwavelet method, *International Journal of Nonlinear Science*, 11(1), 109–113, 2011.
127. A. K. Gupta, and S. Saha Ray, Travelling wave solution of fractional KdV-Burger-Kuramoto equation describing nonlinear physical phenomena, *AIP Advances*, 4, 097120-(1–11), 2014. DOI: 10.1063/1.4895910.
128. A. K. Gupta, and S. Saha Ray, Comparison between homotopy perturbation method and optimal homotopy asymptotic method for the soliton solution of boussinesq-burgers equation, *Computers and Fluids*, 103, 34–41, 2014.
129. B. Zheng, Exact solutions for some fractional partial differential equations by the (G′/G) method, *Mathematical Problems in Engineering*, 2013, 1–13, Article ID 826369, 2013.
130. H. Naher, F. A. Abdullah, and S. T. Mohyud-Din, Extended generalized Riccati equation mapping method for the fifth-order Sawada-Kotera equation, *AIP Advances*, 3, 052104(1–14), 2013.
131. S. Dinarvand, S. Khosravi, A. Doosthoseini, and M. M. Rashidi, The homotopy analysis method for solving the Sawada-Kotera and Lax's fifth-order KdV equations, *Advances in Theoretical and Applied Mechanics*, 1(7), 327–335, 2008.
132. L. Debnath, *Nonlinear Partial Differential Equations for Scientists and Engineers*. Birkhauser/Springer, New York, 2012.
133. Y. Zhang, Time-fractional Camassa-Holm equation: Formulation and solution using variational methods, *Journal of Computational and Nonlinear Dynamics*, 8, 041020(1–7), 2013.
134. F. Guo, and W. Peng, Blowup solutions for the generalized two-component Camassa-Holm system on the circle, *Nonlinear Analysis*, 105, 120–133, 2014.
135. T. Rehman, G. Gambino, and S. Roy Choudhury, Smooth and non-smooth travelling wave solutions of some generalized Camassa-Holm equations, *Communications in Nonlinear Science and Numerical Simulation*, 19, 1746–1769, 2014.
136. R. Camassa, D. Holm, and J. Hyman, A new integrable shallow water equation, *Advances in Applied Mechanics*, 31, 1–33, 1994.
137. R. S. Johnson, Camassa–Holm, Korteweg-de Vries and related models for water waves, *Journal of Fluid Mechanics*, 455, 63–82, 2002.

138. A. Fokas, and B. Fuchssteiner, Symplectic structures, their Backlund transformation and hereditary symmetries, *Physica D: Nonlinear Phenomena*, 4(1), 47–66, 1981.
139. J. Lenells, Conservation laws of the Camassa–Holm equation, *Journal of Physics A: Mathematical and General*, 38(4), 869–880, 2005.
140. R. Camassa, and D. Holm, An integrable shallow water equation with peaked solutions, *Physical Review Letters*, 71, 1661–1664, 1993.
141. S. Saha Ray, Soliton solutions of nonlinear and nonlocal Sine-Gordon equation involving Riesz space fractional derivative, *Zeitschrift für Naturforschung A*, 70(8), 659–667, 2015.
142. G. Alfimov, T. Pierantozzi, and L. Vázquez, Numerical study of a fractional sine-Gordon equation. Fractional differentiation and its applications, *FDA*, 4, 644–649, 2004.
143. R. K. Dodd, J. C. Eilbeck, J. D. Gibbon, and H. C. Morris, *Solitons and Nonlinear Wave Solutions*. Academic, London, U.K., 1982.
144. U. Yücel, Homotopy analysis method for the sine-Gordon equation with initial conditions, *Applied Mathematics and Computation*, 203, 387–395, 2008.
145. D. Kaya, A numerical solution of the sine-Gordon equation using the modified decomposition method, *Applied Mathematics and Computation*, 143, 309–317, 2003.
146. B. Batiha, M. S. M. Noorani, and I. Hashim, Numerical solution of sine-Gordon equation by variational iteration method, *Physics Letters A*, 370, 437–440, 2007.
147. A. M. Wazwaz, The tanh method: Exact solutions of the sine-Gordon and the sinh-Gordon equations, *Applied Mathematics and Computation*, 167, 1196–1210, 2005.
148. M. J. Ablowitz, B. M. Herbst, and C. Schober, On the numerical solution of the sine-Gordon equation. I: Integrable discretizations and homoclinic manifolds, *Journal of Computational Physics*, 126, 299–314, 1996.
149. B. M. Herbst, and M. J. Ablowitz, Numerical homoclinic instabilities in the sine-Gordon equation, *Quaestiones Mathematicae*, 15(3), 345–363, 1992.
150. O. S. Iyiola, A numerical study of Ito equation and Sawada-Kotera equation both of time-fractional type, *Advances in Mathematics: Scientific Journal*, 2(2), 71–79, 2013.
151. A. Saadatmandi, and M. Dehghan, A new operational matrix for solving fractional-order differential equations, *Computers and Mathematics with Applications*, 59, 1326–1336, 2010.
152. M. Dehghan, J. Manafian, and A. Saadatmandi, Solving nonlinear fractional partial differential equations using the homotopy analysis method, *Numerical Methods for Partial Differential Equations*, 26, 448–479, 2009.
153. M. Dehghan, J. Manafian, and A. Saadatmandi, The solution of the linear fractional partial differential equations using the homotopy analysis method, *Zeitschrift für Naturforschung A*, 65a(11), 935–949, 2010.
154. B. He, Q. Meng, and S. Li, Explicit peakon and solitary wave solutions for the modified Fornberg-Whitham equation, *Applied Mathematics and Computation*, 217, 1976–1982, 2010.
155. G. B. Whitham, Variational methods and applications to water wave, *Proceedings of the Royal Society of London A: Mathematical, Physical and Engineering Sciences*, 299, 6–25, 1967.
156. B. Fornberg, and G. B. Whitham, A numerical and theoretical study of certain nonlinear wave phenomena, *Philosophical Transactions of the Royal Society of London A: Mathematical, Physical and Engineering Sciences*, 289, 373–404, 1978.
157. S. Hesam, A. Nazemi, and A. Haghbin, Reduced differential transform method for solving the Fornberg-Whitham type equation, *International Journal of Nonlinear Science*, 13(2), 158–162, 2012.
158. J. Lu, An analytical approach to the Fornberg–Whitham type equations by using the variational iteration method, *Computers and Mathematics with Applications*, 61(8), 2010–2013, 2011.
159. D. Kaya, and S. M. El-Sayed, A numerical method for solving Jaulent–Miodek equation, *Physics Letters A*, 318, 345–353, 2003.
160. A. M. Wazwaz, The tanh-coth and the sech methods for exact solutions of the Jaulent–Miodek equation, *Physics Letters A*, 366, 85–90, 2007.
161. A. Yildirim, and A. Kelleci, Numerical simulation of the Jaulent-miodek equation by He's Homotopy perturbation method, *World Applied Sciences Journal*, 7, 84–89, 2009.

162. J. H. He, and L. N. Zhang, Generalized solitary solution and compacton-like solution of the Jaulent-Miodek equations using the Exp-function method, *Physics Letters A*, 372(7), 1044–1047, 2008.
163. M. M. Rashidi, G. Domairry, and S. Dinarvand, The Homotopy analysis method for explicit analytical solutions of Jaulent–Miodek equations, *Numerical Methods for Partial Differential Equations*, 25(2), 430–439, 2009.
164. H. T. Ozer, and S. Salihoglu, Nonlinear Schrödinger equations and $N = 1$ super-conformal algebra, *Chaos, Solitons and Fractals*, 33, 1417–1423, 2007.
165. S. Y. Lou, A direct perturbation method: Nonlinear Schrodinger equation with loss, *Chinese Physics Letters*, 16, 659–661, 1999.
166. A. Atangana, and D. Baleanu, Nonlinear fractional Jaulent-Miodek and Whitham-Broer-Kaup equations within Sumudu transform, *Abstract and Applied Analysis*, 2013, 1–8, Article ID 160681, 2013.
167. C. H. Su, and C. S. Gardner, Korteweg-de Vries equation and generalizations. III. Derivation of the Korteweg-de-Vries and Burgers' equation, *Journal of Mathematical Physics*, 10(3), 536–539, 1969.
168. R. S. Jonson, A nonlinear equation incorporating damping and dispersion, *Journal of Fluid Mechanics*, 42(1), 49–60, 1970.
169. R. S. Jonson, Shallow water waves in a viscous fluid—The undular bore, *Physics of Fluids*, 15(10), 1693–1699, 1972.
170. H. Grad, and P. N. Hu, Unified shock profile in a plasma, *Physics of Fluids*, 10(12), 2596–2602, 1967.
171. S. Saha Ray, and A. K. Gupta, Application of novel schemes based on Haar wavelet collocation method for Burger and Boussinesq-Burger equations, *Applied Mathematics and Information Sciences*, 10(4), 1513–1524, 2016.
172. D. J. Korteweg, and G. de Vries, On the change of form of long waves advancing in a rectangular Canal and on a new type of long stationary waves, *Philosophical Magazine Series 5*, 39(240), 422–443, 1895.
173. D. Kaya, An application of the decomposition method for the KdVB equation, *Applied Mathematics and Computation*, 152, 279–288, 2004.
174. B. Sahu, and R. Roychoudhury, Travelling wave solution of Korteweg-de Vries–Burger's equation, *Czechoslovak Journal of Physics*, 53(6), 517–527, 2003.
175. M. A. Helal, and M. S. Mehanna, A comparison between two different methods for solving KdV-Burgers equation, *Chaos, Solitons and Fractals*, 28, 320–326, 2006.
176. T. S. A. El-Danaf, Septic B-spline method of the Korteweg-de Vries-Burger's equation, *Communication of Nonlinear Science and Numerical Simulation*, 13, 554–566, 2008.
177. S. Haq, S. U. Islam, and M. Uddin, A mesh-free method for the numerical solution of the KdV-Burgers equation, *Applied Mathematical Modelling*, 33, 3442–3449, 2009.
178. B. Saka, and I. Dag, Quartic B-spline Galerkin approach to the numerical solutions of the KdVB equation, *Applied Mathematics and Computation*, 215, 746–758, 2009.
179. S. I. Zaki, A quintic B-spline finite elements scheme for the KdVB equation, *Computer Methods in Applied Mechanics and Engineering*, 188, 121–134, 2000.
180. Q. Wang, Numerical solution for fractional KdV-Burgers equation by Adomian decomposition method, *Applied Mathematics and Computation*, 182, 1048–1055, 2006.
181. Q. Wang, Homotopy perturbation method for fractional KdV-Burgers equation, *Chaos, Solitons and Fractals*, 35, 843–850, 2008.
182. S. T. Demiray, Y. Pandir, and H. Bulut, Generalized Kudryashov method for time-fractional differential equations, *Abstract and Applied Analysis*, 2014, 1–13, Article ID 901540, 2014.
183. A. El-Ajou, O. A. Arqub, and S. Momani, Approximate analytical solution of the nonlinear fractional KdV–Burgers equation: A new iterative algorithm, *Journal of Computational Physics*, 293, 81–95, 2015.
184. A. G. Johnpillai, and C. M. Khalique, On the solutions and conservation laws for the Sharma-Tasso-Olver equation, *Science Asia*, 40(6), 451–455, 2014.

185. Y. Shang, Y. Huang, and W. Yuan, Bäcklund transformations and abundant exact and explicit solutions of the Sharma-Tasso-Olver equation, *Applied Mathematics and Computation*, 217, 7172–7183, 2011.
186. L. Song, Q. Wang, and H. Zhang, Rational approximation solution of the fractional Sharma–Tasso–Olever equation, *Journal of Computational and Applied Mathematics*, 224, 210–218, 2009.
187. H. Tasso, Cole's ansatz and extensions of Burgers' equation, Max Planck Institute für Plasmaphysik, Report number IPP-6-142, Garching, Germany, 1976.
188. A. S. Sharma, and H. Tasso, Connection between wave envelope and explicit solution of a nonlinear dispersive equation, Max Planck Institute für Plasmaphysik, Report number IPP-6-158, Garching, Germany, 1977.
189. P. J. Olver, Evolution equation possessing infinite many symmetries, *Journal of Mathematical Physics*, 18(6), 1212–1215, 1977.
190. Z. Yan, Integrability of two types of the (2+1)-dimensional generalized Sharma–Tasso–Olver integro-differential equations, *MM Research*, 22, 302–324, 2003.
191. B. Erbas, and E. Yusufoglu, Exp-function method for constructing exact solutions of Sharma-Tasso-Olver equation, *Chaos, Solitons and Fractals*, 41, 2326–2330, 2009.
192. Y. He, S. Li, and Y. Long, Exact solutions to the Sharma-Tasso-Olver equation by using Improved (G/G) expansion method, *Journal of Applied Mathematics*, 2013, 1–6, Article ID 247234, 2013.
193. A. M. Wazwaz, New solitons and kinks solutions to the Sharma-Tasso-Olver equation, *Applied Mathematics and Computation*, 188(2), 1205–1213, 2007.
194. A. Esen, O. Tasbozan, and N. M. Yagmurlu, Approximate analytical solutions of the fractional Sharma-Tasso-Olver equation using Homotopy Analysis Method and a comparison with other methods, *Cankaya University Journal of Science and Engineering*, 9(2), 139–147, 2012.
195. A. Cesar, and S. Gómez, A nonlinear fractional Sharma–Tasso–Olver equation: New exact solutions, *Applied Mathematics and Computation*, 266, 385–389, 2015.
196. S. Sahoo, and S. Saha Ray, A new method for exact solutions of variant types of time fractional Korteweg–de Vries equations in shallow water waves, *Mathematical Methods in Applied Sciences*, 40(1), 106–114, 2017.
197. I. Christie, D. F. Griffths, A. R. Mitchell, and J. M. Sanz-Serna, Product approximation for nonlinear problems in finite element methods, *IMA Journal of Numerical Analysis*, 1, 253–266, 1981.

Index

M

N

S

T

V

W

Adhya AP

28/10/20

£ 130